GENERAL TOPOLOGY

Wacław Sierpiński

Translated and Revised by
C. Cecilia Krieger

DOVER PUBLICATIONS, INC.
Mineola, New York

Bibliographical Note

This Dover edition, first published in 2020, is an unabridged republication of the work originally published in 1956 by the University of Toronto Press, Toronto, Canada.

Library of Congress Cataloging-in-Publication Data

Names: Sierpiński, Wacław, 1882-1969, author. | Krieger, C. Cecilia, 1894-1974, translator.
Title: General topology / Wacław Sierpiński ; translated and revised by C. Cecilia Krieger.
Description: Dover edition [2020 edition]. | Mineola, New York : Dover Publications, Inc., 2020. | Originally published as 2nd edition, 1956: Toronto, Canada : University of Toronto Press. Republished by Dover Publications, 2000. | Includes bibliographical references and index. | Summary: "This critically acclaimed text by a major 20th-century mathematician presents a detailed theory of Frechet (V) spaces and a comprehensive examination of their relevance to topological spaces, plus in-depth discussions of metric and complete spaces. Numerous exercises supplement each chapter"— Provided by publisher.
Identifiers: LCCN 2019043389 | ISBN 9780486842547 (trade paperback)
Subjects: LCSH: Topology. | Set theory.
Classification: LCC QA611 .S47 2020 | DDC 514—dc23
LC record available at https://lccn.loc.gov/2019043389

Manufactured in the United States by LSC Communications
84254101
www.doverpublications.com

2 4 6 8 10 9 7 5 3 1

2020

AUTHOR'S PREFACE

THE theorems of any geometry (e.g., Euclidean) follow, as is well known, from a number of axioms, i.e., hypotheses about the space considered, and from accepted definitions. A given theorem may be a consequence of some of the axioms and may not require all of them. Such a theorem will be true not only in the space defined by all the axioms, but also in more general spaces. It will, therefore, be of importance to introduce axioms gradually and to deduce from them as many conclusions as possible.

We thus arrive at the concept of an abstract space (Fréchet). Theorems obtained for a given abstract space are true for each set of elements which satisfies the axioms of that space; however, the set may also satisfy other axioms. Herein lies the practical advantage of the study of abstract spaces. For, with a suitable choice of axioms for such a space, the theorems obtained from that space may be applied to different branches of mathematics, e.g., to various types of geometry, to the theory of functions, etc.

In the first chapter we develop a fairly detailed theory of the so-called Fréchet (V)spaces. A Fréchet (V)space is a set K whose elements are subject to only one condition, namely, that with each element p of K there is associated at least one subset of K called a neighbourhood of the element p. In chapter II we investigate (V)spaces which satisfy additional axioms, i.e., the so-called topological spaces; in chapters III, IV, and V we study topological spaces satisfying various additional axioms. Chapter VI is devoted to the study of very important particular topological spaces, namely, the so-called metric spaces, which find numerous applications, and chapter VII deals with the so-called complete metric spaces.

It may be said about chapters I, II, V, VI, and VII that in each of them new axioms are introduced about the space under consideration and theorems are derived from them. In general, the theorems of each of these chapters are not true in a space satisfying only the axioms of the preceding chapters.

Such an axiomatic treatment of the theory of point sets, apart from its logical simplicity, has also an advantage in that it supplies excellent material for exercise in abstract thinking and logical argument in the deduction of theorems from stated suppositions alone; i.e., in proving the theorems by drawing logical conclusions only and without any appeal to intuition, which is so apt to mislead one in the theory of sets.

The book differs to quite an extent from the *Introduction to General Topology* (Toronto, 1934). Apart from a different axiomatic treatment, which seems to us much more advantageous, the subject matter has been considerably enlarged and numerous problems added.

In conclusion I wish to express my thanks to the University of Toronto for making the publication of this book possible, and to Dr. Cecilia Krieger for translating it from the Polish manuscript.

WACLAW SIERPIŃSKI

Warsaw, October 1948

TRANSLATOR'S PREFACE

WHEN a new edition of *Introduction to General Topology* was being considered, Professor Sierpiński informed me that he had prepared a new manuscript on "General Topology" differing from the "*Introduction*" in both content and treatment. He expressed the hope that the University of Toronto Press would publish a translation of the new manuscript in preference to a revised edition of the "*Introduction*."

The appendix appearing at the end of the *Introduction* is reprinted here with very slight changes. Material that can be omitted on first reading appears between asterisks. The numerous footnotes have, for economy in printing, been placed at the end of the book. For the same reason, the usual notation for analytic sets was changed. It is hoped that this change will not place any serious difficulties in the way of the reader.

I wish to take this opportunity to express my deep gratitude to all those who with their discussion and criticism contributed to the enjoyment of a task which might easily have proved tedious. My special thanks are due to Mr. L. W. Crompton and Mr. W. T. Sharp who read part of the manuscript and to Dr. R. G. Stanton who read all of it and offered valuable suggestions.

C. CECILIA KRIEGER

Toronto, February 1952

CONTENTS

I. FRECHET (*V*)SPACES

II. TOPOLOGICAL SPACES

III. TOPOLOGICAL SPACES WITH A COUNTABLE BASIS

VII. COMPLETE SPACES

GENERAL TOPOLOGY

CHAPTER I

FRECHET (V) SPACES

1. Fréchet (V)spaces. A Fréchet (V)space, or briefly a (V)space, is a set K of elements in which with each element a there is associated a certain class of subsets of K called *neighbourhoods* of a.

Thus the set of points in the plane is a (V)space if a neighbourhood of a point p is taken to be, e.g., the interior of an arbitrary circle with centre at p. Clearly, a neighbourhood in this case can be defined in many ways as, for instance, the interior and boundary of any square with centre at p. It would also be consistent with the definition to assume that each point p of the plane possesses only one neighbourhood, e.g., the set consisting of p itself.

The set of all real functions of a real variable is a (V)space, if a neighbourhood of $f(x)$ is defined to be the set of all functions $g(x)$ which, for a given positive c and for all values of x, satisfy the inequality

$$|f(x) - g(x)| < c.$$

In particular, an arbitrary set K is a (V)space if each element of K possesses only one neighbourhood, for instance, the set K itself, or if every subset of K is a neighbourhood of each element of K.

A Fréchet (V)space is thus defined by its system of neighbourhoods. A given set K for which there are defined two different systems of neighbourhoods gives rise to two different corresponding (V)spaces. It might seem that the concept of a (V)space without additional assumptions is too general to embrace many properties. It will be seen however that, with suitable definitions, a whole theory of (V)spaces can be developed and that certain of its results find an application in various branches of topology and of the theory of functions.

2. Limit elements and derived sets. Let K be a given (V)space. An element p of K is said to be a *limit element* of a set $E \subset K$ if every neighbourhood of p contains at least one element of E different from p. The set of all limit elements of a set E is called the *derived set* of E and is denoted by E'. It is clear that if $p \in E'$ then $p \in (E - \{p\})'$, and $p \in A'$, where $E - \{p\} \subset A \subset K$.[1]

If a set E has no limit elements its derived set is the null set. In particular, the derived set of the null set is empty. We thus have the following properties of the derived set:

(1) $$E' = 0, \qquad E = 0,$$

$$E'_1 \subset E', \qquad E_1 \subset E \subset K, \tag{2}$$

$$a \in (E - \{a\})', \qquad a \in E'. \tag{3}$$

*Thus the function $f(E) = E'$ assigns to each set $E \subset K$ a set $f(E) \subset K$ which is subject to the following conditions:

(i) If $E = 0$, then $f(E) = 0$;

(ii) If $E_1 \subset E \subset K$, then $f(E_1) \subset f(E)$;

(iii) If $a \in f(E)$, then $a \in f(E - \{a\})$.

Suppose now that K is a given set and $f(E)$ a function which assigns to each set $E \subset K$ a set $f(E) \subset K$ which is subject to conditions (i), (ii), (iii). It is then possible to define neighbourhoods of the elements of K so that K is a (V)space in which

$$E' = f(E), \qquad E \subset K. \tag{4}$$

For let a subset $H \subset K$ be a neighbourhood of the element $a \in K$ if and only if $a \in K - f(K - H)$. This condition is certainly satisfied by $H = K$ for then, from (i), $f(K - H) = 0$; consequently every element of K has at least one neighbourhood.

Suppose that E is a given subset of K and $a \in E'$. Every neighbourhood of a contains at least one element of E; consequently, the set $H = K - E$ cannot be a neighbourhood of a, i.e., $a \notin K - f(K - H)$. But $a \in K$; hence $a \in f(K - H) = f(E)$. This gives

$$E' \subset f(E). \tag{5}$$

Next assume that $a \notin E'$. Then there exists a neighbourhood H of a such that $H(E - \{a\}) = 0$ and therefore $E - \{a\} \subset K - H$. By (ii)

$$f(E - \{a\}) \subset f(K - H). \tag{6}$$

Since H is a neighbourhood of a we have $a \in K - f(K - H)$; hence $a \notin f(K - H)$ and therefore, from (6), $a \notin f(E - \{a\})$; by (iii) $a \notin f(E)$. This gives

$$f(E) \subset E'. \tag{7}$$

Combining (5) and (7), we obtain (4).

It follows from the above argument that all properties of the derived set which can be proved to hold in every (V)space can be deduced from the properties (1), (2), and (3).*

3. Topological equivalence of (V)spaces. Two (V)spaces consisting of the same elements are said to be *topologically equivalent* if the derived set of each subset in one space is the same as the derived set of the same subset in the other space. They are also said to possess the same topological structure or, more briefly, the same topology.

It is easily seen that every (V)space may be associated with a topologically equivalent (V)space in which each element is contained in each one of its own neighbourhoods. It may, therefore, be assumed without any loss of generality that, whatever the definition of neighbourhoods, each element is contained in each one of its neighbourhoods.

THEOREM 1. *Two (V)spaces K_1 and K_2 consisting of the same elements are topologically equivalent (we assume that each element is contained in each one of its neighbourhoods) if and only if to every neighbourhood U of an element in K_1 there exists a neighbourhood of that element in K_2 which is contained in U, and vice versa.*

Proof. Let K_1 and K_2 be two topologically equivalent (V)spaces consisting of the same elements. Let a be a given element of $K = K_1 = K_2$ and U_1 a neighbourhood of a in K_1. Put $E = K - U_1$; hence $E \,.\, U_1 = 0$ and so $a \notin E'$ and, of course, $a \notin E$. Since K_1 and K_2 are topologically equivalent the derived sets of E are the same in both spaces. There exists, therefore, a neighbourhood $U_2 \subset K_2$ of a such that $U_2(E - \{a\}) = U_2 \,.\, E = 0$; hence $U_2 \subset K - E = U_1$. Similarly, because of the symmetry of the conditions, to every neighbourhood $U_2 \subset K_2$ of a there exists a neighbourhood $U_1 \subset K_1$ of a such that $U_1 \subset U_2$. The condition of the theorem is therefore necessary.

Suppose the condition of the theorem satisfied and let E be a set contained in $K_1 = K_2$. If an element $a \notin E' \subset K_1$ there exists a neighbourhood U_1 such that $U_1(E - \{a\}) = 0$; but, by the condition of the theorem, there exists a neighbourhood $U_2 \subset K_2$ such that $U_2 \subset U_1$; hence $U_2(E - \{a\}) = 0$ and therefore $a \notin E' \subset K_2$. Thus every element of a derived set in K_2 is an element of the corresponding derived set in K_1 and conversely, because of the symmetry of the conditions. As a consequence we see that derived sets of a given set in the two spaces are identical and therefore the two spaces are topologically equivalent.

Examples

*1. Given two elements a and b obtain all (V)spaces consisting of these two elements (assuming that each element is contained in each one of its neighbourhoods) and determine which of them are topologically equivalent.

	Neighbourhoods of a:	Neighbourhoods of b:
K_1	$\{a\}$	$\{b\}$
K_2	$\{a\}$	$\{a, b\}$
K_3	$\{a\}$	$\{b\}, \{a, b\}$
K_4	$\{a, b\}$	$\{b\}$
K_5	$\{a, b\}$	$\{a, b\}$
K_6	$\{a, b\}$	$\{b\}, \{a, b\}$
K_7	$\{a\}, \{a, b\}$	$\{b\}$
K_8	$\{a\}, \{a, b\}$	$\{a, b\}$
K_9	$\{a\}, \{a, b\}$	$\{b\}, \{a, b\}$

The following spaces are topologically equivalent: K_1, K_3, K_7, and K_9; K_2 and K_8; K_4 and K_6. But no two of K_1, K_2, K_4, and K_5 are topologically equivalent.

2. Show that the number of topologically non-equivalent (V)spaces consisting of three elements is 125.

Let $K = \{a, b, c\}$; there are 15 different sets of neighbourhoods of the element a:

1. $\{a\}$; 2. $\{a, b\}$; 3. $\{a, c\}$; 4. $\{a, b, c\}$; 5. $\{a\}$, $\{a, b\}$; 6. $\{a\}$, $\{a, c\}$; 7. $\{a\}$, $\{a, b, c\}$; 8. $\{a, b\}$, $\{a, c\}$; 9. $\{a, b\}$, $\{a, b, c\}$; 10. $\{a, c\}$, $\{a, b, c\}$; 11. $\{a\}$, $\{a, b\}$, $\{a, c\}$; 12. $\{a\}$, $\{a, c\}$, $\{a, b, c\}$; 13. $\{a, b\}$, $\{a, c\}$, $\{a, b, c\}$; 14. $\{a\}$, $\{a, b\}$, $\{a, b, c\}$; 15. $\{a\}$, $\{a, b\}$, $\{a, c\}$, $\{a, b, c\}$.

Of these the systems 1, 5, 6, 7, 11, 12, 14, and 15 are topologically equivalent and so are the systems 2 and 9, 3 and 10, 8 and 13; but no two of 1, 2, 3, 4, and 8 are topologically equivalent. Corresponding to each element of K there are 5 topologically non-equivalent systems of neighbourhoods; consequently there are 5^3 topologically non-equivalent (V)spaces each consisting of the same three elements.

3. Show that there are 19^4 topologically non-equivalent (V)spaces each consisting of the same 4 elements.

4. Show that the number of different (V)spaces consisting of the same n elements is

$$(2^{2^{n-1}} - 1)^n.$$

5. Determine the number of different topologies in a (V)space consisting of (a) two elements, (b) three elements (see examples 1 and 2).*

Given a set M of cardinal $\mathbf{m}$ one may divide all (V)spaces obtained from M into disjoint classes assigning two (V)spaces to the same class if and only if they are topologically equivalent. How many of these classes are there? In other words, how many different topological structures can be induced into a space of cardinal $\mathbf{m}$?

It can be shown that in an infinite space of cardinal $\mathbf{m}$ there can be defined $2^{2^{\mathbf{m}}}$ different topologies (hence as many as there are different (V)spaces obtained from a given set of cardinal $\mathbf{m}$).

4. Closed sets. A set which contains all its limit elements is called *closed*. Thus E is closed if and only if $E' \subset E$.

THEOREM 2. *The intersection of any aggregate of closed sets is closed.*

Proof. Let $P = \prod E$ be the intersection of the closed sets E. Hence $P \subset E$ for every E of the aggregate; by property (2) of derived sets

$P' \subset E' \subset E$ since E is closed. Therefore, $P' \subset \prod E = P$; consequently P is closed.

Since, in a given (V)space, the derived set is uniquely defined it follows that the family Φ of all closed sets of this (V)space is also uniquely defined. Thus the families of closed sets in two topologically equivalent spaces are identical. But, as is shown in § **5**, there are topologically non-equivalent spaces consisting of the same elements and having all closed sets in common.[2] Hence the family of all closed sets of a (V)space does not determine the topology of this space.

Theorem 3. *If a set E is closed then every set contained in E and containing E′ is closed.*

Proof. Let T be a set such that $E' \subset T \subset E$; then $T' \subset E' \subset T$ and therefore T is closed.

In particular, the derived set of a closed set is closed. However, the derived set of a set which is not closed may not be closed.[3]

5. The closure of a set. It follows from the definition of a closed set that *the null set is closed* and *the whole (V)space is closed.* Thus for every set $E \subset K$ there exist closed sets containing E (e.g., the set K). Denote by $\bar{E}$ the intersection of all closed sets containing E. By Theorem 2, $\bar{E}$ is closed; it is called the *closure* of the set E. Hence the closure of every set is a closed set. Moreover, it is the smallest closed set containing E, that is to say, it is contained in every closed set containing E. Consequently *E is closed if and only if $E = \bar{E}$.* In particular,

$$\bar{\bar{E}} = \bar{E} \qquad \text{(where } \bar{\bar{E}} = (\bar{\bar{E}})\text{)}.$$

From $E \subset \bar{E}$, since $\bar{E}$ is closed, we obtain at once that $E' \subset \bar{E}$ and so $E + E' \subset \bar{E}$ for every set $E \subset K$.

It follows immediately that the closure of a set possesses the following properties:[4]

1) $\bar{E} = 0, \qquad E = 0;$

2) $\bar{E}_1 \subset \bar{E} \qquad E_1 \subset E \subset K;$

3) $E \subset \bar{E}, \qquad E \subset K;$

4) $\bar{\bar{E}} = \bar{E}. \qquad E \subset K.$

We have already defined the function $f(E) = E'$ for every $E \subset K$; we can now define the function $\phi(E) = \bar{E}$ in terms of the function f. For $\phi(E)$ is the intersection of all sets $F \subset K$ such that $E + f(F) \subset F$. But we cannot define the function f in terms of the function ϕ. Two (V)spaces with the same elements and having two different functions $f(E) = E'$ defined in them may have the same function $\phi(E) = \bar{E}$, as can be seen from the following example: Let V_1 be a (V)space with three elements a, b, c, each element having only one neighbourhood, namely:

$$U_1(a) = \{a, c\}, U_1(b) = \{b, a\}, U_1(c) = \{c, b\}.$$

The set $\{a\}$ has a single limit element b, hence $\{a\}' = \{b\}$; similarly, $\{b\}' = \{c\}$ and $\{c\}' = \{a\}$. These sets are obviously not closed; this proves incidentally that, in a (V)space, derived sets need not be closed. Nor are the sets consisting of two elements closed. For $\{a, b\}' = \{b, c\}$ which is obviously not contained in $\{a, b\}$. The only closed sets of V_1 are the null set and the set V_1. Hence for $E \subset V_1$ and $E \neq 0$, we have

$$\phi_1(E) = \bar{E} = V_1$$

while for $E = \{a\}$, we have

$$f_1(E) = E' = \{b\}.$$

Next, let V_2 be a (V)space with the same three elements a, b, c, each element having the same neighbourhood, namely,

$$U_2(a) = U_2(b) = U_2(c) = \{a, b, c\}.$$

Here $\{a\}' = \{b, c\}, \{b\}' = \{c, a\}, \{c\}' = \{a, b\}, \{b, c\}' = \{c, a\}' = \{a, b\}' = \{a, b, c\}$; hence the only closed sets of V_2 are the null set and V_2. Thus for $E \subset V_2$, $E \neq 0$, we find that

$$\phi_2(E) = V_2 = V_1 = \phi_1(E)$$

but for $E = \{a\}$ we have

$$f_2(E) = \{b, c\} \neq \{b\} = f_1(E).$$

It is thus seen that, even if the closures of a given set in two (V)spaces with the same elements be the same, the derived sets of that set may be different. Hence, if in a given (V)space the derived set is known, then the closure also is known, but not conversely. The function $\phi(E) = \bar{E}$ does not, therefore, define the topology of a (V)space.

The function $\phi(E) = \bar{E}$ associates with each set $E \subset K$ a definite set $\phi(E) \subset K$ subject to the conditions:

1. $$\phi(E) = 0, \qquad E = 0;$$
2. $$\phi(E_1) \subset \phi(E), \qquad E_1 \subset E;$$
3. $$E \subset \phi(E);$$
4. $$\phi(\phi(E)) = \phi(E).$$

Let now K be a given set, $\phi(E)$ a function defined for every $E \subset K$ and subject to the conditions 1, 2, 3, and 4; it is then possible to define neighbourhoods in K so that K becomes a (V)space in which

(8) $$\bar{E} = \phi(E) \qquad \text{for all } E \subset K.$$

Thus, for example, let a set $H \subset K$ be a neighbourhood of $a \in K$ if and only if

(9) $$a \in K - \phi((K - H) - \{a\}).$$

The set $H = K$ satisfies (9) hence every element of K has at least one neighbourhood. We first show that for every $E \subset K$ we have

(10) $$\phi(E) = E + E'.$$

In fact, if $a \notin \phi(E)$ then, from 3, $a \notin E$; since $E - \{a\} \subset E$ we get from condition 2, $\phi(E - \{a\}) \subset \phi(E)$ and so $a \notin \phi(E - \{a\})$. But this gives $a \in K - \phi(E - \{a\})$. Let $H = K - E$, then $a \in K - \phi((K - H) - \{a\})$; consequently H is a neighbourhood of a and H contains no elements of E. Hence $a \notin E'$. This gives

(11) $$E + E' \subset \phi(E).$$

Next, suppose that $a \notin E + E'$. Then $a \notin E'$; hence there exists a neighbourhood H of a such that $H \,.\, E = 0$. Therefore, $E \subset K - H$ and, since $a \notin E$, $E \subset (K - H) - \{a\}$; hence, from condition 2, we have

(12) $$\phi(E) \subset \phi((K - H) - \{a\}).$$

But $a \in H$, i.e., $a \in K - \phi((K - H) - \{a\})$; hence, from (12), $a \notin \phi(E)$; this gives

(13) $$\phi(E) \subset E + E'.$$

Relations (11) and (13) give (10).

From (10) and condition 4 we obtain, for every $E \subset K$, the relation

$$\phi(E + E') = \phi(\phi(E)) = \phi(E) = E + E',$$

that is,

(14) $$\phi(E + E') = E + E';$$

since, by (10), $E' \subset \phi(E)$ for every $E \subset K$, (14) implies that

$$(E + E')' \subset \phi(E + E') = E + E'.$$

Hence the set $E + E'$ is closed and since it contains E it must contain $\bar{E}$. Therefore,

(15) $$\bar{E} \subset \phi(E).$$

On the other hand, we have $E + E' \subset \bar{E}$ and so by (10)

(16) $$E + E' = \phi(E) \subset \bar{E}.$$

From (15) and (16), we obtain

(17) $$\bar{E} = \phi(E) = E + E' \qquad \text{for every } E \subset K.$$

We have thus proved that for every function $\phi(E)$ defined for $E \subset K$ and subject to the conditions 1, 2, 3, and 4, neighbourhoods can be so defined that K becomes a (V)space in which (17) holds.

The relation $\bar{E} = E + E'$ holds in many important (V)spaces (§ **19**), but need not be true in general, as may be seen from the example of the space $V_1 = \{a, b, c\}$ given in this section, where $E = \{a\}$ and $E + E' = \{a, b\} \neq \{a, b, c\} = \bar{E}$. It follows from the above established properties of the functions $\phi(E)$ that every property of the closure of a set which holds in all (V)spaces must result from conditions 1, 2, 3, and 4. It can be easily shown that these conditions are independent.

*The closure $\bar{E}$ of a set $E \subset K$ (K a (V)space) can be obtained by means of transfinite construction as follows:

Let $E_0 = E$; for every ordinal number $\alpha > 0$ define by transfinite induction the set

(18) $$E_\alpha = \Big(\sum_{0 \leqslant \xi < \alpha} E_\xi\Big)'.$$

Since

$$\sum_{0 \leqslant \xi < \alpha} E_\xi \subset \sum_{0 \leqslant \xi < \beta} E_\xi, \qquad 0 < \alpha < \beta,$$

we have

(19) $$E_\alpha \subset E_\beta.$$

Suppose that the cardinal of K is $\aleph_\mu$; then there exists an ordinal number ν, where $0 < \nu < \omega_{\mu+1}$, such that

(20) $$E_\nu = E_{\nu+1}.$$

For if not, assume that

(21) $$E_\alpha \neq E_{\alpha+1}, \qquad 0 \leqslant \alpha < \omega_{\mu+1}.$$

By (19), $E_\alpha \subset E_{\alpha+1}$ and therefore for every ordinal α satisfying the inequality $0 < \alpha < \omega_{\mu+1}$, there exists, by (21), an element p_α such that $p_\alpha \in E_{\alpha+1}$ but $p_\alpha \notin E_\alpha$. Consequently $p_\alpha \notin E_{\xi+1}$ for $\xi < \alpha$. But $p_\xi \in E_{\xi+1}$; hence $p_\alpha \neq p_\xi$, for $\xi < \alpha$.

The transfinite sequence $\{p_\alpha\}$, $\alpha < \omega_{\mu+1}$, consisting of different elements has cardinal $\aleph_{\mu+1}$, contrary to the fact that it is a subset of K whose cardinal is $\aleph_\mu$. The existence of an ordinal number ν, where $0 < \nu < \omega_{\mu+1}$, such that (20) is true is thus established.

Furthermore,

(22) $$\bar{E} = \sum_{0 \leqslant \xi < \nu} E_\xi.$$

For, from (18) and (20), we have for $\alpha = \nu + 1$,

$$(\sum_{0 \leqslant \xi < \nu} E_\xi)' = E_{\nu+1} = E_\nu \subset \sum_{0 \leqslant \xi < \nu} E_\xi;$$

this proves that the set $F = \sum_{0 \leqslant \xi < \nu} E_\xi$ is a closed set.

On the other hand, let α be an ordinal number, where $0 \leqslant \alpha \leqslant \nu$. For $\alpha = 0$ we have $E_0 \subset \bar{E}$; assume that $E_\xi \subset \bar{E}$, for $\xi < \alpha$; then $\sum_{0 \leqslant \xi < \alpha} E_\xi \subset \bar{E}$ and therefore, from (18) and the fact that $\bar{E}$ is closed, we obtain

$$E_\alpha = (\sum_{0 \leqslant \xi < \alpha} E_\xi)' \subset \bar{E}' \subset \bar{E};$$

consequently $E_\alpha \subset \bar{E}$. Hence, by transfinite induction,

$$E_\alpha \subset \bar{E}, \qquad 0 \leqslant \alpha \leqslant \nu$$

and this gives

$$\sum_{0 \leqslant \xi < \nu} E_\xi = F \subset \bar{E}.$$

Since F is closed and $E \subset F \subset \bar{E}$ we have $F = \bar{E}$; this proves (22).

Example. Prove that the $*$ operation on sets A and B (contained in a (V)space) defined by

$$A * B = A.\bar{B} + \bar{A}.B$$

has the following properties

$$A * A = A, \quad A * B = B * A, \quad A * (A * B) = (A * A) * B,$$

but that the $*$ operation need not be associative.*

6. Open sets. The interior of a set. The complement of a closed set with respect to the space K in which it is contained is called an *open set*. Hence the complement of an open set is a closed set.

Let S denote the sum of a given aggregate of open sets G. By De Morgan's rule the complement of a sum of sets is the intersection of the complements of these sets. Hence

$$S = \sum G \quad \text{implies that} \quad CS = \prod CG.$$

Since G is open CG is closed and so, by Theorem 2, $\prod CG$ is closed. Thus CS is closed and therefore S is open. This gives:

THEOREM 4. *The sum of any aggregate of open sets is an open set.*

The intersection of two open sets contained in a (V)space need not be open. For instance, let K be a (V)space consisting of three elements a, b, c, where a and b have one neighbourhood each, $V(a) = \{a\}$, $V(b) = \{b\}$, while c has two neighbourhoods, $V_1(c) = \{a, c\}$, $V_2(c) = \{b, c\}$. The sets $E_1 = \{a, c\}$ and $E_2 = \{b, c\}$ are open since their complements are closed. But the set $E_1 . E_2 = \{c\}$ is not open since its complement $C\{c\} = \{a, b\}$ is not closed. This leads at once to the result that *in a (V)space the sum of two closed sets need not be closed.*

Let U be a given open set and let $p \in U$; then there exists a neighbourhood V_p of p which is contained in U. For, if there is no such neighbourhood, every neighbourhood of p must contain at least one element of CU and so $p \in (CU)' \subset CU$, since CU is closed; but this is impossible since $p \in U$. Thus, if U is open and $p \in U$, there exists at least one neighbourhood of p which is contained in U.

Conversely, if a set U contains a neighbourhood of each of its elements then U is open. To prove this, it will be sufficient to show that the set $E = K - U$ is closed. Let $q \in E'$; then $q \notin U$. For if q were in U there would be a neighbourhood $V_q \subset U$; this would imply that $V_q \,.\, E = 0$, and hence that $q \notin E'$ which is impossible. Consequently $q \in E$ and E is closed. This proves the following theorem:

THEOREM 5. *In order that a set U be open it is necessary and sufficient that it contain at least one neighbourhood of each of its elements.*

The sum of all open sets contained in a given set E is called the *interior* of E and is denoted by $I(E)$. In particular, the set $I(E)$ may be empty. By Theorem 4, the interior of a set E is open. It is the largest open set contained in E.

Clearly, if

$$E_1 \subset E \subset K, \text{ then } I(E_1) \subset I(E). \tag{23}$$

We shall show that

$$I(E) = K - \overline{K - E}. \tag{24}$$

In fact, the set

$$W = K - \overline{K - E} = C\overline{CE} = C\prod F,$$

where F is closed and $CE \subset F$; hence $E \supset CF = U$, where U is open. Therefore $W = C\prod F = \sum U = I(E)$ (since $\sum U$ is the sum of all open sets contained in E). This establishes (24).

It is clear that a set E is open if and only if $E = I(E)$. Furthermore, for every $E \subset K$,

$$I(I(E)) = I(E).$$

An element of the interior of a set E is called an *interior element* of E. A given element of a (V)space is clearly an interior element of a set E if and only if it is not an element of the closure of the complement of E.

Certain authors define interior elements of a set E differently, e.g. as elements which are not limit elements of the complement of E. The interior of E is then defined as the set of all interior elements of E. According to this definition, the interior of a set E is the set

$$E \,.\, C(CE)' = E - (K - E)'.$$

The two definitions are obviously equivalent in all (V)spaces in which $\bar{E} = E + E'$ for all $E \subset K$.

7. Sets dense-in-themselves. The nucleus of a set. Scattered sets. A set every element of which is one of its limit elements is called *dense-in-itself*. It is consistent with this definition to consider the null set as dense-in-itself.

Thus a set E is dense-in-itself if and only if $E \subset E'$.

A set which is closed and dense-in-itself is called *perfect*. Hence a set E is perfect if and only if $E = E'$. In particular, the null set is perfect.

THEOREM 6. *The closure of a set which is dense-in-itself is dense-in-itself.*

Proof. Let E be a set which is dense-in-itself; then $E \subset E'$. Since $E \subset \bar{E}$ we get, from (2), $E' \subset \bar{E}'$ and therefore $E \subset \bar{E}'$. But $\bar{E}$ is closed; hence, by the Corollary to Theorem 3, $\bar{E}'$ is closed and, since $\bar{E}'$ contains E, it must contain $\bar{E}$. Since $\bar{E} \subset \bar{E}'$, it follows that $\bar{E}$ is dense-in-itself.

THEOREM 7. *The sum of any aggregate of sets which are dense-in-themselves is dense-in-itself.*

Proof. Let $S = \sum E$ denote the sum of sets E each of which is dense-in-itself. Then $E \subset E'$ for each $E \subset S$. Hence $E' \subset S'$; consequently $E \subset S'$ for every E. Therefore $S \subset S'$; hence S is dense-in-itself.

THEOREM 8. *If a set E is dense-in-itself, then every set containing E and contained in E' is dense-in-itself.*

Proof. If E is dense-in-itself, then $E \subset E'$. Let T be a set such that $E \subset T \subset E'$; then $E' \subset T'$ and so $T \subset T'$; therefore T is dense-in-itself.

In particular, *the derived set of a set which is dense-in-itself is dense-in-itself.*

We note in connection with Theorems 6 and 8 that, if $E \subset E'$ and $E \subset T \subset \bar{E}$, then T need not be dense-in-itself. For example, in the space K with the 4 elements a, b, c, d, each having only one neighbourhood, namely, $V(a) = \{a, b\}$, $V(b) = \{a, b\}$, $V(c) = \{b, c\}$, $V(d) = \{c, d\}$, let $E = \{a, b\}$ and $T = \{a, b, d\}$; then $E' = \{a, b, c\}$ and $\bar{E} = \{a, b, c, d\} = K$. Hence $E \subset T \subset \bar{E}$ but T is not dense-in-itself since $d \in T - T'$.

For a given set E, denote by N the sum of all sets which are dense-in-themselves and contained in E. The set N is dense-in-itself by Theorem 7. It is clearly the largest subset of E which is dense-in-itself. The set N is called the *nucleus* of E. A set whose nucleus is empty is called *scattered*. In particular, the null set is scattered.

THEOREM 9. *Every set E can be represented in the form $E = N + R$, where N is the nucleus of E and R is scattered.*

Proof. Let E be a given set and N its nucleus; then $N \subset E$. Write $E - N = R$, and suppose R is not scattered. There exists then a subset of

R which is dense-in-itself and therefore contained in N; this is impossible. The theorem is therefore proved.

*The nucleus N of a given set E contained in a (V)space K may be obtained by a transfinite process as follows:

Put $E_0 = E$ and, for ordinal numbers $\alpha > 0$, define by transfinite induction sets E_α by the relation

$$(25) \qquad E_{\alpha+1} = E_\alpha \,.\, E'_\alpha$$

and, where α is an ordinal of the second kind, by

$$(26) \qquad E_\alpha = \prod_{\xi<\alpha} E_\xi.$$

It follows immediately from (25) and (26) that

$$(27) \qquad E_\alpha \supset E_\beta, \qquad 0 \leqslant \alpha < \beta.$$

We shall show that, if the cardinal of K is $\aleph_\mu$, then there exists an ordinal number ν, satisfying the inequality $0 < \nu < \omega_{\mu+1}$, and such that

$$(28) \qquad E_\nu = E_{\nu+1}.$$

Suppose that such a number does not exist; we then have

$$(29) \qquad E_\alpha \neq E_{\alpha+1}, \qquad 0 \leqslant \alpha < \omega_{\mu+1}.$$

Since, by (27), $E_\alpha \supset E_{\alpha+1}$, (29) will ensure, for every ordinal α, the existence of an element p_α such that

$$p_\alpha \in E_\alpha \text{ but } p_\alpha \notin E_{\alpha+1};$$

whence, and from (27), $p_\alpha \in E_{\xi+1}$ for $\xi + 1 \leqslant \alpha$. But $p_\xi \in E_\xi - E_{\xi+1}$; therefore $p_\xi \neq p_\alpha$ for $\xi < \alpha$. The transfinite sequence $\{p_\alpha\}$, $\alpha < \omega_{\mu+1}$, contains $\aleph_{\mu+1}$ different elements of K; this is impossible, since the cardinal of K is $\aleph_\mu$. Hence ν exists.

We next show that

$$(30) \qquad N = E_\nu.$$

From (28) and (25) we have $E_\nu = E_{\nu+1} = E_\nu \,.\, E'_\nu$; hence $E_\nu \subset E'_\nu$ and therefore E_ν is dense-in-itself. Moreover $E_\nu \subset E$; consequently $E_\nu \subset N$.

Let now T be a subset of E which is dense-in-itself; then

$$(31) \qquad T \subset T'.$$

We wish to show that

$$(32) \qquad T \subset E_\alpha, \qquad 0 \leqslant \alpha < \omega_{\mu+1}.$$

This result is true for $\alpha = 0$ since $T \subset E = E_0$. Let α be an ordinal number greater than 0 and suppose that $T \subset E_\xi$ for $\xi < \alpha$. If $\alpha = \beta + 1$, then $T \subset E_\beta$; hence $T' \subset E'_\beta$ and so, from (31), $T \subset E'_\beta$ which gives

$T \subset E_\beta \,.\, E'_\beta = E_{\beta+1} = E_\alpha$. If α is a number of the second kind it follows immediately, from (26) and the assumption for $\xi < \alpha$, that $T \subset E_\alpha$. This proves (32).

For $\alpha = \nu$, we get $T \subset E_\nu$. The set E_ν contains every dense-in-itself subset of E and, since it is dense-in-itself and contained in E, it must be the nucleus of E.

Furthermore, we can show that

$$E_\alpha = E_\nu, \qquad \alpha > \nu. \tag{33}$$

Statement (33) is true for $\alpha = \nu$; let β be a number greater than ν and assume (33) is true for α, where $\nu \leqslant \alpha < \beta$. If β is a number of the first kind, say $\beta = \alpha + 1$, then $E_\alpha = E_\nu$ and so from (25) and (28) we have

$$E_\beta = E_\alpha \,.\, E'_\alpha = E_\nu \,.\, E'_\nu = E_{\nu+1} = E_\nu;$$

if β is a number of the second kind, we obtain from (26)

$$E_\beta = \prod_{0 \leqslant \alpha < \beta} E_\alpha = \prod_{\nu \leqslant \alpha < \beta} E_\alpha = \prod E_\nu = E_\nu.$$

This proves (33).

It follows from (30) that if $N = 0$ then $E_\nu = 0$. Suppose there exists an ordinal number α such that $E_\alpha = 0$. If $\alpha \leqslant \nu$ then $E_\nu = 0$ by (27), and therefore $N = 0$ by (30). If $\alpha > \nu$, then $E_\alpha = E_\nu = 0$, by (33), and so again $N = 0$. Thus *a set E is scattered if and only if there exists an ordinal number ν such that $E_\nu = 0$.*

It can also be shown that

$$R = E - N = \sum_{0 \leqslant \xi < \nu} (E_\xi - E_{\xi+1}).$$

There exists a duality of a certain kind in a (V)space between open sets and sets which are dense-in-themselves. The following theorem can be established:

If K is a Fréchet (V)space which is dense-in-itself, then there exists a (V)space K^*, with the same elements as K, such that every open set in K is dense-in-itself in K^* and conversely, and every set which is dense-in-itself in K is open in K^* and conversely.

To prove this theorem, it is sufficient to define neighbourhoods in K^* as follows: a subset $U \subset K^*$ is a neighbourhood of $a \in K^*$ if $a \in U$ and $U(V - \{a\}) \neq 0$ for every set V which is a neighbourhood of a in K.*

8. Sets closed in a given set. A set E_1 is said to be *closed in E* if

$$E'_1 \,.\, E \subset E_1, \tag{34}$$

that is to say, if E_1 contains all those limit elements of E_1 which belong to E.

If E_1 is closed in E then it is closed in every subset of E. For, from $E'_1 \,.\, E \subset E_1$ and $E_2 \subset E$, it follows that

$$E'_1 \,.\, E_2 \subset E'_1 \,.\, E \subset E_1; \text{ hence } E'_1 \,.\, E_2 \subset E_1.$$

However, E_1 need not be closed in a set T which contains E.

A set contained in a closed set and closed in that set is itself closed. Here $E_1 \subset E$ gives $E'_1 \subset E$, since E is closed, and so $E'_1 = E'_1 . E \subset E_1$ by (34); consequently E_1 is closed.

A closed set is closed in every subset of a (V)space. For the relation $E'_1 \subset E_1$ implies (34) for every set E.

The intersection of a set E and a closed set is closed in E. To prove this, let $E_1 = E . F$, F closed; then $E_1 \subset F$ and $E'_1 \subset F' \subset F$ since F is closed. Hence

$$E'_1 . E \subset F . E = E_1.$$

However, a set E_1 which is contained and closed in E is not necessarily the intersection of E and a closed set in every (V)space.

In fact, let $K = \{a, b, c\}$, and let $V(a) = \{a, b\}$, $V(b) = \{b, a\}$, $V(c) = \{b, c\}$ be the only neighbourhoods of a, b, c, respectively; put $E = \{a, c\}$, $E_1 = \{a\}$. Then $E'_1 = \{b\}$, $E''_1 = \{b\}' = \{a, c\} = E$ and $E'_1 . E = 0 \subset E_1$; hence E_1 is contained and closed in E. Let F be a closed set containing E_1; then $E'_1 \subset F$ and $E''_1 \subset F$ and therefore $E \subset F$; hence $E . F = E \neq E_1$.

THEOREM 10. *The intersection of any aggregate of sets closed in a set E_0 is closed in E_0.*

Proof. Let $P = \prod E$ denote the intersection of the given sets E which are closed in E_0; then $P \subset E$; hence $P' \subset E'$ for every E of the aggregate. Therefore $P' . E_0 \subset E' . E_0 \subset E$, since E is closed in E_0; consequently $P' . E_0 \subset \prod E = P$.

THEOREM 11. *The nucleus of a set is closed in the set.*

Proof. Let E_1 be the nucleus of E; then E_1 is dense-in-itself and contained in E. We therefore have $E_1 \subset E'_1 . E \subset E'_1$; hence, by Theorem 8, $E'_1 . E$ is dense-in-itself and, since it is a subset of E, it must be contained in the nucleus of E. This gives $E'_1 . E \subset E_1$; hence E_1 is closed in E.

It follows that *the nucleus of a closed set is perfect.*

9. Separated sets. Connected sets. Two sets A and B are said to be *separated* if

$$A \neq 0, \quad B \neq 0, \quad A . B = A . B' = A' . B = 0. \tag{35}$$

*Two (V)spaces consisting of the same elements but not topologically equivalent may have the same pairs of separated sets (i.e., if the sets A and B are separated in one space, they are separated in the other and vice versa).

In fact, let V_1 be the (V)space with two elements a and b, each having the single neighbourhood $V_1(a) = V_1(b) = \{a, b\}$, and let V_2 be the (V)space consisting of the same elements a and b, with the neighbourhoods $V_2(a) = \{a, b\}$ and $V_2(b) = \{b\}$. It is easily seen that there are no separated sets in either V_1 or V_2 although the spaces are not topologically equivalent. In V_1 we have $\{a\}' = b$, while in V_2, $\{a\}' = 0$.

We may mention here that Szymański[5] has employed separated sets as the primitive concept in topology.*

THEOREM 12. *If A and B are separated sets and A_1 and B_1 satisfy the conditions*

$$(36) \qquad A_1 \neq 0, \quad B_1 \neq 0, \quad A_1 \subset A, \quad B_1 \subset B,$$

then A_1 and B_1 are separated sets.

Proof. From (35), (36), and (2), we have $A_1 . B_1 \subset A . B = 0$, $A_1 . B'_1 \subset A . B' = 0$, and $A'_1 . B_1 \subset A' . B = 0$; hence A_1 and B_1 are separated.

A set which is not the sum of two separated sets is called *connected* (Hausdorff). In accordance with this definition, the null set is connected and any set consisting of only one element is connected.

THEOREM 13. *A set E is connected if and only if it is not the sum of two non-empty disjoint sets each closed in E.*

Proof. Assume $E = A + B$, where A and B are non-empty disjoint sets closed in E. Hence

$$(37) \qquad A \neq 0, \; B \neq 0, \; A . B = 0, \; A' . E \subset A, \; B' . E \subset B.$$

Since $E = A + B$, we have $A' . B = A' . E . B \subset A . B = 0$, and $A . B' = A . E . B' \subset A . B = 0$; it follows that A and B are separated and therefore E is not connected. Hence the condition of the theorem is necessary.

Suppose next that E is not connected. Then $E = A + B$, where A and B are separated sets and so satisfy (35). We have therefore $A' . E = A'(A + B) = A' . A + A' . B = A' . A \subset A$; hence A is closed in E. Similarly, B is closed in E. This proves the sufficiency of the condition.

From § 8 and Theorem 13, we obtain

THEOREM 14. *A closed set is connected if and only if it is not the sum of two non-empty disjoint and closed sets.*

A set $E_1 \subset E$ is said to be *open in E if $E - E_1$ is closed in E.*

Theorem 13 leads to the following

COROLLARY. *A set is connected if and only if it does not possess a non-empty proper subset which is both closed and open in the set.*

*Also, as can easily be seen, a set E is connected if and only if it satisfies the following condition W:

If A is a non-empty subset of E with the property that for every element of the set $A + A' . E$ there exists a neighbourhood U of this element such that $U . E \subset A$, then $A = E$.

Saks[6] refers to sets satisfying condition W as sets for which *the principle of generalized induction* holds. It may therefore be stated that the principle of generalized induction applies to a set E (contained in a (V)space) if and only if E is connected.

We remark further that the principle of induction for the set of all real numbers, given by Khintchine,[7] may be considered a special case of the principle of generalized induction; the same is true of the principle of induction for natural numbers. The set of natural numbers is connected if a neighbourhood of a number n consists of n and all the natural numbers smaller than n.*

THEOREM 15. *The closure of a connected set is connected.*

Suppose the contrary to be true. Since the closure $\bar{E}$ of a set E is closed, if not connected, it is, by Theorem 14, the sum of two disjoint non-empty closed sets A and B. Put $A_1 = E . A$, $B_1 = E . B$; then $E = A_1 + B_1$. If $A_1 = 0$, then $E = B_1 \subset B$ and, since B is closed, $\bar{E} \subset B$, which is impossible since $A \neq 0$ and $A . B = 0$. Hence $A_1 \neq 0$ and, by a similar argument, $B_1 \neq 0$. Thus, by Theorem 12, A_1 and B_1 are separated sets which implies that E is not connected, contrary to hypothesis. This establishes Theorem 15.

In connection with the above theorem it should be noted that a set containing a connected set and contained in the closure of that set need not be connected.

For instance, let K be a (V)space with three elements a, b, c, each having only one neighbourhood, namely, $V(a) = \{a\}$, $V(b) = \{a, b\}$, $V(c) = \{b, c\}$. Put $E = \{a\}$; E is connected and $\bar{E} = \{a, b, c\}$. Let $T = \{a, c\}$; then $E \subset T \subset \bar{E}$ but $T = \{a\} + \{c\}$ is obviously the sum of two separated sets and therefore not connected.

THEOREM 16. *A connected set which is contained in the sum of two separated sets is contained in one of the two sets.*

Proof. Let E be a connected set contained in the sum of two separated sets A and B; hence (35) holds and, since $E \subset A + B$, we have $E = A . E + B . E$. Let $A_1 = A . E$ and $B_1 = B . E$. If $A_1 \neq 0$ and $B_1 \neq 0$ the sets A_1 and B_1 would be separated which is impossible since E is connected. Hence, either $A_1 = 0$, i.e. $E \subset B$, or $B_1 = 0$, in which case $E \subset A$.

COROLLARY. *If E is connected then every set containing E and contained in $E + E'$ is connected.*

Let E denote a connected set and T a set such that $E \subset T \subset E + E'$. If T is not connected, then $T = A + B$, where A and B are separated sets. Since E is connected, and $E \subset A + B$, by Theorem 16, E is contained in one of the sets, say A. Hence, from (2), $E' \subset A'$ and therefore $E' . B \subset A' . B = 0$; hence $E' . B = 0$. Also $E . B = 0$; thus

$$B = (A + B) . B = T . B \subset (E + E') . B = E . B + E' . B = 0$$

which is impossible. Similarly $E \subset B$ leads to a contradiction. The set T is therefore connected. In particular, if E is connected then the set $E + E'$ is also connected.

THEOREM 17. *If a set S (contained in a (V)space K) is connected and E is a set contained in K such that*

$$S . E \neq 0, \text{ and } S - E \neq 0, \tag{38}$$

then

$$S . E(S - E)' + (S - E)(S . E)' \neq 0. \tag{39}$$

Proof. Assume the conditions of the theorem satisfied and

$$S . E(S - E)' + (S - E)(S . E)' = 0; \tag{40}$$

since

$$S = S . E + (S - E),$$

it follows from (38) and (40) that S is the sum of two separated sets which is impossible. This proves the theorem.

The set

$$F(E) = E(K - E)' + (K - E)E' \tag{41}$$

is called the *frontier* of the set E. Obviously

$$F(K - E) = F(E).$$

From Theorem 17 we obtain the following:

COROLLARY. *A connected set which has elements in common with two complementary sets contains at least one element of their frontier.*

For let S denote a connected set which has elements in common with the sets E and $K - E$; then (38) holds and so, from Theorem 17, (39) follows. But

$$S . E(S-E)' + (S - E)(S . E)' \subset S . E(K - E)' + S(K - E)E' = S . F(E);$$

hence $S . F(E) \neq 0$.

THEOREM 18 (Hausdorff). *If every two elements of a set E belong to some connected subset of E, then E is connected.*

Proof. Suppose that the set E is not connected, i.e. $E = A + B$, where A and B are separated sets. Let a denote any element of A, b any element of B, E_1 any subset of E containing the elements a and b. Letting $A_1 = A \,.\, E_1$, $B_1 = B \,.\, E_1$, we shall have $A_1 \neq 0$, $B_1 \neq 0$, since $a \in A \,.\, E_1$, $b \in B \,.\, E_1$, and $A_1 \subset A$, $B_1 \subset B$; by Theorem 12, the sets A_1 and B_1 are separated. The set $E_1 = (A + B)E_1 = A \,.\, E_1 + B \,.\, E_1 = A_1 + B_1$ is therefore not connected. It follows that none of the subsets of E containing the elements a and b is connected, contrary to hypothesis; the set E is therefore connected.

THEOREM 19. *Two connected sets whose sum is not connected are separated.*

Proof. Let E_1 and E_2 be two connected sets whose sum $E = E_1 + E_2$ is not connected; hence $E = A + B$, where A and B are separated. Since $E_1 \subset A + B$, by Theorem 16 the set E_1 must be contained in one of the sets A and B, say $E_1 \subset A$. The same is true of E_2. If $E_2 \subset A$, then $E = E_1 + E_2 \subset A$, which is impossible since $E = A + B$, $B \neq 0$, and $A \,.\, B = 0$. Hence $E_2 \subset B$. The sets E_1 and E_2 are not empty; hence, by Theorem 12, they are separated.

COROLLARY. *The sum of two connected sets which have an element in common is connected.*

Otherwise, the sets would be separated and could not have an element in common, contrary to hypothesis.

THEOREM 20 (Hausdorff). *The sum of any aggregate of connected sets, every pair of which has an element in common, is connected.*

Proof. Let S denote the sum of the connected sets E, every two of which have at least one element in common, and let a and b be any two elements of S. There exist, therefore, two sets E_1 and E_2, members of the aggregate ($E_1 \neq E_2$ or $E_1 = E_2$) contained in S and such that $a \in E_1$, $b \in E_2$. By hypothesis, $E_1 \,.\, E_2 \neq 0$ and so, since E_1 and E_2 are connected, the set $E_1 + E_2$ is connected by the corollary to Theorem 19. Moreover, the set $E_1 + E_2$ contains the elements a and b; hence S satisfies the conditions of Theorem 18 and is therefore connected.

*We remark here that Knaster and Kuratowski[8] constructed a plane connected set S consisting of more than one element (in fact, a bi-connected set, i.e., a set which is not the sum of two disjoint non-empty connected sets), which contains an element a such that the set $S - \{a\}$ does not contain any connected subset consisting of more than one element.

An example of a bi-connected (V)space is supplied by the set K consisting of the three elements a, b, c, each element having only two neighbourhoods, namely, $V_1(a) = \{a, b\}$, $V_2(a) = \{a, c\}$, $V_1(b) = \{a, b\}$, $V_2(b) = \{b, c\}$,

$V_1(c) = \{a, c\}$, $V_2(c) = \{b, c\}$. It is easy to see that K is bi-connected.*

Two elements a and b of a set E are said to be *separated* in E if E is the sum of two separated sets one of which contains a and the other b. It is easily seen that a set in which every two elements are separated, does not possess any connected subsets consisting of more than one element (i.e. it is *totally disconnected*). The converse, however, need not be true.

*For example, let K be a (V)space consisting of three elements a, b, c, where a has the two neighbourhoods $V_1(a) = \{a, b\}$ and $V_2(a) = \{a, c\}$, b has the two neighbourhoods $V_1(b) = \{a, b\}$, $V_2(b) = \{a, c\}$, while c has only one neighbourhood $V(c) = \{c\}$. It is easily seen that no subset of K, consisting of more than one element, is connected, since each such set can be expressed as the sum of two separated sets. But the elements a and b are not separated in K since neither the sets $\{a\}$ and $\{b, c\}$, nor the sets $\{a, c\}$ and $\{b\}$ are separated.

Furthermore, an example of a (non-countable) totally disconnected plane set was constructed[9] in which two particular elements are not separated but this example is much more complicated.*

Let E denote a given set $\subset K$ and $a \in E$. There exist connected subsets of E containing a, e.g., the set consisting of a itself. Denote by $C(a)$ the sum of all connected sets containing a and contained in E; $C(a)$ is connected, by Theorem 20.

The set $C(a)$, which is the largest connected subset of E containing a, is called the *component* of E corresponding to a. In particular, the component $C(a)$ may reduce to the element a itself. It follows from the definition of components and Theorem 20 that the components corresponding to two different elements of E are either identical or have no elements in common in which case they are separated, by Theorem 19.

*Thus every set (in a (V)space) can be decomposed into disjoint connected subsets, any two of which are separated (but not necessarily each from the sum of the remaining ones). Such a decomposition is not necessarily unique, as is seen in the case of a set E consisting of two disjoint open intervals; E can be decomposed into connected subsets, each consisting of only one point, any two of which are separated.

Menger introduced the concept of connected sets of various orders (of connectedness). Sets consisting of single elements are connected sets of order 0. Connectedness of order $\geqslant n$ is defined by induction. A set E is said to be connected of order $\geqslant n$ if under every partition $E = A + B$, where A and B are closed in E, the set $A \, . \, B$ contains a connected subset of order $\geqslant n - 1$. A set is connected of order n if it is connected of order $\geqslant n$ but not connected of order $\geqslant n + 1$.*

10. Images and inverse images of sets. Biuniform functions. Let P and Q be any two sets (not necessarily in a Fréchet (V)space). Suppose that to each element of P there corresponds a unique element of Q (to different elements of P may correspond the same element of Q). Such a correspondence is said to establish a one-valued mapping of the set P into the set Q or to define a function of the elements of P. (Clearly, a function defined in a set is always one-valued.) If q denotes the element of Q which corresponds to the element p of the set P we write

$$q = f(p)$$

and call q the *image* of the element p.

The set of all elements $f(p)$ which correspond to the elements p of P is called the *image* of the set P and is denoted by $f(P)$. In particular, the sets P and Q may be identical; we then have a mapping of the set P into itself.

If f is a function defined in the set P and E is any subset of P then the set of all elements $f(p)$ corresponding to the elements p of E is denoted by $f(E)$, the image of E.

We have clearly, for every function f defined in the set P,

$$f(E_1) \subset f(E_2), \qquad E_1 \subset E_2 \subset P,$$

i.e., *the image of a subset of a set is a subset of the image of that set;* furthermore, it can be proved that

$$f(E_1 + E_2) = f(E_1) + f(E_2), \qquad E_1 \subset P, E_2 \subset P$$

and, more generally, for every sum $S = \sum E$ of sets $E \subset P$,

$$f(\sum E) = \sum f(E),$$

where the summation extends over all sets E of the sum S. Hence the image of a sum of sets is the sum of the images of these sets.

All that can be said about the image of the difference of two sets is that, in general,

$$f(E_1 - E_2) \supset f(E_1) - f(E_2), \qquad E_1 \subset P, E_2 \subset P$$

(i.e., *the image of the difference contains the difference of the images*).

For the intersection of any aggregate of sets E we have, as is easily seen,

$$f(\prod E) \subset \prod f(E), \tag{42}$$

i.e., *the image of the intersection of sets is contained in the intersection of the images of these sets.*

If P is a given set then the set of all elements p of P which satisfy a given condition $W(p)$ is denoted (after Lebesgue) by $\mathrm{E}_{p\epsilon P}[W(p)]$ or by $\mathrm{E}_p[p \in P, W(p)]$. Similarly, the set of all elements p of P which satisfy conditions $W_1(p), W_2(p), \ldots, W_n(p)$, is denoted by $\mathrm{E}_{p\epsilon P}[W_1(p), W_2(p), \ldots, W_n(p)]$.

Clearly

$$\mathop{\mathrm{E}}_{p \epsilon P}[W_1(p), W_2(p)] = \mathop{\mathrm{E}}_{p \epsilon P}[W_1(p)] \, . \, \mathop{\mathrm{E}}_{p \epsilon P}[W_2(p)]$$

and, more generally,

$$\mathop{\mathrm{E}}_{p \epsilon P}[W_1(p), W_2(p), \ldots, W_n(p)] = \prod_{k=1}^{n} \mathop{\mathrm{E}}_{p \epsilon P}[W_k(p)].$$

Thus, for example, if N is the set of all natural numbers, then $\mathrm{E}_{x \epsilon N}[x > 10]$ is the set of all natural numbers > 10, $\mathrm{E}_x[x \in N, 3 \leqslant x < 6]$ is the set consisting of the numbers 3, 4, and 5, $\mathrm{E}_{x \epsilon N}[x^2 + 6 = 5x]$ is the set $\{2, 3\}$, $\mathrm{E}_x[x \in N, 2x = 3]$ is the null set, $\mathrm{E}_{x \epsilon N}[x \equiv 1 \pmod 2]$ is the set of all odd natural numbers, $\mathrm{E}_{p \epsilon N}[p > 1, (p-1)! \equiv -1 \pmod p]$ denotes the set of all primes (a result well known in the theory of numbers).

If $f(p)$ is a function defined in P then, as is easily seen,

$$\mathop{\mathrm{E}}_{p \epsilon P}[f(p) \in Q_1 + Q_2] = \mathop{\mathrm{E}}_{p \epsilon P}[f(p) \in Q_1] + \mathop{\mathrm{E}}_{p \epsilon P}[f(p) \in Q_2]$$

and

$$\mathop{\mathrm{E}}_{p \epsilon P}[f(p) \in Q_1 \, . \, Q_2] = \mathop{\mathrm{E}}_{p \epsilon P}[f(p) \in Q_1] \, . \, \mathop{\mathrm{E}}_{p \epsilon P}[f(p) \in Q_2]$$

for $Q_1 \subset f(P)$ and $Q_2 \subset f(P)$; more generally, for every sum $\sum Q$ and every intersection $\prod Q$ of sets $Q \subset f(P)$,

$$\mathop{\mathrm{E}}_{p \epsilon P}[f(p) \in \sum Q] = \sum \mathop{\mathrm{E}}_{p \epsilon P}[f(p) \in Q],$$

$$\mathop{\mathrm{E}}_{p \epsilon P}[f(p) \in \prod Q] = \prod \mathop{\mathrm{E}}_{p \epsilon P}[f(p) \in Q],$$

and

$$\mathop{\mathrm{E}}_{p \epsilon P}[f(p) \in Q_1 - Q_2] = \mathop{\mathrm{E}}_{p \epsilon P}[f(p) \in Q_1] - \mathop{\mathrm{E}}_{p \epsilon P}[f(p) \in Q_2],$$
$$Q_1 \subset f(P), \, Q_2 \subset f(P).$$

The set $\mathrm{E}_{p \epsilon P}[f(p) \in Q]$, where $Q \subset f(P)$, is called the *inverse image* of the set Q (obtained by means of the function f) and is denoted by $f^{-1}(Q)$.

If $f(p)$ is a function defined in a set P and if $f(p_1) \neq f(p_2)$ for $p_1 \in P, p_2 \in P$, where $p_1 \neq p_2$, then the function f is said to be *biuniform* or briefly (1,1), since it obviously establishes a (1,1) correspondence between the elements of the sets P and $f(P)$. Corresponding to each element q of the set $f(P)$ there exists one and only one element p of P such that $f(p) = q$; this element is denoted by $f^{-1}(q)$. The function $f^{-1}(q)$ defined in the set $f(P)$ establishes a mapping of the set $Q = f(P)$ on the set P; thus $f^{-1}(Q) = P$. The function f is said to have an inverse in P and the function f^{-1} is called the inverse function of f.

It is easy to see that if f is (1,1) in P then

$$f^{-1}(f(p)) = p, \qquad p \in P,$$

$$f(f^{-1}(q)) = q, \qquad q \in f(P),$$

$$f^{-1}(f(E)) = E, \qquad E \subset P,$$

$$f(f^{-1}(T)) = T, \qquad T \subset f(P).$$

Clearly, the function f defined in P is the inverse of the function f^{-1} defined in the set $f(P)$.

If f is a (1,1) function defined in a set P, then for every aggregate of sets $E \subset P$ we have

$$f(\prod E) = \prod f(E). \tag{43}$$

From (42)

$$f(\prod E) \subset \prod f(E);$$

applying (42) to the function f^{-1} and the sets $f(E)$, we obtain

$$f^{-1}(\prod f(E)) \subset \prod f^{-1}(f(E)) = \prod E$$

and

$$f(f^{-1}(\prod f(E))) \subset f(\prod E)$$

or

$$\prod f(E) \subset f(\prod E).$$

This and (42) give (43).

Similarly, it can be easily proved that, if a function f is (1,1) in a set P, then

$$f(E_1 - E_2) = f(E_1) - f(E_2), \qquad E_1 \subset P \text{ and } E_2 \subset P.$$

11. Continuity. Continuous images. Let K and K_1 be two (V)spaces (which may be identical). We shall assume, without any loss of generality (§ 3), that every element is contained in each of its neighbourhoods. Let E denote a given set $\subset K$ and f a function defined in E which maps E into K_1.

A function f is said to be continuous in the set E at an element p_0 of that set, if for every neighbourhood V (in K_1) of the element $f(p_0)$, there exists a neighbourhood U (in K) of p_0 such that

$$f(p) \in V, \qquad p \in U.E. \tag{44}$$

*We note that continuity of a function could be defined differently, for example, as the so-called "limit continuity" (*Limesstetigkeit*).[10] An infinite sequence $p_1, p_2, \ldots$ of elements of a space K is said to have for its limit the element p of this space (written $\lim_{n\to\infty} p_n = p$), if for every neighbourhood U of p there exists a natural number μ such that $p_n \in U$ for $n > \mu$ (§ **35**).

A function $f(p)$ is said to have limit continuity in a set E at the element $p \in E$ if, for every infinite sequence $p_1, p_2, \ldots$ for which $\lim_{n\to\infty} p_n = p$, we have $\lim_{n\to\infty} f(p_n) = f(p)$. It is easily seen that a function continuous at an element according to our definition has also limit continuity at that element. The converse, however, need not be true unless certain assumptions be made about the space K (see § **37**).*

It follows from the definition of continuity that a function $f(p)$ which is constant in E is continuous at every element of E. In particular, every function is continuous in a set E consisting of a single element.

The set of all real numbers is a (V)space if by a neighbourhood of the number x_0 we mean the interior of any interval containing x_0. Here, as is easily seen, our definition of continuity is identical with that used by Cauchy in defining a continuous (real) function of a real variable.

If a function f is continuous at every element of a set E, it is said to be continuous in E, and the set $T = f(E)$ is called a *continuous image* of E obtained by means of the function f. Obviously, if $p_0 \in E_1 \subset E$ and if f is continuous in E at p_0, then f is also continuous in E_1 at p_0.

THEOREM 21. *If the function f is continuous in E at p_0 and the function g is continuous in $T = f(E)$ at $q_0 = f(p_0)$, then the function $\phi(p) = g(f(p))$ is continuous in E at p_0.*

Proof. Let W be a neighbourhood of $\phi(p_0)$ (in the space to which this element belongs); since $\phi(p_0) = g(f(p_0))$ and g is continuous in T at $f(p_0)$, there exists a neighbourhood V (corresponding to W) of $f(p_0)$ (in the space to which $f(p_0)$ belongs) such that

$$g(q) \in W \qquad \text{for } q \in V.T \tag{45}$$

Since $f(p)$ is continuous in E at p_0 there exists a neighbourhood U (corresponding to V) of p_0 such that (44) is true. From (45) and (44) we obtain at once

$$g(f(p)) \in W \qquad \text{for } p \in U.E. \tag{46}$$

Thus for every neighbourhood W of $\phi(p_0)$ there exists a neighbourhood U of p_0 for which (46) holds; since $\phi(p) = g(f(p))$, this proves that the function ϕ is continuous in E at p_0.

In particular, if f is continuous in the whole set E and g is continuous in the whole set $T = f(E)$, then the function ϕ is continuous in the whole set E. In other words, *a continuous image of a continuous image of a given set is a continuous image of that set.*

12. Conditions for continuity in a set.

THEOREM 22. *A function f defined in a set E is continuous in E if and only if the image of any element of E which is a limit element of any subset of E is an element or limit element of the image of that subset.*

In other words, for f to be continuous it is necessary and sufficient that

$$f(E'_1 \,.\, E) \subset f(E_1) + (f(E_1))', \qquad E_1 \subset E. \tag{47}$$

Proof. Let f be a function defined and continuous in E, E_1 a subset of E, and p_0 an element of $E'_1 \,.\, E$. Assume that $f(p_0) \notin f(E_1)$ and let V denote a neighbourhood of $f(p_0)$. Since f is continuous in E at p_0 there exists a neighbourhood U of p_0 for which (44) holds. From $p_0 \in E'_1 \,.\, E \subset E'_1$, it follows that $U \,.\, E_1 \neq 0$; there exists, therefore, at least one element $p \in U \,.\, E_1 \subset U \,.\, E$ such that $f(p) \in f(E_1)$ and, from (44), $f(p) \in V$; since $f(p_0) \notin f(E_1)$, we have $f(p) \neq f(p_0)$. Thus every neighbourhood of $f(p_0)$ contains an element of $f(E_1)$ different from $f(p_0)$; consequently $f(p_0) \in (f(E_1))'$. This proves the necessity of the condition.

Next assume that f is a function defined in E but not continuous at $p_0 \in E$. Hence there exists a neighbourhood V of $f(p_0)$ such that every neighbourhood U of p_0 contains at least one element $p \in U \,.\, E$ for which $f(p) \notin V$. Let E_1 denote the set of all elements p of E for which $f(p) \notin V$; then $p_0 \notin E_1$ since $f(p_0) \in V$, and therefore $f(p_0) \notin f(E_1)$; also $f(p_0) \notin (f(E_1))'$ since $f(E_1)(V - f(p_0)) = 0$. Consequently (47) is not true. The condition of the theorem is therefore sufficient and the theorem is proved.

Let f be a (1,1) function, continuous in a set E. Let E_1 be a subset of E, p_0 an element of $E \,.\, E'_1$, and V a neighbourhood of $f(p_0)$. Since f is continuous in E, there exists a neighbourhood U of p_0 for which (44) holds. But $p_0 \in E \,.\, E'_1 \subset E'_1$; hence there exists an element $p \in U \,.\, E_1$ such that $p \neq p_0$; then $f(p) \in f(E_1)$ and, since f is (1,1), $f(p) \neq f(p_0)$. Thus every neighbourhood V of $f(p_0)$ contains an element of $f(E_1)$ different from $f(p_0)$; consequently $f(p_0) \in (f(E_1))'$.

We have therefore

$$f(E'_1 \,.\, E) \subset (f(E_1))' \qquad \text{for all } E_1 \subset E. \tag{48}$$

On the other hand, if a function f defined in a set E satisfies (48), it also satisfies (47) and so, by Theorem 22, it is continuous in E. We thus obtain:

THEOREM 23. *In order that a function f, defined and* (1,1) *in a set E, be continuous in E, it is necessary and sufficient that*

$$f(E \,.\, E'_1) \subset (f(E_1))' \qquad \textit{for all } E_1 \subset E.$$

From Theorem 22 we obtain the following

COROLLARY. *If a function f is continuous in a set E and T_1 is a set closed in*

$T = f(E)$, *then the set* $E_1 = \mathrm{E}p[f(p) \in T_1]$, *that is, the set of all elements* p *of* E *for which* $f(p) \in T_1$, *is closed in* E.

Proof. Let $p_0 \in E \, . \, E'_1$; since f is continuous, by Theorem 22, $f(p_0) \in f(E_1) + (f(E_1))'$. If $f(p_0) \in f(E_1)$, then $p_0 \in E_1$; and if $f(p_0) \in (f(E_1))'$, since $p_0 \in E$, $f(p_0) \in T(f(E_1))' \subset T \, . \, T'_1 \subset T_1$, because T_1 is closed in T. Hence in this case too, $p_0 \in E_1$. Consequently $E \, . \, E'_1 \subset E_1$, that is, E_1 is closed in E.

In particular, if $T_1 \subset T$, we can state

THEOREM 24. *The inverse image of a set contained and closed in a continuous image* T *of a set* E *is closed in* E

13. A continuous image of a connected set.

THEOREM 25. *A continuous image of a connected set is connected.*

Proof. Let E denote a connected set, $T = f(E)$ its continuous image, and assume that T is not connected. By Theorem 13, $T = A_1 + B_1$, where A_1 and B_1 are non-empty disjoint sets, each closed in T. Put $A = \mathrm{E}_p[f(p) \in A_1]$, $B = \mathrm{E}_p[f(p) \in B_1]$; the sets A and B are obviously non-empty, disjoint, and, by Theorem 24, each closed in E; moreover, $E = A + B$ and, since E is connected, this is impossible. Hence T is connected.

Theorem 25 may be restated as follows: *connectedness of sets is an invariant under continuous transformations.*

The set of all real numbers is a (V)space if by a neighbourhood of a number x we mean any interval[11] $< a, b >$ where a and b are any numbers such that $a < x < b$. A connected set of real numbers has the property that, if it contains the numbers a and b, it contains all numbers between a and b; consequently, *if a connected set of real numbers contains more than one number, it must be an interval* $< a, b >$ *, finite or infinite.* For suppose E is a connected set of real numbers $a \in E$, $b \in E$, $a < c < b$, but $c \notin E$. Let U_1 and U_2 be the sets of real numbers $< c$ and $> c$ respectively and put $E_1 = U_1 \, . \, E$, $E_2 = U_2 \, . \, E$; then $E = E_1 + E_2$ is the sum of two separated sets. (This follows from the fact that U_1 is a neighbourhood of every element of E_1 and $U_1 \, . \, E_2 = 0$; hence $E_1 \, . \, E'_2 = 0$; similarly $E_2 \, . \, E'_1 = 0$ while $E_1 \neq 0$, $E_2 \neq 0$ and $E_1 \, . \, E_2 = 0$.)

That an interval of real numbers is a connected set follows from Dedekind's postulate. For suppose that an interval E is the sum of two separated sets E_1 and E_2, where $a \in E_1$, $b \in E_2$, and $a < b$. Let $[A, B]$ denote a cut of the set of all real numbers, where A consists of all real numbers $x \leqslant a$ and all real numbers x_0 such that, for $x \in E_1$ and $x < b$, $a < x_0 \leqslant x$; the class B consists of the remaining real numbers. $B \, . \, E_2 \neq 0$ since $b \in E_2$. There exists, by Dedekind's postulate, a number c which is the greatest number in the class A or the least in the class B. In the first case, we have, as is easily seen, $c \in E_1 . E'_2$; in the second, $c \in E'_1 . E_2$, which is impossible since E_1 and E_2 are separated. Consequently E is connected. Since connected linear[12] sets are necessarily intervals it follows that *there are altogether* $\mathbf{c}$ *connected linear sets.*

There are, however, $2^{\mathbf{c}}$ different plane connected sets. To prove this, let P denote the set of points in the plane. P is a (V)space if by a neighbourhood of a point we understand the interior of any circle with the point as centre. Let Q be the set of points in the interval $< 0, 1 >$ of the x-axis and, for any $H \subset Q$, let $E = H + (P - Q)$. The set E is connected, because any two points of E can be joined by two line segments having a point of H in common and therefore are contained in a connected set. But to different subsets H correspond different sets E and, since Q has $2^{\mathbf{c}}$ different subsets, there are at least $2^{\mathbf{c}}$ different connected plane sets; there cannot be more than $2^{\mathbf{c}}$ of them because there are altogether $2^{\mathbf{c}}$ different subsets of P. Plane connected sets are much more complicated than linear connected sets.

If a function f defined in a set E contained in a space K takes on values in the set of real numbers, then it is called a real-valued function defined in E.

From Theorem 25 and the property of connected sets deduced above, we obtain immediately the following

COROLLARY. *If a real-valued function f defined and continuous in a connected set E (contained in a (V)space) takes on the values y_1 and y_2 respectively at two elements of E, then f takes on in E every value between y_1 and y_2.*

This theorem is a generalization of a property well known in analysis, namely, the property of real functions continuous in an interval (the so-called Darboux property).

Conversely: *If a set E is such that every real function defined and continuous in E assumes in E every value that is intermediate to any two values it attains in E, then E is connected.*

For if E were the sum of two separated sets A and B, then the function $f(p)$ defined to be zero for $p \in A$ and one for $p \in B$ would obviously be continuous in E but would not assume any value between 0 and 1.

14. Homeomorphic sets. If a function f establishes a (1,1) mapping of a set E on a set T, if further f is continuous in the set E and its inverse f^{-1} is continuous in the set T, then T is said to be a (1,1) and bicontinuous image of E. In this case E is clearly also a (1,1) and bicontinuous image of T.

Two sets E and T which are (1,1) and bicontinuous images of each other are called *homeomorphic*, in symbols, $E\ h\ T$. To indicate that the function f establishes this mapping we write $E\ h_f\ T$. We have clearly $T\ h_{f^{-1}} E$; the relation of homeomorphism is obviously symmetric. It is also reflexive, since $E\ h_f\ E$ for every E, if we define $f(p)$ so that $f(p) = p$ when $p \in E$.

It is clear that, if

$$E\,h_f\,T,\quad E_1 \subset E,\quad \text{and}\, f(E_1) = T_1$$

then

$$E_1\,h_f\,T_1.$$

Thus, if two sets are homeomorphic, their corresponding subsets are homeomorphic. It follows from Theorem 23 that, if

$$E\,h_f\,T \text{ and } T\,h_g\,Y$$

and if $\phi(p) = g[f(p)]$ for $p \in E$, then

$$E\,h_\phi\,Y;$$

the relation of homeomorphism is therefore transitive.

*It follows readily from the definition of homeomorphism that any two sets, each consisting of one element, are homeomorphic. Two sets, each consisting of two elements, are not, however, necessarily homeomorphic. For let K be a (V)space with three elements a, b, c, where each element has only one neighbourhood, namely, $V(a) = \{a\}$, $V(b) = V(c) = \{b, c\}$. Put $E = \{b, c\}$ and $T = \{a, b\}$. Let f be a (1,1) function which maps E on T. Then either $f(b) = a$ and $f(c) = b$, or $f(b) = b$ and $f(c) = a$. In the first case, the only neighbourhood $\{b, c\}$ of b contains c but $f(c) = b \notin V(f(b)) = V(a) = \{a\}$; in the second case, $b \in V(c)$ but $f(b) = b \notin V(f(c)) = V(a) = \{a\}$. In neither case does the function f satisfy the condition (44) of continuity; consequently, the sets E and T are not homeomorphic.

We note further that, *if in a* (V)*space a set* E_2 *is a continuous and* (1,1) *image of a set* E_1 *while* E_1 *is a continuous and* (1,1) *image of* E_2, *then the sets* E_1 *and* E_2 *are not necessarily homeomorphic.*

Kuratowski[13] gives the following example of two such linear sets. Let E_1 be the set of all integers of the form $3n + 2$ and all numbers in the open interval $(3n, 3n + 1)$, for $n = 0, 1, 2, \ldots$. Let $E_2 = (E_1 + \{1\}) - \{2\}$. It is clear that the sets E_1 and E_2 are not homeomorphic since there is no point in E_1 which corresponds to the point 1 in E_2. Nevertheless, if for $x \in E_1$ we put $f(x) = x$, $x \neq 2$, and $f(2) = 1$, the function f so defined maps E_1 continuously and biuniformly on E_2. For $x \in E_2$, put $g(x) = x/2$ if $x \leqslant 1$, $g(x) = x/2 - 1$, if $3 < x < 4$, and $g(x) = x - 3$, for $x \geqslant 5$; it is easily seen that the function $g(x)$ maps the set E_2 continuously and biuniformly on the set E_1.*

THEOREM 26.[14] *A function* f *defined in a set* E *establishes a homeomorphic mapping of* E *on the set* $T = f(E)$ *if and only if it is* (1,1) *and satisfies the condition*

$$f(E \,.\, E'_1) = f(E) \,.\, (f(E_1))' \qquad \textit{for all } E_1 \subset E. \tag{49}$$

Proof. Suppose that the function f establishes a homeomorphic mapping of the set E on the set $T = f(E)$; it is therefore (1,1) and continuous in E, Since $f(E \,.\, E'_1) \subset f(E)$, we have, by Theorem 23,

$$f(E \,.\, E'_1) \subset f(E)(f(E_1))' \qquad \text{for any } E_1 \subset E. \tag{50}$$

Similarly, we obtain for the inverse function f^{-1},

$$f^{-1}(T \,.\, T'_1) \subset f^{-1}(T)(f^{-1}(T_1))' \qquad \text{for any } T_1 \subset T. \tag{51}$$

Let E_1 be any subset of E and take $T_1 = f(E_1)$; (51) gives

$$f(f^{-1}(T \,.\, T'_1)) \subset f(f^{-1}(T)(f^{-1}(T_1))')$$

and so, since $f^{-1}(T) = E, f^{-1}(T_1) = E_1$, we obtain

$$T \,.\, T'_1 \subset f(E \,.\, E'_1),$$

that is,

$$f(E) \,.\, (f(E_1))' \subset f(E \,.\, E'_1), \qquad E_1 \subset E. \tag{52}$$

Relations (50) and (52) establish (49). Hence the condition of the theorem is necessary.

Next, suppose that the function f, defined and (1,1) in E, satisfies (49). It must also satisfy the conditions of Theorem 23 and is, therefore, continuous. Since f is (1,1) in E, it possesses an inverse function f^{-1} defined in the set $T = f(E)$. Let T_1 be an arbitrary subset of T and take $E_1 = f^{-1}(T_1)$. From (49) we obtain at once

$$f^{-1}(f(E \,.\, E'_1)) = f^{-1}(f(E) \,.\, (f(E_1))'),$$

which gives

$$E \,.\, E'_1 = f^{-1}(T \,.\, T'_1)$$

or

$$f^{-1}(T) \,.\, (f^{-1}(T_1))' = f^{-1}(T \,.\, T'_1) \qquad \text{for any } T_1 \subset T;$$

hence, by Theorem 23, the function f^{-1} is continuous. Thus the sets E and T are homeomorphic and the condition of the theorem is therefore sufficient.

Assume, in particular, that $E = K$ and $T = K_1$, where K and K_1 are two given (V)spaces. For $E_1 \subset K$, we have $E \,.\, E'_1 = K \,.\, E'_1 = E'_1$; hence relation (49) takes the form

$$f(E'_1) = (f(E_1))' \qquad \text{for any } E_1 \subset K.$$

From Theorem 26 we therefore obtain

THEOREM 27. *Two (V)spaces K and K_1 are homeomorphic if and only if there exists a* (1,1) *correspondence between their elements such that the derived set of the image of any subset of K is the image of the derived set of that subset (in other words, if and only if derived sets of corresponding sets are corresponding sets).*

It follows at once from Theorem 27 that, if two (V)spaces are homeomorphic, there exists a (1,1) correspondence between their elements such that the image of a closed set of one space is always a closed set of the other. The converse, however, is not necessarily true.

*For let K_1 be a (V)space with three elements a, b, c, each element having only one neighbourhood, namely,

$$V_1(a) = \{a, c\},\ V_1(b) = \{b, a\},\ V_1(c) = \{c, b\},$$

and let K_2 be a (V)space with three elements p, q, r, each element having only one neighbourhood, namely,

$$V_2(p) = V_2(q) = V_2(r) = \{p, q, r\}.$$

If to the elements a, b, c, be assigned the elements p, q, r, respectively, then closed sets of K_1 will map on closed sets of K_2 and conversely, as was shown in § **5**. The spaces K_1 and K_2 are not, however, homeomorphic. For suppose that $K_1 \, h_f \, K_2$. Since all the elements of K_2 have the same neighbourhood, it is immaterial which of these elements is denoted by p, which by q, and which by r; it may therefore be supposed that $f(a) = p, f(b) = q$, and $f(c) = r$. For $E_1 = \{a\}$ we have $E'_1 = \{b\}$; hence $f(E'_1) = \{q\}$; but in K_2, $(f(E_1))' = \{q, r\}$. Thus $f(E'_1) \neq (f(E_1))'$; hence, by Theorem 26, the function f, contrary to assumption, does not establish a homeomorphism between K_1 and K_2.*

Theorem 27 leads to the following

COROLLARY. *Two (V)spaces are homeomorphic if and only if there exists a (1,1) correspondence between their elements such that the image of a limit element of any set contained in one space is a limit element of the image of that set in the other space.*

15. Topological properties. A property of a set E possessed by every set homeomorphic with E is called a *topological property*. The subject of Topology is the study of topological properties of sets, that is, of invariants under (1,1) and bicontinuous transformations.

Examples of topological properties:

It follows from Theorem 25 that *connectedness of a set* is a topological property (since, by Theorem 25, connectedness of a set is invariant under every continuous transformation).

The property of *being dense-in-itself* is a topological property of a set. In fact, we shall prove

THEOREM 28. *A (1,1) and continuous image of a set which is dense-in-itself is dense-in-itself.*

Proof. Let E denote a set which is dense-in-itself, f a (1,1) function continuous in E and $T = f(E)$. Let q_0 be an element of T; hence there exists an element p_0 of E such that $f(p_0) = q_0$.

Let V denote a neighbourhood of q_0; since f is continuous at p_0 and $f(p_0) = q_0 \in V$, there exists a neighbourhood U of p_0 such that $p \in U . E$ implies $f(p) \in V$. Since E is dense-in-itself, $E \subset E'$; consequently $p_0 \in E'$ and therefore there exists in U an element p_1 of E different from p_0. Since f is (1,1) in E, we have $q_1 = f(p_1) \neq f(p_0) = q_0$ and, since $p_1 \in U . E$, we have $q_1 = f(p_1) \in V$; but $q_1 = f(p_1) \in f(E) = T$; hence $q_1 \in T$. This gives $q_0 \in T'$ and, since q_0 is any element of T, $T \subset T'$; hence T is dense-in-itself.

We note here that, although there is a certain duality (§ 7) between the properties of open sets and sets which are dense-in-themselves, sets which are dense-in-themselves cannot be replaced by open sets in Theorem 28. In fact, *the property of being open is not, in general, a topological property of a set.*

For let K denote a (V)space with two elements a and b, each having a single neighbourhood, namely, $V(a) = \{a\}$, $V(b) = \{a, b\}$.

The set $E = \{a\}$ is obviously open in K but the set $T = \{b\}$, homeomorphic with E, is not open in K. Moreover, T is closed in K while E is not. Hence *the property of being closed is not, in general, a topological property of a set.*

In linear spaces the property of being open is topological but the property of being closed is not. In a (V)space, where every element a has the single neighbourhood $V(a) = \{a\}$, both properties, that of being open and that of being closed, are invariant under continuous transformations.

From the fact that the property of a set being dense-in-itself is topological, it follows that the property of being scattered is topological. Scatteredness is not, however, invariant under all (1,1) continuous transformations; a (1,1) and continuous image of a scattered, or even, as we shall show later, an isolated set, may be dense-in-itself. However, it follows at once from Theorem 28 that the property of being scattered is invariant under those (1,1) transformations (not necessarily continuous) which have a continuous inverse.

It follows readily from Theorem 28 and the definition of the nucleus of a set that under a homeomorphic transformation the nucleus of a set is mapped into the nucleus of the image of that set whereas, under a (1,1) and continuous transformation, the nucleus is mapped into a subset of the nucleus of the image.

A set which does not contain any of its limit elements is called *isolated.* Thus a set E is isolated if and only if $E . E' = 0$. It is clear from the Corollary to Theorem 27 that *the property of being isolated is a topological invariant of a set.* It is not, however, an invariant under all (1,1) and continuous transformations.

For let K_1 be a (V)space consisting of two elements a and b, each having the single neighbourhoods $V_1(a) = \{a\}$, $V_1(b) = \{b\}$, and let K_2 be a (V)space consisting of the same elements each having the single neighbourhood $V_2(a) = V_2(b) = \{a, b\}$. The function $f(x) = x$ maps K_1 on K_2 in a (1,1) and continuous manner but K_1 is clearly isolated while K_2 is dense-in-itself.

However, it follows readily from Theorem 28 that the property of being isolated is invariant under those (1,1) transformations which have a continuous inverse.

16. Limit elements of order $\mathfrak{m}$. Elements of condensation. $\mathfrak{m}$-compact sets. Let K be a given (V)space, E a given set $\subset K$, a a given element $\in K$, and $\mathfrak{m}$ a given cardinal number. We shall say that a is a limit element of the set E of order $\geqslant \mathfrak{m}$ if every neighbourhood of a contains at least $\mathfrak{m}$ elements of E different from a. If the cardinal $\mathfrak{m}$ is transfinite the expression "different from a" may be omitted. According to this definition, limit elements as previously defined are of order $\geqslant 1$.

If a is a limit element of E of order $\geqslant \mathfrak{m}$, then obviously $\mathfrak{m} \leqslant \bar{\bar{E}}$ (since a neighbourhood of an element of E cannot contain more elements of E than the set E itself).[15]

The set M of cardinal numbers $\mathfrak{m}$ such that a is a limit element of E of order $\geqslant \mathfrak{m}$ contains a greatest number. For let $\mathfrak{n}$ be the smallest cardinal number such that a is not a limit element of order $\geqslant \mathfrak{n}$ (such cardinal numbers exist, for example, every cardinal number $> \bar{\bar{E}}$). Suppose there is no greatest number in the set M of all cardinal numbers $< \mathfrak{n}$; this implies that $\mathfrak{n} \geqslant \aleph_0$. Let V be a neighbourhood of a and $\mathfrak{m}$ a cardinal number $< \mathfrak{n}$. It follows from the definition of $\mathfrak{n}$ that a is a limit element of E of order $\geqslant \mathfrak{m}$; hence V contains at least $\mathfrak{m}$ elements of E. We have therefore

$$\overline{\overline{V \, . \, E}} \geqslant \mathfrak{m}$$

for every cardinal number $\mathfrak{m} < \mathfrak{n}$. Moreover, equality is impossible. For if

$$\overline{\overline{V \, . \, E}} = \mathfrak{m}_1 < \mathfrak{n},$$

since there is no greatest $\mathfrak{m} < \mathfrak{n}$, there exists a number $\mathfrak{m}_2$ such that $\mathfrak{m}_1 < \mathfrak{m}_2 < \mathfrak{n}$. Hence a is a limit element of E of order $\geqslant \mathfrak{m}_2$ and so

$$\overline{\overline{V \, . \, E}} \geqslant \mathfrak{m}_2 > \mathfrak{m}_1,$$

contrary to assumption. Consequently,

$$\overline{\overline{V \, . \, E}} \geqslant \mathfrak{n}.$$

Thus every neighbourhood V of a contains at least $\mathfrak{n}$ elements of E and so (since $\mathfrak{n} \geqslant \aleph_0$) a is a limit element of E of order $\geqslant \mathfrak{n}$, contrary to the definition of $\mathfrak{n}$.

Hence the set of cardinal numbers $< \mathfrak{n}$ contains a greatest number; denote it by $\mathfrak{m}_0$. It follows from the definition of $\mathfrak{n}$ that a is a limit element of E of order $\geqslant \mathfrak{m}_0$ but not of order $\geqslant \mathfrak{m}$ for any cardinal number $\mathfrak{m} > \mathfrak{m}_0$ (for in that case $\mathbf{m} \geqslant \mathbf{n}$); the number $\mathbf{m}_0$ is therefore the greatest cardinal number in the set M.

The element a is then said to be a *limit element of E of order* $\mathbf{m}_0$.

Thus every limit element of E is a limit element of a well defined order which is a cardinal number $\geqslant 1$ and $\leqslant \bar{\bar{E}}$.

An element $a \in E$ is a limit element of E of order $\mathbf{m}$ if and only if every neighbourhood of a contains at least $\mathbf{m}$ elements of E and there exists at least one neighbourhood of a which contains precisely $\mathbf{m}$ elements of the set E.

Limit elements of a set E of order $> \aleph_0$ (i.e. of non-countable order) are called *elements of condensation* of E. Hence a is an element of condensation of a set E (E non-countable) if and only if every neighbourhood of a contains a non-countable subset of E. This is Lindelöf's definition.

Fréchet calls an element a (belonging to E or not) an element of condensation of E if a is a limit element of every set obtained by removing from E any finite or countable subset.

The definitions of Lindelöf and Fréchet are equivalent. For let a be an element of condensation of the set E according to Lindelöf's definition, P a finite or countable subset of E, and V a neighbourhood of a, then V contains a non-countable subset of E and therefore also of $E - P$. Hence a is an element of condensation according to the definition of Fréchet.

Next suppose that a is not an element of condensation of E according to Lindelöf; there exists, therefore, a neighbourhood V of a such that the set $P = V \,.\, E$ is finite or countable. Hence the set $V(E - P) = 0$, and so a is not a limit element of $E - P$ and, therefore, not an element of condensation of E according to Fréchet's definition. The two definitions are therefore equivalent.

A set E is called $\mathbf{m}$-*compact* if it is finite or if every infinite subset of E possesses at least one limit element (belonging to E or not) of order $\geqslant \mathbf{m}$. A 1-compact set is called *compact*. Thus a set is compact if and only if it is finite or if every infinite subset of the set has a non-empty derived set.

Clearly a subset of an $\mathbf{m}$-compact set is $\mathbf{m}$-compact. Furthermore, *the sum of a finite number of* $\mathbf{m}$-*compact sets is* $\mathbf{m}$-*compact.*

For if $E = E_1 + E_2 + \ldots + E_n$, where each E_k ($k = 1, 2, \ldots, n$) is an $\mathbf{m}$-compact set, and if T is an infinite subset of E, then at least one (say $T \,.\, E_i$) of the sets $T \,.\, E_1, T \,.\, E_2, \ldots, T \,.\, E_n$ is infinite. Since $T \,.\, E_i \subset E_i$ and E_i is $\mathbf{m}$-compact, the set $T \,.\, E_i$ possesses a limit element of order $\geqslant \mathbf{m}$; hence T possesses a limit element of order $\geqslant \mathbf{m}$; consequently E is $\mathbf{m}$-compact.

17. Cantor's theorem.

THEOREM 29. *If*

$$E_1 \supset E_2 \supset E_3 \supset \ldots \tag{53}$$

is an infinite descending sequence[16] *of non-empty closed sets (contained in a*

(V)space) at least one of which is $\aleph_0$-compact, then the intersection of these sets is not empty.

Proof. Let E_p be the first $\aleph_0$-compact set in the sequence. Let a_n denote an element of E_n for $n = 1, 2, \ldots$, and let P be the set of all a_n for $n \geqslant p$. From (53), $P \subset E_p$. If P is finite, one of the elements of the sequence $a_p, a_{p+1}, \ldots$ is repeated from a certain stage onwards; this element is obviously contained in E_n for $n = 1, 2, \ldots$ and, therefore, in the intersection $\prod_1^\infty E_n$.

On the other hand, if P is infinite, it possesses a limit element, say a, of order $\geqslant \aleph_0$, since it is a subset of an $\aleph_0$-compact set. Let V be a neighbourhood of a; it therefore contains an infinite number of elements of P and, since $P \subset E_p \subset E_n$, $n < p$, V contains an infinite number of elements of E_n, $n \leqslant p$. But only a finite number of elements of P may be outside the set E_n for $n > p$; hence V contains an infinite number of elements of E_n for $n = 1, 2, \ldots$ Therefore $a \in E'_n$ and, since E_n is closed, $a \in E_n$ for $n = 1, 2, \ldots$; hence

$$a \in \prod_1^\infty E_n.$$

This proves the theorem.

The sequence of sets (53) in Theorem 29 cannot be replaced by a transfinite sequence (unless certain assumptions be made about the (V)space in which the sets are contained). For example, let K denote the set of all ordinal numbers of the first and second classes. K becomes a (V)space if the interior of any segment of ordinal numbers (i.e., all ordinal numbers ξ satisfying the relation $\alpha < \xi < \beta$) containing the number $\eta \neq 1$ is regarded as the neighbourhood of η, and if the neighbourhood of 1 is the set of all ordinal numbers $< \alpha$, where α is any ordinal such that $1 < \alpha < \Omega$. For any $\alpha < \Omega$, denote by E_α the set of all ordinals ξ such that $\alpha \leqslant \xi < \Omega$. It is easily seen that each of the sets E_α is closed and $\aleph_0$-compact, and the transfinite sequence $\{E_\alpha\}$, $\alpha < \Omega$, is descending but the intersection $\prod_{\alpha<\Omega} E_\alpha = 0$.

Theorem 29 may be easily generalized as follows:

THEOREM 30. *The infinite sequence*

$$E_1, E_2, E_3, \ldots \tag{54}$$

of closed sets, at least one of which is $\aleph_0$-compact, has a non-empty intersection if and only if the intersection of any finite aggregate of these sets is not empty.

That the condition of Theorem 30 is necessary is obvious. To prove its sufficiency consider the sets

$$F_n = \prod_1^n E_k,$$

where $n = 1, 2, \ldots$. The sets F_n form a descending sequence, they are non-empty and closed, and for sufficiently large n, as subsets of an $\aleph_0$-compact set, they are $\aleph_0$-compact. Hence, by Theorem 29, their intersection is not empty.

THEOREM 31 (Borel). *Let E denote a closed and $\aleph_0$-compact set contained in a (V)space K. If*

$$Q_1, Q_2, Q_3, \dots \tag{55}$$

is an infinite sequence of open sets such that

$$E \subset Q_1 + Q_2 + \dots, \tag{56}$$

then there exists a natural number N such that

$$E \subset Q_1 + Q_2 + \dots + Q_N. \tag{57}$$

(In other words, *a closed and $\aleph_0$-compact set which is covered by a countable aggregate of open sets can be covered by a finite number of these sets.*)

Proof. Put, for $n = 1, 2, \dots$,

$$Q_1 + Q_2 + \dots + Q_n = S_n, \qquad K - S_n = F_n, \qquad E \,.\, F_n = E_n.$$

Clearly $S_n \subset S_{n+1}$; hence $F_n \supset F_{n+1}$ and $E_n \supset E_{n+1}$, that is, $\{E_n\}$ is a descending sequence. The sets S_n are open (as sums of open sets); hence their complements F_n are closed and so are the sets E_n (as intersections of closed sets). Moreover, the sets E_n are also $\aleph_0$-compact (as subsets of E which is $\aleph_0$-compact). If none of the sets E_n were empty, then, by Theorem 29, their intersection P would be non-empty and so we would have

$$P \subset E_n \subset F_n = K - S_n, \qquad n = 1, 2, \dots;$$

consequently $P \,.\, S_n = 0$ and, since $Q_n \subset S_n$, $P \,.\, Q_n = 0$ for $n = 1, 2, \dots$. Hence, from (56), $P \,.\, E = 0$; this is impossible since $P \neq 0$ and $P \subset E_n \subset E$.

There exists, therefore, a natural number N such that $E_N = 0$ i.e. $E \,.\, F_N = 0$ and, since $F_N = K - S_N$ and $E \subset K$, we have $E \subset S_N$; this proves (57).

18. Topological limits of a sequence of sets. Let E_n $(n = 1, 2, \dots)$ be a given sequence of sets contained in a (V)space K. The *lower topological limit*, Lt inf E_n, of this sequence is the set of all elements p of K such that, for every neighbourhood V of p and for all sufficiently large values of n, the set $V \,.\, E_n$ is not empty.

The *upper topological limit*, Lt sup E_n, of the sequence $\{E_n\}$ is the set of all elements p of K such that, for every neighbourhood V of p, the set $V \,.\, E_n$ is non-empty for an infinite number of different values of n.

It is obvious that for every infinite sequence $\{E_n\}$ of sets $\subset K$ we have

$$\text{Lt inf } E_n \subset \text{Lt sup } E_n.$$

It also follows readily that for all sequences $\{A_n\}$ and $\{B_n\}$ of sets $\subset K$ we have

$$\begin{aligned}
&\text{Lt inf } A_n + \text{Lt inf } B_n \subset \text{Lt inf } (A_n + B_n),\\
&\text{Lt sup } A_n + \text{Lt sup } B_n = \text{Lt sup } (A_n + B_n),\\
&\text{Lt inf } (A_n \,.\, B_n) \subset (\text{Lt inf } A_n)(\text{Lt inf } B_n),\\
&\text{Lt sup } (A_n \,.\, B_n) \subset \text{Lt sup } A_n)(\text{Lt sup } B_n),\\
&\text{Lt inf } A_n - \text{Lt sup } B_n \subset \text{Lt inf } (A_n - B_n).
\end{aligned}$$

If for a given infinite sequence $\{E_n\}$ of sets $\subset K$

$$\text{Lt sup } E_n = \text{Lt inf } E_n,$$

then the sequence is said to be *topologically convergent* and the set

$$\text{Lt inf } E_n = \text{Lt sup } E_n$$

is called the *topological limit* of the sequence $\{E_n\}$; it is denoted by Lt E_n.

It is easily seen that, if the topological limits Lt A_n and Lt B_n exist, then Lt $(A_n + B_n)$ also exists and we have Lt $(A_n + B_n) =$ Lt $A_n +$ Lt B_n, but Lt $(A_n - B_n)$ need not exist. (For example, in linear space, if $A_n = \{1/(2n-1)\}$, $B_n = \{1/2n\}$ or $B_n = \{2^{\frac{1}{2}}/(2n-1)\}$, for $n = 1, 2, \ldots.$)

Similarly, even if Lt A_n and Lt B_n exist, Lt $(A_n \,.\, B_n)$ need not exist as seen from the above example.

CHAPTER II

TOPOLOGICAL SPACES

19. Topological spaces. In chapter I we investigated (V)spaces which were not subjected to any conditions. In this chapter we shall consider (V)spaces K whose neighbourhoods satisfy the following four conditions:

α. *Every element of K possesses at least one neighbourhood. Every element is contained in all its neighbourhoods.*

β. *If V_1 and V_2 are two neighbourhoods of an element a, there exists a neighbourhood V of a such that $V \subset V_1 \cdot V_2$.*

γ. *If b is an element of K different from a, there exists a neighbourhood V of a which does not contain b.*

δ. *If V is a neighbourhood of a and $b \in V$, there exists a neighbourhood W of b such that $W \subset V$.*

A space K which satisfies conditions α, β, γ, and δ is called a *topological space.* Conditions α and γ are, as can easily be seen, equivalent to the condition:

Every element of K is the intersection of all its neighbourhoods.

If for any set K we define the only neighbourhood of an element $a \in K$ to be the element a itself, then K becomes a topological space. Thus there exist topological spaces of any given cardinal number.

It follows from Theorem 5 that condition δ is equivalent to the condition that *every neighbourhood of an element of K is an open set.*

*Let K denote a given topological space. We obtain a topological space K_1, topologically equivalent to K, if each open set E of K be taken to be a neighbourhood in K_1 of every element contained in E. It follows from the above remark about condition δ that every neighbourhood of p in K is a neighbourhood of p in K_1. Conversely, if U is a neighbourhood of p in K_1 and hence an open set of K containing p, there exists a neighbourhood of p in K contained in U. Hence, by Theorem 1, K and K_1 are topologically equivalent.

Thus *every topological space K can be changed into a topologically equivalent space K_1 if every open set of K containing $p \in K$ be taken to be a neighbourhood of p in K_1.*

Let Φ denote the family of all open sets of a topological space K. The family Φ obviously satisfies the two conditions:

(a) If $p \in U_1 \in \Phi$, and $p \in U_2 \in \Phi$, then there exists a set $U_3 \in \Phi$ such that $p \in U_3 \subset U_1 . U_2$.

(b) If $p \in K, q \in K$, and $p \neq q$, there exists a set $U \in \Phi$ such that $p \in U$ and $q \notin U$.

Let K be any set containing more than one element and Φ a family of all its subsets which satisfy conditions (a) and (b). If every set of Φ be taken to be a neighbourhood of each element it contains, then K becomes a topological space. It can easily be proved that K satisfies conditions α, β, γ, and δ; consequently the sets of the family Φ are open sets of K.

But, as previously shown, every topological space K contains a family Φ of sets, namely, the family of all open sets $\subset K$, which satisfy conditions (a) and (b) and are such that, if they be chosen for neighbourhoods of every element they contain, we obtain a space which is topologically equivalent to K. Thus an investigation[1] of topological spaces is topologically equivalent to an investigation of spaces K in which we have defined a certain family Φ of sets which satisfy conditions (a) and (b).*

Condition δ leads to the following result:

$$\bar{E} = E + E' \qquad \text{for all } E \subset K. \tag{1}$$

For, let E be a given set $\subset K$, a a limit element of the set $T = E + E'$, and suppose that $a \notin T$. Let V be any neighbourhood of a; since $a \in T'$, there exists an element $b \neq a$ such that $b \in T . V$. If $b \in E$, then V contains an element of E different from a. If $b \notin E$, then $b \in E'$, since $b \in T = E + E'$. Condition δ implies the existence of a neighbourhood W of b such that $W \subset V$. Since $b \in E'$, there exists an element $c \in W . E$, where $c \neq a$, since $a \notin T$. Hence, in either case, there exists in V an element of E different from a; so $a \in E'$ contrary to the assumption that $a \notin T$. Consequently $a \in T$ and so $T' \subset T$, i.e., the set $E + E'$ is closed. Since $E \subset E + E'$, we obtain from the definition of $\bar{E}$ (§5)

$$\bar{E} \subset E + E'. \tag{2}$$

On the other hand, we have $E + E' \subset \bar{E}$ (§5); combining this result with (2), we obtain (1). We have therefore proved that condition δ implies relation (1).

Suppose now that K is a (V)space in which relation (1) is true. We may assume without changing the topology of K (§3) that every element of K is contained in all its neighbourhoods. We shall show that every neighbourhood V of $a \in K$ (where $a \in V$) contains an open set U which contains a.

Put

$$U = K - \bar{T}, \text{ where } T = K - V. \tag{3}$$

The set U is obviously open (as the complement of the closed set $\bar{T}$) and it contains a. For $a \in V$ implies $a \notin T$, and $T . V = 0$ gives $a \notin T'$. Hence from (1), $a \notin \bar{T}$, and so from (3), $a \in U$. Furthermore, $U \cdot \bar{T} = 0$ from (3); hence $U \cdot T = 0$, i.e., $U(K - V) = 0$, and so $U \subset V$.

Thus every neighbourhood of an element $a \in K$ contains an open set in which a is contained. On the other hand, by Theorem 5, every open set containing an element a contains a neighbourhood of that element. Hence, by Theorem 1, K is topologically equivalent to the space K_1 obtained from K by assigning to every element $a \in K$ as its neighbourhood in K_1 all open sets $\subset K$ and containing a.

Clearly the neighbourhoods in K_1 satisfy condition δ; we have therefore the result that a (V)space satisfying condition (1) is topologically equivalent to a (V)space satisfying condition δ. Since we have previously shown that condition δ implies (1), it follows that *in every* (V)*space conditions* (1) *and* δ *are equivalent.*

If K is a topological space and K_1 is a given subset of K, then K_1 becomes a topological space if every set $V \cdot K_1$, where V is a neighbourhood of $a \in K$, be taken to be a neighbourhood of $a \in K_1$. This means that *a subset of a topological space is a topological space.*

If $K_1 \subset K$, then obviously closed sets of K_1 are identical with the sets of K which are closed in K_1. The open sets of K_1 are identical with those sets of K whose complements with respect to K are closed in K_1.

LetK_1 and K_2 be two topological spaces. The set K of all ordered pairs $a = (a_1, a_2)$, where $a_1 \in K_1$ and $a_2 \in K_2$, is called the *cartesian* or *combinatorial product*, or simply the *product*, of the sets K_1 and K_2, and is denoted by $K_1 \times K_2$.

The product K will become a topological space if the set $V = V_1 \times V_2$ be taken as a neighbourhood of the element $a = (a_1, a_2)$, where V_1 is a neighbourhood of $a_1 \in K_1$ and V_2 is a neighbourhood of $a_2 \in K_2$. The proof is left to the reader.

The definition of a product of two topological spaces may be extended at once to a finite number of factors. We may also consider the product of an infinite sequence of topological spaces, i.e., the set $K = K_1 \times K_2 \times K_3 \times \ldots$, consisting of all arrangements $(a_1, a_2, \ldots)$, where $a_i \in K_i$ for $i = 1, 2, \ldots$. If the element $(a_1, a_2, \ldots) \in K$ be assigned as neighbourhoods the sets $V = V_1 \times V_2 \times \ldots$, where V_i is a neighbourhood of $a_i \in K_i$, then *the product of an infinite sequence of topological spaces becomes a topological space.*

20. Properties of derived sets. We shall deduce in this section a number of theorems which hold in topological spaces. None of these theorems is true in all (V)spaces.

THEOREM 32. *The derived set of a sum of two sets is the sum of the derived sets of these sets.*

Proof. Let E_1 and E_2 be two given sets contained in a topological space K.

Property (2) of derived sets gives

$$(4) \qquad E'_1 + E'_2 \subset (E_1 + E_2)'.$$

We suppose next that

$$(5) \qquad a \notin E'_1 + E'_2.$$

Hence $a \notin E'_1$ and $a \notin E'_2$; there exist, therefore, neighbourhoods V_1 and V_2 of a such that $V_1(E_1 - \{a\}) = 0$ and $V_2(E_2 - \{a\}) = 0$. By β, there exists a neighbourhood $V \subset V_1 . V_2$; hence $V(E_1 - \{a\}) = 0$ and $V(E_2 - \{a\}) = 0$, and so $V(E_1 + E_2 - \{a\}) = V(E_1 - \{a\}) + V(E_2 - \{a\}) = 0$. Consequently

$$(6) \qquad a \notin (E_1 + E_2)'.$$

Thus (5) implies (6) which gives

$$(E_1 + E_2)' \subset E'_1 + E'_2$$

and this, in combination with (4), gives the relation

$$(E_1 + E_2)' = E'_1 + E'_2.$$

Theorem 32 is therefore proved.

It follows by induction that Theorem 32 holds for any *finite* number of sets. It can, however, be easily shown that it need not be true for an infinite sequence of sets.

Relation (1) and Theorem 32 give, for $E_1 \subset K$ and $E_2 \subset K$,

$$\begin{aligned} \overline{E_1 + E_2} &= E_1 + E_2 + (E_1 + E_2)' = E_1 + E_2 + E'_1 + E'_2 \\ &= (E_1 + E'_1) + (E_2 + E'_2) = \bar{E}_1 + \bar{E}_2. \end{aligned}$$

We have therefore the following

COROLLARY.

$$\overline{E_1 + E_2} = \bar{E}_1 + \bar{E}_2$$

for all $E_1 \subset K$ *and all* $E_2 \subset K$, *i.e., the closure of the sum of two sets is the sum of the closures of the sets.* This result may be extended by induction to any finite number of sets.

All that can be said about the closure of the intersection of two sets is that it is contained in the intersection of their closures but the converse is not necessarily true. For example, if in linear space $A = \mathrm{E}_x[x < 0]$, and $B = \mathrm{E}_x[x > 0]$, then $A . B = 0$, $\overline{A . B} = 0$, but $\bar{A} . \bar{B} = \{0\} \neq 0$.

THEOREM 33. *The derived set of a finite set is empty.*

Proof. By Theorem 32 (generalized to a finite number of terms), it is sufficient to prove that the derived set of a set consisting of a single element is empty.

Hence let $E = \{a\}$ and let p be any element of K. If $p = a$, then no neighbourhood of p contains an element of E different from a; hence $p \notin E'$. If $p \neq a$, then, by γ, there exists a neighbourhood V of p such that $a \notin V$ and so, $V . E = 0$; consequently $p \notin E'$. Hence $E' = 0$ and this establishes Theorem 33.

From Theorems 32 and 33 we obtain

COROLLARY 1. *The derived set of a given set remains unchanged if a finite number of elements be removed from the set.*

The proof is left to the reader. We obtain furthermore

COROLLARY 2. *If a is a limit element of a set E, then every neighbourhood of a contains infinitely many elements of E.*

For suppose there is a neighbourhood V of a which contains only a finite number of elements of E. Put $V . E = E_1$, $E - E_1 = E_2$; the set E_1 is finite; therefore, by Corollary 1, $E' = E'_2$. But $V . E_2 = V(E - E_1) = V . E - V . E_1 = E_1 - E_1 = 0$; hence $a \notin E'_2$, contrary to assumption. This proves the corollary.

It follows from Corollary 2 that *every limit element of a set E is a limit element of E of order $\geqslant \aleph_0$* and that therefore *every compact set is $\aleph_0$-compact.* Hence, for topological spaces, we may replace in Theorems 29, 30, and 31, the expression *$\aleph_0$-compact* by *compact.*

From Theorem 33 (and the fact that the null set is a subset of every set) we obtain

COROLLARY 3. *Every finite set is closed or, in other words, every finite set is identical with its closure.*

COROLLARY 4. *Every connected set containing more than one element is dense-in-itself.*

To prove this, let E denote a connected set containing more than one element and let p be an element of E. Put $A = \{p\}$ and $B = E - A$; then $A' = 0$, by Theorem 33. If we had $A . B' = 0$, the sets A and B would be separated, contrary to assumption that E is connected. Consequently $A . B' \neq 0$; hence $p \in B'$ and so certainly $p \in E'$. This gives $E \subset E'$; hence E is dense-in-itself.

From Corollary 4 we conclude further that *a connected and closed set containing more than one element is perfect.*

A set $E_1 \subset E$ is said to be *compact in the set* if every infinite subset of E_1 possesses at least one limit element contained in E, in other words, if for every infinite subset $E_2 \subset E_1$ we have $E'_2 . E \neq 0$. In particular, a set E is said to be *compact-in-itself* if every infinite subset of E possesses at least one limit element which is contained in E. A set which is compact-in-itself is obviously compact. It need not, however, be closed as is illustrated by the following example.

Let K denote the set of all ordinal numbers $\leqslant \Omega$. For any ordinal α, $0 < \alpha \leqslant \Omega$, and any ordinal $\beta < \alpha$, we define a neighbourhood of α to be every set of ordinals ξ such that $\beta < \xi \leqslant \alpha$; the number 0 has only one neighbourhood, itself. It follows readily that K is a topological space and that the set E of all ordinal numbers $< \Omega$ is compact-in-itself but is not closed.

THEOREM 34. *The derived set of every set is closed.*

Proof. Let E be a given set contained in a topological space. Let $a \in (E')'$ and let V denote a neighbourhood of a. Hence V contains an element b of E'. By δ, there exists a neighbourhood W of b such that $W \subset V$. It follows from Corollary 2 to Theorem 33 that W and, therefore, also V, contains infinitely many elements of E; hence $a \in E'$. Consequently $(E')' \subset E'$ and so E' is closed.

21. Properties of families of closed sets. We shall show here that *the family of all closed sets of a* topological space determines the topology of that space (which is not so in the most general (V)space, see § **4**); in other words, *given the family Φ of all closed sets of a topological space K, the derived set E' of every set $E \subset K$ is determined.*

To this end we shall show that if Φ is the family of all closed sets of a topological space K and E any set $\subset K$, then E' is the set $f(E) \subset K$ consisting of all elements a of K such that conditions

$$F \in \Phi \text{ and } F \supset E - \{a\} \tag{7}$$

imply

$$a \in F. \tag{8}$$

To prove this suppose that $a \in E'$; hence $a \in (E - \{a\})'$. If, therefore, F satisfies conditions (7), then $F' \supset (E - \{a\})'$ and so $a \in F'$; but $F \in \Phi$ and so F is closed; hence $a \in F$. Therefore conditions (7) imply (8) and so, from the definition of the set $f(E)$, it follows that $a \in f(E)$.

Next, suppose that $a \in f(E)$. Hence conditions (7) imply (8); but conditions (7) are certainly satisfied by the set $F = \overline{E - \{a\}}$. We have therefore $a \in \overline{E - \{a\}}$ and so, from (1) § **19**,

$$a \in (E - \{a\}) + (E - \{a\})'$$

and since $a \notin E - \{a\}$, we must have $a \in (E - \{a\})'$; consequently $a \in E'$. Thus for every set $E \subset K$ we have $E' = f(E)$.

It follows from the above that *in a topological space K the function $\phi(E) = \bar{E}$* (defined for all $E \subset K$) *determines the topology of that space.* (This is not the case in general (V)spaces, see § 5.) For if the function $\phi(E) = \bar{E}$ is known for every $E \subset K$, the family of all closed sets contained in K can be defined as the family of all sets $E \subset K$ for which $\phi(E) = E$.

THEOREM 35. *The sum of a finite number of closed sets is a closed set.*

Proof. It is obviously sufficient to prove the theorem for two sets and then to generalize the result by induction. Hence let E_1 and E_2 be two closed sets. Therefore

$$E'_1 \subset E_1 \text{ and } E'_2 \subset E_2 \tag{9}$$

and so, by Theorem 32,

$$(E_1 + E_2)' = E'_1 + E'_2 \subset E_1 + E_2;$$

consequently, the set $E_1 + E_2$ is closed.

Passing to complements we obtain from Theorem 35

THEOREM 36. *The intersection of a finite number of open sets is an open set.*

Let Φ denote the family of all closed sets of a topological space K. We have shown that the family Φ possesses the following three properties:

1. *The intersection of any aggregate of sets of* Φ *belongs to* Φ.
2. *The sum of every finite number of sets of* Φ *belongs to* Φ.
3. *The null set, the set consisting of a single element of* K*, and the set* K *itself belong to* Φ.

THEOREM 37. *If* K *is any set and* Φ *a family of subsets of* K *which satisfy conditions* 1, 2, *and* 3, *then, taking for neighbourhoods of* $a \in K$ *all sets* V *of* K *containing* a *and whose complements are sets of* Φ, *we obtain a topological space in which the family of all closed sets is identical with the family* Φ.

Proof. Let K be a given set and Φ a family of its subsets which satisfy conditions 1, 2, and 3. Let a be any element of K; then $V \subset K$ will be a neighbourhood of a if $a \in V$ and $K - V \in \Phi$. It will be shown that the neighbourhoods of K so defined satisfy conditions $\alpha, \beta, \gamma, \delta$.

The null set is contained in Φ, by condition 3; hence $V = K$ satisfies the condition $K - V \in \Phi$ and $a \in V$ for every element of K. Thus every element of K has at least one neighbourhood and, from the definition adopted for neighbourhoods in K, it follows that every element is contained in all its neighbourhoods. Condition α is therefore satisfied.

If V_1 and V_2 are two neighbourhoods of a, then, by the definition of neighbourhoods in K, $a \in V_1$, $a \in V_2$, $K - V_1 \in \Phi$, and $K - V_2 \in \Phi$; but from condition 2, $K - V_1 . V_2 = (K - V_1) + (K - V_2) \in \Phi$. Hence $V_1 . V_2$ is a neighbourhood of a. Thus the intersection of any two neighbourhoods of a is a neighbourhood of a (for all $a \in K$); consequently condition β is satisfied.

Let a and b be two different elements of K. The set $\{b\} \in \Phi$, by condition 3; hence $V = K - \{b\}$ is a neighbourhood of a which does not contain b. Condition γ is therefore satisfied.

Finally, let V be a neighbourhood of a and $b \in V$. Hence $K - V \in \Phi$. But $b \in V$; therefore, from the definition of neighbourhoods in K, V is a neighbourhood of b. This proves that condition δ is satisfied in K.

We have thus proved that the set K, with neighbourhoods as defined above, is a topological space. We shall now show that the family of all closed sets of K is identical with the family Φ.

For let E be a given set of Φ. If $E = K$, then (§ 5) E is closed. If $E \neq K$, then there exists an element $a \in K - E$ and so, since $E \in \Phi$, it follows from the definition of neighbourhoods that $K - E$ is a neighbourhood of a; hence, by δ, $K - E$ is open and therefore E is closed.

On the other hand, let E denote any closed set contained in K. The set $K - E$ is open and so, by Theorem 5, there exists for every element $a \in K - E$, a neighbourhood $V(a) \subset K - E$; this gives

$$K - E = \sum_{a \in K-E} V(a)$$

and so

$$E = K - \sum_{a \in K-E} V(a) = \prod_{a \in K-E} (K - V(a)).$$

But $K - V(a) \in \Phi$, by the definition of neighbourhoods of K; hence E is an intersection of sets of the family Φ and so, by 1, $E \in \Phi$.

This completes the proof of Theorem 37. This theorem implies that every property of the family Φ of all closed sets which holds in every topological space is a consequence of conditions[2] 1, 2, and 3.

22. Properties of closure. The closure of a set contained in a topological space K was shown to possess the following three properties:

I. $\overline{E_1 + E_2} = \bar{E}_1 + \bar{E}_2$ *for all* $E_1 \subset K$ *and all* $E_2 \subset K$.

II. $\bar{E} = E$, *if E is the null set or if E consists of a single element.*

III. $\bar{\bar{E}} = \bar{E}$ *for all* $E \subset K$.

Let K denote any given set and suppose that each set $E \subset K$ is associated with a certain set $\phi(E) \subset K$, where $\phi(E)$ satisfies the following three conditions:

(I) $\phi(E_1 + E_2) = \phi(E_1) + \phi(E_2)$, $E_1 \subset K, E_2 \subset K$.

(II) $\phi(E) = E$, if E is the null set or consists of a single element.

(III) $\phi(\phi(E)) = \phi(E)$, $E \subset K$.

If $E_1 \subset E_2 \subset K$, then $E_2 = E_1 + E_2$ and so, from (I),

$$\phi(E_2) = \phi(E_1) + \phi(E_2) \supset \phi(E_1).$$

Hence condition (I) leads to

$$\text{(IV)} \qquad \phi(E_1) \subset \phi(E_2), \qquad E_1 \subset E_2 \subset K.$$

Furthermore, if $E \subset K$ and $p \in E$, then, from (II) and (IV), $p = \phi(\{p\}) \subset \phi(E)$ and so

$$\text{(V)} \qquad E \subset \phi(E) \qquad \text{for all } E \subset K.$$

In particular, $K \subset \phi(K)$ and, since $\phi(E) \subset K$ for all $E \subset K$, $\phi(K) \subset K$ and so

$$\text{(VI)} \qquad \phi(K) = K.$$

Next, denote by Φ the family of sets $\phi(E)$ for all $E \subset K$; this will be a certain family of subsets of K. We shall show that the family Φ satisfies conditions 1, 2, and 3 of Theorem 37.

Let Φ_1 be a subset of Φ and put $P = \prod_{T \epsilon \Phi_1} T$. Corresponding to every $T \in \Phi_1$ there exists, by the definition of the family $\Phi \supset \Phi_1$, a set $E \subset K$ such that $\phi(E) = T$. Let Φ_2 denote the family of all sets $E \subset K$ such that $\phi(E) \in \Phi_1$; hence $P = \prod_{E \epsilon \Phi_2} \phi(E)$ and so $P \subset \phi(E)$ for all $E \in \Phi_2$. But, from (IV) and (III), $\phi(P) \subset \phi(\phi(E)) = \phi(E)$ for all $E \in \Phi_2$; hence $\phi(P) \subset P$. However, from (V), $P \subset \phi(P)$; consequently $P = \phi(P)$ and so $P \in \Phi$. The family Φ satisfies, therefore, condition 1.

If $T_1 \in \Phi$ and $T_2 \in \Phi$, then there exist sets $E_1 \subset K$ and $E_2 \subset K$ such that $T_1 = \phi(E_1)$, $T_2 = \phi(E_2)$ and so, by (I), $T_1 + T_2 = \phi(E_1) + \phi(E_2) = \phi(E_1 + E_2)$; since $E_1 + E_2 \subset K$, it follows that $T_1 + T_2 \in \Phi$. Thus the sum of two sets of Φ belongs to Φ; consequently, by induction, condition 2 is satisfied by Φ. That Φ satisfies condition 3 is evident from conditions (II) and (VI) of the function ϕ.

The family Φ of subsets of K therefore satisfies conditions 1, 2, and 3. Thus if we define neighbourhoods of elements of K to be the sets V such that $a \in V$ and $K - V \in \Phi$ then, by Theorem 37, we obtain a topological space in which all closed sets are identical with the sets of the family Φ.

Let now E denote any set $\subset K$. From (V), $E \subset \phi(E)$ and, since $\phi(E) \in \Phi$, $\phi(E)$ is closed; hence $\bar{E} \subset \phi(E)$. On the other hand, since $\bar{E}$ is closed, $\bar{E} \in \Phi$ and so there exists a set $X \subset K$ such that $\bar{E} = \phi(X)$; since $E \subset \bar{E}$ we obtain, from (IV) and (III), $\phi(E) \subset \phi(\bar{E}) = \phi(\phi(X)) = \phi(X) = \bar{E}$.

We have therefore proved

THEOREM 38. *If a function ϕ associates with each subset E of a set K a set $\phi(E) \subset K$ subject to conditions* (I), (II), *and* (III), *and if neighbourhoods of a are taken to be all sets V containing a which are such that for some set $X \subset K$*

we have $K - V = \phi(X)$, *then K becomes a topological space in which*

$$\phi(E) = \bar{E} \qquad \textit{for all } E \subset K.$$

It follows from Theorem 38 (and from the fact that conditions I, II, and III, are satisfied in every topological space) that *the investigation of topological spaces is equivalent to the investigation of spaces K in which to every set* $E \subset K$ *corresponds a set* $\bar{E} \subset K$ *subject to conditions* I, II, *and* III.

Taking conditions $\alpha, \beta, \gamma, \delta$ as a starting point, we obtain an axiomatic basis for topological spaces founded on the concept of neighbourhood (following Fréchet and Hausdorff).[3] On the other hand, assuming conditions I, II, and III, we obtain an axiomatic basis for topological spaces founded on the concept of closure (Kuratowski[4] and later Hopf and Alexandroff[5]). Hausdorff also investigated spaces which satisfy conditions I and II and called them "Gestufte Räume."[6]

Theorem 38 establishes the fact that all properties of the closure of a set which hold in every topological space result from the conditions I, II, and III.

In **§21** we have shown that in a topological space the function $\phi(E) = \bar{E}$ (the closure) defines the function $f(E) = E'$ (the derived set). We shall show that in a topological space K,

$$E' = \underset{p}{K}[p \in \overline{E - \{p\}}] \qquad \text{for all } E \subset K. \tag{10}$$

For assume that $p \in E'$; then $p \in (E - \{p\})'$, and so $p \in \overline{E - \{p\}}$. On the other hand, if

$$p \in \overline{E - \{p\}},$$

then, putting $E_1 = E - \{p\}$, we have $p \in \bar{E}_1 = E_1 + E'_1$ and, since $p \notin E_1$, we must have $p \in E'_1 \subset E'$. This proves relation (10).

23. Examples of topological spaces.

1. Let K denote a topological space and K_1 a given subset of K. The set K_1 becomes a topological space if in K_1 the closure $\phi(E)$ of each set $E \subset K_1$ be defined by the relation

$$\phi(E) = \bar{E} \,.\, K_1,$$

where $\bar{E}$ is the closure of E in K. We show that the function ϕ as defined above satisfies (I), (II), and (III). The first two conditions follow immediately. Also, from the definition of the function ϕ,

$$\phi(\phi(E)) = \overline{\bar{E} \,.\, K_1} \,.\, K_1 \qquad \text{for all } E \subset K_1.$$

But, for $E \subset K_1$, we have $E \subset \bar{E} \,.\, K_1$; hence

$$\bar{E} \subset \overline{\bar{E} \,.\, K_1} \subset \bar{\bar{E}} = \bar{E}$$

and so

$$\overline{\bar{E} \,.\, K_1} = \bar{E}.$$

Thus $\phi(\phi(E)) = \bar{E} \,.\, K_1 = \phi(E)$ which proves (III).

2. Let K be any set. K becomes a topological space if we define

$$\bar{E} = E \qquad \text{for all } E \subset K.$$

For it is obvious that the function $\phi(E) = \bar{E}$, as defined above, satisfies the conditions (I), (II), and (III).

It follows from the above that *there exist topological spaces of any given cardinal* (cf. § **19**). It can be shown that a given infinite set of cardinal $\mathfrak{m}$ gives rise to $2^{2^{\mathfrak{m}}}$ different topological spaces.

It is easy to see that the topological space K in example 2 has the property that each of its subsets is both closed and open; hence K is isolated. As a consequence, every mapping of K into itself is continuous. Thus, if K has cardinal $\mathfrak{m}$, it contains $2^{\mathfrak{m}}$ closed sets (and the same number of open sets) and every two subsets of K of the same power are homeomorphic. And if the cardinal of K is $\aleph_\alpha$, then K contains $\aleph_\alpha + \bar{\alpha}$ sets,[7] no two of which are homeomorphic.

3. Let K be a given infinite set of cardinal $\mathfrak{m}$. Let the function $\bar{E}$ for all $E \subset K$ be defined by the conditions:

(i) $\bar{E} = E$, if E is finite (or empty),

(ii) $\bar{E} = K$, if E is infinite.

It is easily verified that the function $\bar{E}$ satisfies the conditions (I), (II), and (III); consequently K is a topological space with closure in K defined as above.

It follows from conditions (i) and (ii) that $E = \bar{E}$ for $E \subset K$ only when E is finite or empty or $E = K$. These are, therefore, the only closed sets of K and there are evidently $\mathfrak{m}$ of them. Moreover, every (1,1) mapping of K on itself transforms closed sets into closed sets from which it follows readily that the mapping is homeomorphic. Furthermore, if K is an infinite set, there are $2^{\mathfrak{m}}$ such (1,1) mappings of K on itself. Hence there are $2^{\mathfrak{m}}$ homeomorphic transformations of K into itself.

Consider sets closed in a subset K_1 of K. The sets closed in K_1 will obviously be all finite sets, the null set, and the subset K_1 itself. Consequently, every two subsets of K_1 of the same power are homeomorphic. It can also be shown that every infinite subset of K is connected.

A space which is the closure of a certain finite or countable subset of itself is called *separable*. If every subset of a space K is contained in the closure of a certain finite or countable subset of itself, then K is called *hereditarily separable*.

The space K of this example is hereditarily separable, since every set $E \subset K$ contains a finite or countable subset P such that $E \subset \bar{P}$. For if E is finite, for $P = E$ we have $E = \bar{P}$; if E is infinite, then, denoting by P any countable subset of E, we have $\bar{P} = K$ and so $E \subset \bar{P}$. Hence there exist topological and hereditarily separable spaces of any given cardinal.

4. Let K denote a topological space and let the closure of a set $E \subset K$ be denoted by $\phi(E)$. Let $P = \{p_1, p_2, \ldots\}$ denote an infinite sequence of distinct elements not belonging to K. We define the closure $\bar{E}$ of a set $E \subset K_1 = K + P$ as follows:

If $E \subset K_1$ and E contains a finite number (> 0) of elements of P we put

$$\bar{E} = \phi(K \,.\, E) + P \,.\, E.$$

Hence, in particular, $\bar{E} = \phi(E)$ for all $E \subset K$. If, on the other hand, E contains an infinity of elements of P, we put

$$\bar{E} = K_1.$$

It can easily be shown that the closure as defined in K_1 satisfies the conditions (I), (II), and (III). K_1 is therefore a topological space.

It is evident from the definition of closure in K_1 that, in particular,

$$\bar{P} = K_1.$$

Hence K_1 is separable. The above example shows that *every topological space K is a subset of a separable topological space obtained from K by adjoining a countable set.*

If K is a non-countable set in example 2 then $\bar{P} = K$ is impossible for any finite or countable subset $P \subset K$ (for $\bar{P} = P$ and P has cardinal $\leq \aleph_0$ whereas K is non-countable). But, as shown above, K is a subset of a certain topological and separable space K_1; hence K_1 is not hereditarily separable (cf. example 3). There exist, therefore, separable topological spaces which are not hereditarily separable.

5. Let K denote any topological space in which the closure of a set E is denoted by $\phi(E)$. Let a be a given element not belonging to K. We define the closure $\bar{E}$ in the set $K_1 = K + \{a\}$ as follows:

If E is finite, we put $\bar{E} = E$; if E is an infinite subset of K_1, we put

$$\bar{E} = \phi(E - \{a\}) + \{a\}.$$

As a consequence of the definition of closure in K_1, we have

$$\phi(E) = \bar{E} \,.\, K \qquad \text{for all } E \subset K$$

(which is consistent with the definition of closure in a subspace).

We see that the closure in K_1, as defined above, satisfies conditions (I), (II), and (III); K_1 is therefore a topological space. We shall now show that K_1 is compact.

Let E denote an infinite subset of K_1. Put $E_1 = E - \{a\}$; hence E_1 is infinite and so, by the definition of closure in K_1, $a \in \bar{E}_1$. But $\bar{E}_1 = E_1 + E'_1$ and $a \notin E_1$; hence $a \in E'_1 \subset E'$; consequently $E' \neq 0$. The space K_1 is therefore compact.

Thus *every topological space K is a subset of a compact topological space obtained from K by adjoining a single element.*

24. Properties of relatively closed sets.

THEOREM 39. *A subset of $E \subset K$ (K a topological space) which is closed in E is the intersection of E and a closed set of K and conversely.*

Proof. If $E_1 \subset E$ and E_1 is closed in E, then $E \, . \, E'_1 \subset E_1$ (§ **8**); so $E_1 = (E_1 + E'_1)E = \bar{E}_1 \, . \, E$. Hence E_1 is the intersection of E and the closed set $\bar{E}_1$.

On the other hand, if $E_1 = F \, . \, E$, where F is closed, then $E_1 \subset F$ and so $E'_1 \subset F$; therefore $E'_1 \, . \, E \subset F \, . \, E = E_1$. Hence E_1 is closed in E and this proves the theorem. In the general (V)space only the second part of Theorem 39 holds.

25. Homeomorphism in topological spaces. By Theorem 27, two (V)spaces K and K_1 are homeomorphic if and only if there exists a (1,1) mapping f of K on K_1 such that

$$(f(E))' = f(E') \qquad \text{for all } E \subset K. \tag{11}$$

We now show that *if K and K_1 are topological spaces* (11) *is equivalent to*

$$\overline{f(E)} = f(\bar{E}) \qquad \textit{for all } E \subset K. \tag{12}$$

Proof. If we assume (11), then (1) gives

$$\overline{f(E)} = f(E) + (f(E))' = f(E) + f(E') = f(E + E') = f(\bar{E})$$

which proves (12).

Assume that (12) is true. Take any $b \in f(E')$. We then have $b = f(a)$ for some element $a \in E'$. Put $E_1 = E - \{a\}$. Since $E'_1 = E'$, we see that $a \in E'_1 \subset \bar{E}_1$. Hence $b = f(a) \in f(\bar{E}_1)$ and so, by (12),

$$b \in \overline{f(E_1)} = f(E_1) + (f(E_1))'.$$

Since f is (1,1) in K, and $a \notin E_1$, we see that $b = f(a) \notin f(E_1)$. Thus we have $b \in (f(E_1))'$ and, since $E_1 \subset E$, $b \in (f(E))'$. Therefore relation (12) implies that $f(E') \subset (f(E))'$.

Suppose next that $b \in (f(E))'$. Since $b \in K_1 = f(K)$, there exists some element $a \in K$ such that $b = f(a)$. Put $E_1 = E - \{a\}$. Since f is (1,1) in K, $f(E_1) = f(E) - \{b\}$ and we see that $b \notin f(E_1)$. Thus, since $(f(E_1))' = (f(E))'$, we have

$$b \in (f(E_1))' \subset \overline{f(E_1)}.$$

By (12), $b \in f(\bar{E}_1) = f(E_1 + E'_1) = f(E_1) + f(E'_1)$. But $b \notin f(E_1)$; so we know that $b \in f(E'_1) \subset f(E')$. Therefore (12) implies that $(f(E))' \subset f(E')$. Hence $(f(E))' = f(E')$ and relations (11) and (12) have been shown to be equivalent. We have therefore proved

THEOREM 40. *Two topological spaces K and K_1 are homeomorphic if and only if there exists a* (1,1) *mapping of K on K_1 such that the closure of the image of any set contained in K is the image of the closure of this set.* In other words, *the closures of corresponding sets under a* (1,1) *transformation are corresponding sets.*

It follows from Theorem 40 that if two topological spaces are homeomorphic it is possible to establish a (1,1) correspondence between their elements such that the image of a closed subset of either space is always a closed subset of the other. This is true in every (V)space—see remark to Theorem 27, § **14**. We now prove the converse, which need not hold in the general (V)space.

Suppose that f establishes a (1,1) transformation of the topological space K on the topological space K_1 in such a manner that if E is a closed subset of K then $f(E)$ is a closed subset of K_1 and if T is a closed subset of K_1 then $f^{-1}(T)$ is a closed subset of K. Hence if E_0 be any subset of K, then each closed set E containing E_0 is mapped into a closed subset of K_1 containing $f(E_0)$ and the function f^{-1} transforms every closed set containing $f(E_0)$ into a closed set containing E_0. Since f is (1,1) in K, it transforms the intersection of all closed sets (contained in K) which contain E_0 into the intersection of all closed sets (contained in K_1) which contain $f(E_0)$ i.e., it transforms the set $\bar{E}_0$ into the set $\overline{f(E_0)}$. We have therefore

$$f(\bar{E}) = \overline{f(E)}$$

for all $E \subset K$; hence, by Theorem 40, K and K_1 are homeomorphic. We thus obtain the

COROLLARY. *Two topological spaces K and K_1 are homeomorphic if and only if there exists a* (1,1) *correspondence between their elements such that closed sets of K correspond to closed sets of K_1 and vice versa.*

It should be noted, however, that in order that two topological spaces K and K_1 be homeomorphic it is not sufficient that there exist between their elements a (1,1) correspondence under which closed sets of K correspond to closed sets of K_1. The theorem fails if closed sets of K_1 map on sets of K which are not closed.

For example, if K denotes the set of all real numbers x, where $0 \leqslant x \leqslant 1$, and K_1 is the set consisting of all numbers x, where $0 \leqslant x < 1$, and of the number 2, then the sets K and K_1, as subsets of the set of all real numbers, are topological spaces. Also, the function $f(x) = x$ for $0 \leqslant x < 1$ and $f(1) = 2$ transforms every closed subset of K into a closed subset of K_1; but, since K is connected and K_1 is not, the spaces are not homeomorphic. This is because the inverse function transforms the set $0 \leqslant x < 1$, closed in K_1, into a set not closed in K.

Note that in Theorem 40 it is sufficient to state the condition in one direction only because the condition in the other direction follows immediately. In connection with the corollary, we may add that the expression "closed" may be replaced everywhere by the expression "open."

*Furthermore, an example can be given of two topological spaces which are not homeomorphic although there exists a (1,1) mapping of each on the other such that the image of a set closed in one space is a set closed in the other space. Consider, for example, the sets E_1 and E_2 constructed by Kuratowski and discussed in § **14**, in which the function g^{-1} transforms closed sets of E_1 into closed sets of E_2 and the function f^{-1} transforms closed sets of E_2 into closed sets of E_1.*

26. The border of a set. Nowhere-dense sets. If E is a set contained in a topological space K, then the set

$$(13) \qquad B(E) = E\overline{(K - E)}$$

is called the *border* of E and the elements of $B(E)$ are called *border elements* of E.

Given the function $B(E)$ for all $E \subset K$, the function $\bar{E}$ is determined because $\bar{E} = E + B(K - E)$ for all $E \subset K$, as can easily be verified; hence the topology of the space is determined.

If every element of E is a border element of E, E is called a *border set*. A set $E \subset K$ is a border set if and only if

$$E = B(E)$$

or else from (13), if and only if

$$(14) \qquad E \subset \overline{K - E}.$$

Since

$$K - E \subset \overline{K - E},$$

(14) gives

$$K = E + (K - E) \subset \overline{K - E},$$

and, since

$$\overline{K - E} \subset K,$$

we obtain

$$(15) \qquad \overline{K - E} = K.$$

On the other hand, since $E \subset K$, (15) gives (14). Relations (14) and (15) are therefore equivalent. Hence a set $E \subset K$ is a border set if and only if (15) holds.

In § 6 we have deduced the relation

$$I(E) = E - \overline{K - E}$$

for the interior $I(E)$ of a set E (see (24)). As a consequence, condition (15) is equivalent to the condition

$$I(E) = 0.$$

Hence a set E is a border set if and only if its interior is empty.

If $E_1 \subset E$, then

$$I(E_1) \subset I(E)$$

and so, if E is a border set, then

$$I(E) = 0;$$

hence

$$I(E_1) = 0$$

and therefore E_1 is a border set. Thus *a subset of a border set is a border set.*

It follows from (13) that for every $E \subset K$ we have

$$B(B(E)) = B(E) . \overline{K - B(E)} = E . \overline{K - E}(\overline{K - E . \overline{K - E}}). \tag{16}$$

But

$$E . \overline{K - E} \subset E;$$

therefore

$$\overline{K - E} \subset \overline{K - E . \overline{K - E}}$$

and so (16) gives

$$B(B(E)) = E . \overline{K - E} = B(E).$$

Hence *the border of every set is a border set.*

A set whose closure is a border set is called *nowhere dense.* Hence a set $E \subset K$ is nowhere dense if and only if

$$\bar{E} \subset \overline{K - \bar{E}}, \tag{17}$$

which, as we have seen (from the equivalence of (14) and (15)), is equivalent to

$$\overline{K - \bar{E}} = K, \tag{18}$$

and also to

$$E \subset \overline{K - \bar{E}},$$

since $E \subset \bar{E}$ and $\overline{K - \bar{E}}$ is closed.

If $E_1 \subset E$ then

$$\overline{K - \bar{E}} \subset \overline{K - \bar{E}_1}$$

and so, if E is nowhere dense, then from (18)

$$E \subset \overline{K - \bar{E}};$$

hence

$$E_1 \subset \overline{K - \bar{E}_1}.$$

Consequently, E_1 is nowhere dense. Thus *a subset of a set which is nowhere dense is nowhere dense.* In other words, *the property of being nowhere dense is a hereditary property of sets.*

Let E_1 and E_2 be two nowhere-dense sets $\subset K$. We have therefore

(19) $$K = \overline{K - \bar{E}_1} = \overline{K - \bar{E}_2}.$$

Now

$$K - \overline{E_1 + E_2} \subset \overline{K - \overline{E_1 + E_2}} \qquad \text{(since } E \subset \bar{E}\text{)}$$

and so

$$K - (\bar{E}_1 + \bar{E}_2) \subset \overline{K - \overline{E_1 + E_2}}.$$

But

$$K - \bar{E}_1 \subset K - (\bar{E}_1 + \bar{E}_2) + \bar{E}_2 = K - \overline{(E_1 + E_2)} + \bar{E}_2;$$

hence (19) gives

$$K = \overline{K - \bar{E}_1} \subset \overline{K - \overline{(E_1 + E_2)} + \bar{E}_2} = \overline{K - \overline{(E_1 + E_2)}} + \bar{E}_2,$$

and so

$$K - \bar{E}_2 \subset \overline{K - \overline{(E_1 + E_2)}};$$

consequently

$$K = \overline{K - \bar{E}_2} \subset \overline{K - \overline{(E_1 + E_2)}}.$$

This proves that the set $E_1 + E_2$ is nowhere dense. Hence *the sum of two* (and therefore of any finite number) *nowhere-dense sets is nowhere dense.*

If E is a border set then

$$K = \overline{K - E};$$

if, in addition, E is closed, then $E = \bar{E}$ and so

$$K = \overline{K - \bar{E}}.$$

Hence *a closed border set is nowhere dense.*

It follows from (13) that *the border of a closed set is closed* and, since it is a border set, *it is nowhere dense;* and from (18) that *the closure of a set which is nowhere dense is nowhere dense.*

We have shown previously that a set is a border set if and only if its interior is the null set. Hence a set E is nowhere dense if and only if $I(\bar{E}) = 0$; in other words, if and only if $\bar{E}$ contains no non-empty open set. But if $\bar{E}$ contains no non-empty open set, then any open set $U \subset K$ contains a non-empty open set free of elements of E. To see this, note that $U - \bar{E} \neq 0$ and that $V = U - \bar{E}$ is an open set $\subset U$ and has no elements in common with E. Conversely, if every open set $U \subset K$ contains a non-empty open set V such that $V . E = 0$, then $E \subset K - V$ and $\bar{E} \subset K - V$ since $K - V$ is closed. Hence $U - \bar{E} \supset U - (K - V) = V \neq 0$ and therefore $\bar{E}$ contains no non-empty open set. Consequently, *a set E is nowhere dense if and only if every open non-empty set contains an open non-empty set which has no elements in common with E.*

A set T is said to be *dense on a set* E if

$$E \subset \overline{T \, . \, E}.$$

If S is contained in T and is dense on T, and if T is contained in E and is dense on E, then S is dense on E. For if

$$S \subset T \subset \overline{S \, . \, T} \text{ and } T \subset E \subset \overline{T \, . \, E},$$

then $T \subset \bar{S}$ and $E \subset \bar{T}$ and so $E \subset \bar{T} \subset \bar{\bar{S}}$. Since $S \subset E$, we have $S = S \, . \, E$; hence

$$S \subset E \subset \bar{S} = \overline{S \, . \, E}$$

and consequently S is dense on E. If, however, the set S is dense on T and T is dense on E, then S is not necessarily dense on E (for example, if S is the set of all rational numbers, T the set of all real numbers, and E the set of all irrational numbers).

In consequence of the property of nowhere-dense sets previously obtained a nowhere-dense set cannot be dense on any non-empty open set. The converse is also true. For if a set E is not dense on a non-empty open set V, then the set

$$U = V - \overline{V \, . \, E} \neq 0;$$

but

$$U = V(K - \overline{V \, . \, E})$$

is the intersection of two open sets; hence U is open and contained in V and, since

$$U \subset K - \overline{V \, . \, E}$$

and $U \subset V$, it follows that $U \, . \, E = 0$ and E is nowhere dense on V. Thus, if a set E is not nowhere dense, there exists an open set on which E is dense.

*For nowhere-dense linear sets we have the following theorem by Scheeffer:[8]

*If E is a nowhere-dense linear set and P a linear set, at most countable, then for a and $b > a$ any two real numbers, there exists a real number c in $a < c < b$ such that the set obtained from P by a translation of length c has no elements in common with E.**

Problems

1. Prove that every infinite connected set S contained in a topological space possesses an infinite connected proper subset.

Solution. Let $p \in S$; if the set $S - \{p\}$ is connected, then it is the required subset of S. If, on the other hand, $S - \{p\}$ is the sum of two separated sets A and B, then at least one of the sets $A + \{p\}$ and $B + \{p\}$ is infinite and neither of them is equal to S. It will be shown that each of these sets is connected. For if $A + \{p\}$ were the sum of two separated sets A_1 and B_1

and $p \in A_1$, then $p \notin B_1$, and so from $A + \{p\} = A_1 + B_1$ we would have $B_1 \subset A$; hence $B'_1 \subset A'$. Since A and B as well as A_1 and B_1 are separated sets, we have

$$(A_1 + B)B_1 \subset A_1 . B_1 + A . B = 0,$$

$$(A_1 + B)' . B_1 \subset A'_1 . B_1 + B' . B_1 \subset A'_1 . B_1 + B' . A = 0,$$

$$(A_1 + B)B'_1 = A_1 . B'_1 + B . B'_1 = B . B'_1 \subset B . A' = 0.$$

Hence the sets $A_1 + B$ and B_1 are separated; but this is impossible since $A_1 + B + B_1 = A_1 + B_1 + B = A + \{p\} + B = S$ a connected set. Similarly, it can be shown that the set $B + \{p\}$ is connected. That one of the two sets which is infinite is a proper subset of S with the required property.

2. Give an example of an infinite connected (V)space which has no infinite proper subset which is connected.

Solution. Let $K = \{p_1, p_2, \ldots\}$ where the element p_1 has only one neighbourhood $\{p\}$ but, for $n > 1$, each of the $n - 1$ sets $\{p_n, p_k\}$, $k = 1, 2, \ldots, n - 1$, is a neighbourhood of p_n. It is easily verified that K has the required properties.

3. Give an example of a connected (V)space consisting of three elements no subset of which consisting of two elements is connected.

Solution. Such a space is $K = \{a, b, c\}$, where a, b, c, have, respectively, the neighbourhoods $\{a, b\}$ and $\{a, c\}$, $\{a, b\}$ and $\{b, c\}$, $\{a, c\}$ and $\{b, c\}$.

4. Give an example of a connected topological space which becomes disconnected when any one of the elements is removed.

Answer. The straight line.

5. Prove that every infinite connected set contained in a topological space is the sum of two infinite connected proper subsets.[9]

6. Prove that, if S and T are two connected sets contained in a topological space, where $S \supset T$ and $S - T$ is the sum of two separated sets A and B, then the sets $T + A$ and $T + B$ are connected.[10]

7. Show that, if in a topological space $P + Q$ and $P . Q$ are connected and P and Q are closed, then the sets P and Q are each connected.[11]

Proof. Let P and Q be two closed sets such that $P + Q$ and $P . Q$ are connected, and suppose that P is not connected; hence it is the sum of two non-empty disjoint closed sets A and B (by Theorem 14). The sets A and B are separated and, since $P . Q \subset A + B$ and is connected, it must be contained in one of the sets A or B (by Theorem 16). Suppose $P . Q \subset A$; hence $P . Q . B = 0$ and $Q . B = Q(A + B)B = Q . P . B = 0$. But $P + Q = A + B + Q = (A + Q) + B$, where $A + Q$ and B are two disjoint, non-empty, closed sets contrary to the assumption that $P + Q$ is connected. Consequently P is connected. It is proved similarly that Q is connected.

8. Give an example of a (V)space in which the result of problem 7 is not true.

Solution. Let $K = \{a, b, c, d\}$, where a has the three neighbourhoods $\{a, b\}$, $\{a, c\}$, and $\{a, d\}$, b has the three neighbourhoods $\{a, b\}$, $\{b, c\}$, and $\{b, d\}$, c has the two neighbourhoods $\{a, c, d\}$ and $\{b, c, d\}$, and d has the single neighbourhood $\{d\}$. It follows that the sets $P = \{a, b, c\}$ and $Q = \{c, d\}$ are closed and that their sum and intersection are connected; but the set P is not connected since it is the sum of the two separated sets $\{a\}$ and $\{b, c\}$.

9. Prove that the result of problem 7 is true in all (V)spaces consisting of less than four elements.

10. Give an example of a countable topological space which is totally disconnected and which possesses two elements which are not separated.

Solution. Let $K = \{p_1, p_2, \ldots\}$ and let $\bar{E}$ be defined as follows:

$$\bar{E} = E \text{ for } E \text{ finite (or empty)},$$

$$\bar{E} = E + \{p_1, p_2\}, \text{ if } E \text{ is infinite.}$$

For if E is finite and contains more than one element it is obviously not connected; and if E is infinite there exists an integer $k > 2$ such that $p_k \in E$ and so the set $E = \{p_k\} + (E - \{p_k\})$ is the sum of two separated sets. It is also easily verified that p_1 and p_2 are not separated in K.

CHAPTER III

TOPOLOGICAL SPACES WITH A COUNTABLE BASIS

27. Topological spaces with countable bases. A topological space K is said to possess a *countable basis* if there exists an infinite sequence

$$(1) \qquad U_1, U_2, U_3, \ldots$$

of open sets such that every open set contained in K is the sum of a certain aggregate of sets belonging to the sequence (1). The sets of the sequence (1) will be called *rational sets.*

We shall deduce in this chapter a number of theorems for topological spaces with a countable basis.

It is evident from the Corollary to Theorem 40 that *the existence of a countable basis in a topological space is a topological property.*

LEMMA 1. *If a is a given element belonging to K, and U an open set containing a, then there exists a rational set containing a and contained in U.*

Proof. The set U, being open, is the sum of a certain aggregate M of rational sets, i.e.,

$$U = \sum_{U_k \in M} U_k;$$

hence if $a \in U$, $a \in U_k \subset U$ for some k. This proves the lemma.

LEMMA 2. *Corresponding to every element a belonging to K there exists an infinite sequence $V_1, V_2, \ldots$ of rational sets such that their intersection consists of the element a only, and every open set V containing a contains all but a finite number of the sets $V_1, V_2, \ldots$.*

Proof. Let a be a given element of the space K and let

$$U_{n_1}, U_{n_2}, \ldots$$

be the sets of the sequence (1) which contain a in the same order as they appear in (1). (It may always be assumed that this sequence is infinite by repeating the last term infinitely many times if necessary.) Let k be a given natural number. The intersection

$$U_{n_1} . U_{n_2} \ldots U_{n_k}$$

is, by Theorem 36, an open set and it contains a; hence, by Lemma 1, there exists a rational set V_k such that

$$a \in V_k \subset U_{n_1} . U_{n_2} \dots U_{n_k}.$$

(It may be assumed that V_k is the first set of the sequence (1) which satisfies the above condition.) We shall show that the sequence $\{V_k\}$, $k = 1, 2 \dots$, satisfies the conditions of Lemma 2. For, on the one hand, $a \in V_k$ for $k = 1, 2, \dots$; and if V is an open set containing a then, by Lemma 1, there exists a rational set, say U_{n_q}, containing a and contained in V. Hence for $k \geqslant q$ we have, from the definition of the sets V_k,

$$V_k \subset U_{n_q} \subset V$$

and so $V_k \subset V$ for $k \geqslant q$.

On the other hand, if b is an element of K different from a, there exists, by condition γ and the fact that in topological spaces neighbourhoods are open sets, an open set V containing a but not containing b. Hence the sets V_k, which are contained in V for all sufficiently large k, do not contain b; consequently the intersection $V_1 . V_2 . V_3 \dots$ contains no element different from a. Lemma 2 is therefore proved. Moreover, it can be proved that the sequence $V_1, V_2, \dots$ of Lemma 2 may be assumed to be descending, that is, $V_1 \supset V_2 \supset V_3 \supset \dots$.

In connection with Lemma 2 we remark further that the following condition W is called *the first axiom of countability* (Hausdorff):

Condition W. *Corresponding to every element a of a space K there exists an infinite sequence* $V_1, V_2, \dots$ *of open sets containing a and such that, for every open set V containing a, all but a finite number of the sets* $V_1, V_2, \dots$ *are contained in V.*

It follows from Lemma 2 that every topological space with a countable basis satisfies the first axiom of countability. There exist, however, topological spaces which satisfy the first axiom of countability but do not possess a countable basis. For example, the non-countable space K defined in example 2 of § **23** is such a space. The condition that a countable basis exist is called by Hausdorff *the second axiom of countability.*

From Lemma 2 we obtain

THEOREM 41. *A topological space with a countable basis has cardinal* $\leqslant \mathbf{c}$.

Proof. It follows from Lemma 2 that every element of the space K is the intersection of all those sets of the sequence (1) in which it is contained. Put

$$N(a) = \mathop{\mathrm{E}}_{n} [a \in V_n] \qquad a \in K$$

(i.e., $N(a)$ is the set of all natural numbers n for which $a \in V_n$); obviously

$$\{a\} = \prod_{n \in N(a)} V_n.$$

Every element of K is therefore completely determined by the set $N(a)$, that is, by a certain set of natural numbers. Since the aggregate of all sets of natural numbers has cardinal **c**, it follows that $\bar{\bar{K}} \leqslant \mathbf{c}$. This proves the theorem.

It is obvious that *every subset of a topological space with a countable basis is a topological space with a countable basis.* For if the sequence (1) is a countable basis for K and E is any subset of K, then the sequence

$$E \,.\, U_1,\ E \,.\, U_2,\ \ldots$$

is a countable basis for E.

The linear space (with the usual definition of neighbourhoods) is a topological space of cardinal **c** with a countable basis consisting of all open intervals with rational end-points. Consequently there exists a topological space with a countable basis of any cardinal $\leqslant \mathbf{c}$.

28. Hereditary separability of topological spaces with countable bases. Let K denote a given topological space with the countable basis (1) and E any set $\subset K$. Form the set P consisting of one element from each $E \,.\, U_n$ which is not empty. Clearly $P \subset E$ and $\bar{\bar{P}} \leqslant \aleph_0$. Further, let a denote any element of E and U any neighbourhood of a; then U is an open set $\subset K$ (§ **19**) and so, from Lemma 1, § **27**, there exists a rational set U_n such that $a \in U_n \subset U$. Since $a \in E$ we have $E \,.\, U_n \neq 0$ and so, from the definition of the set P, there exists an element $p_n \in P$ such that $p_n \in U_n \subset U$. Hence every neighbourhood of a contains at least one element of P; consequently $a \in P + P' = \bar{P}$.

Thus E contains a finite or countable subset P such that $E \subset \bar{P}$. K is therefore hereditarily separable (cf. § **23**, example 3). We thus obtain

THEOREM 42. *Every topological space with a countable basis is hereditarily separable.*

The converse of Theorem 42 is not necessarily true for we have proved in § **23** (example 3) that there exist hereditarily separable topological spaces of any cardinal whereas, by Theorem 41, no such space of cardinal $> \mathbf{c}$ can possess a countable basis. Moreover, there exist countable topological spaces (these are always hereditarily separable) which do not possess a countable basis.

Appert[1] has proved that the set of all natural numbers is such a space, when neighbourhoods are defined as follows: every natural number > 1 has only one neighbourhood consisting of the number itself; the number 1 has for its neighbourhoods all sets V containing 1 and such that

$$\lim_{n \to \infty} \frac{N(n, V)}{n} = 1,$$

where $N(n,V)$ denotes the number of those numbers in the set V which are $\leqslant n$. This space does not even satisfy the first axiom of countability.

*Another such space is the set S of all natural numbers where every number > 1 has only one neighbourhood consisting of the number itself. The number 1 has its neighbourhoods defined as follows: a set V is a neighbourhood of the number 1 if it contains 1 and if there exists an infinite sequence $l_1, l_2, \ldots$ of natural numbers such that V is the set of all natural numbers of the form $2^k(2l - 1)$ where k and l are natural numbers and $l > l_k$.

It is easily verified that S is a topological space. Let now $V_1, V_2, \ldots$ denote any infinite sequence of open sets containing 1 and contained in S and let k be a given natural number. The number 1 is obviously a limit element of the set of numbers $\{2^k(2l - 1)\}$, $l = 1, 2, \ldots$ (since every neighbourhood of 1 contains infinitely many numbers of this set). There exists, therefore, a natural number l_k such that the number $2^k(2l_k - 1) \in V_k$. Let V denote the set of all natural numbers of the form $2^k(2l - 1)$ where $k = 1, 2, \ldots$ and $l > l_k$. Hence V is a neighbourhood of 1 and consequently an open set. But $2^k(2l_k - 1) \notin V$ for $k = 1, 2, \ldots$ and so $V_k - V \neq 0$ for $k = 1, 2, \ldots$. Thus no infinite sequence $V_1, V_2, \ldots$ of open sets satisfies condition W of **§ 27** for the element 1. The space S does not satisfy the first axiom of countability; consequently it does not possess a countable basis.

However, *every countable* (V)*space which satisfies the first axiom of countability possesses a countable basis.*

Proof. Let K denote a given countable (V)space which satisfies the first axiom of countability. Hence, corresponding to every element $a \in K$ there exists an infinite sequence $V_n(a)$, $n = 1, 2, \ldots$, of open sets which satisfies condition W of **§ 27**. Since the cardinal number of K is $\aleph_0$, the family B of the sets $V_n(a)$, $a \in K$, $n = 1, 2, \ldots$, is countable. Moreover, the family B forms a countable basis of the space K. For if U is an open set contained in K, then U is the sum of all sets of B which are contained in U. This follows at once from the fact that if $a \in U$ then there exists a natural number n such that $V_n(a) \subset U$; consequently every element of the set U is contained in some set of the family B which is contained in U.*

29. The power of an aggregate of open sets.

THEOREM 43. *The aggregate of all open (closed) sets in a topological space with a countable basis has cardinal* $\leqslant \mathbf{c}$.

Proof. Let K denote a topological space with the countable basis (1) and let U be an open set $\subset K$. Denote by $N(U)$ the set of all natural numbers n for which $U_n \subset U$. It is evident from the definition of the basis (1) that

$$U = \sum_{n \in N(U)} U_n.$$

Hence every open set $U \subset K$ is uniquely determined by a set of natural numbers i.e., by $N(U)$; consequently the aggregate of all open sets of K has cardinal $\leqslant \mathbf{c}$. The theorem is thus proved for open sets and, since a closed set is the complement of an open set, the aggregate of all closed sets has the same cardinal number as that of all open sets.

Note that there exist infinite topological spaces with a countable basis in which the aggregate of all open (closed) sets has cardinal $< \mathbf{c}$. For instance, the space K in example 3, § **23**, for $\mathbf{m} = \aleph_0$, where the aggregate of all open (closed) sets is countable (cf. Theorem 59, § **41**).

30. The countability of scattered sets.

THEOREM 44. *The set of all elements of a given set contained in a topological space with a countable basis which are not its elements of condensation is at most countable.*

Proof. Let K denote a topological space with the countable basis (1) and E a given set $\subset K$. Denote by E_1 the set of all elements of E which are not elements of condensation of E and by $N(E)$ the set of all natural numbers n for which

$$\overline{\overline{U_n \,.\, E}} \leqslant \aleph_0.$$

We shall show that

$$E_1 = E \,.\sum_{n \in N(E)} U_n. \tag{2}$$

For if $a \in E_1$ then, from the definition of an element of condensation (§ **16**), there exists a neighbourhood U of a such that

$$\overline{\overline{U \,.\, E}} \leqslant \aleph_0.$$

But a neighbourhood in a topological space is an open set (§ **19**); hence, by Lemma **1**, § **27**, there exists a natural number m such that $a \in U_m \subset U$ and so

$$\overline{\overline{U_m \,.\, E}} \leqslant \aleph_0;$$

consequently $m \in N(E)$ and

$$a \in U_m \subset \sum_{n \in N(E)} U_n$$

and clearly $a \in E$, since $a \in E_1 \subset E$.

Next, suppose

$$a \in E \,.\sum_{n \in N(E)} U_n;$$

there exists, therefore, a natural number $m \in N(E)$ such that $a \in U_m$. Consequently,

$$\overline{\overline{E \,.\, U_m}} \leqslant \aleph_0;$$

since U_m is an open set and contains at most a countable subset of E, a is not an element of condensation of E and so $a \in E_1$. This proves (2). Since, for $n \in N(E)$,

$$\overline{\overline{U_n \,.\, E}} \leqslant \aleph_0,$$

relation (2) gives at once

$$\bar{\bar{E}}_1 \leqslant \sum_{n \in N(E)} \overline{\overline{U_n \,.\, E}} \leqslant \aleph_0$$

and this establishes the theorem.

Theorem 44 implies that *every non-countable set $E \subset K$ contains a non-countable subset of elements of condensation of E.*

A set is called *condensed* if every one of its elements is an element of condensation of the set.

THEOREM 45. *The set of all elements of condensation of a given set contained in a topological space with a countable basis is a condensed set.*

Proof. Let E_2 be the set of all elements of condensation of E and E_1 the same as in Theorem 44. Then $E_2 = E - E_1$. Let $a \in E_2$ and let U be any one of its neighbourhoods. Since a is an element of condensation of E, the set $E \,.\, U$ is non-countable. But $E_2 \,.\, U \supset E \,.\, U - E_1 \,.\, U$ and $E_1 \,.\, U \subset E_1$ is, by Theorem 44, at most countable; hence $E_2 \,.\, U$ is non-countable. Consequently, a is an element of condensation of E_2. Since a denotes any element of E_2, Theorem 45 is proved. A condensed set is obviously dense-in-itself.

Theorems 44 and 45 lead to the

COROLLARY. *Every non-countable set contained in a topological space with a countable basis contains a non-countable subset which is dense-in-itself.*

As an immediate consequence of this corollary we obtain

THEOREM 46. *A scattered set contained in a topological space with a countable basis is at most countable.*

Furthermore, it follows immediately from Theorem 44 that *the set of all the elements of a given set contained in a topological space with a countable basis which are not its limit elements is at most countable* (since every element of condensation of a set is also a limit element of the set). Consequently *every isolated set is at most countable.* This result is also a direct consequence of Theorem 46 since an isolated set is scattered (§ 7).

31. The Cantor-Bendixson theorem. Suppose the set E to be closed. We have proved at the end of § 8 that the nucleus N of a closed set is perfect and in Theorem 9 that the set $E = N + R$, where R is scattered and so, by Theorem 46, at most countable. This gives

THEOREM 47 (Cantor-Bendixson). *Every closed set contained in a topological space with a countable basis is the sum of a perfect set and a set at most countable (either of which may be empty).*

Moreover, *the decomposition of a closed set into two disjoint sets, one of which is perfect and the other at most countable, is unique.*

To prove this we shall first show that *if E is dense-in-itself and U is any open set then the set $U \,.\, E$ is dense-in-itself or empty.*

For suppose the contrary; then $U \,.\, E$ contains an isolated element p. There exists, therefore, a neighbourhood V of p which contains no element of $U \,.\, E$ different from p. But V is open (condition δ, § **19**); hence, by Theorem 36, the set $U \,.\, V$ is open. There exists therefore a neighbourhood W of p contained in $U \,.\, V$. Thus $p \in W \subset U \,.\, V$ and so $W \,.\, E - \{p\} = 0$. Hence $p \notin E'$, contrary to the assumption that E is dense-in-itself.

Next, suppose that the set $E = P + S = P_1 + S_1$, where the sets P and P_1 are perfect (or empty), the sets S and S_1 are scattered (or empty), and where $P \,.\, S = 0$, $P_1 \,.\, S_1 = 0$, $P \neq P_1$, and $S \neq S_1$. Hence in at least one of the sets, say in S_1, there is an element p which is not in S. Since $p \notin S$ but $p \in S_1 \subset P_1 + S_1 = P + S$, it follows that $p \in P$. Also, $p \in S_1$ and $P_1 \,.\, S_1 = 0$; hence $p \notin P_1$ and, since P_1 is perfect, and so closed, $p \notin P'_1$. There exists, therefore, a neighbourhood U of p such that $P_1 \,.\, U = 0$. But $U \,.\, E = U(P_1 + S_1) = U \,.\, S_1 \subset S_1$ and so, since S_1 is scattered, $U \,.\, E$ must be scattered. Now $U \,.\, P \subset U \,.\, E$ and U is open, P is dense-in-itself, and $U \,.\, P \neq 0$ (since $p \in U \,.\, P$); hence $U \,.\, P$ is dense-in-itself and is contained in a scattered set; this is a contradiction.

We have therefore proved that the decomposition $E = P + S$ is unique.

Similarly, we can obtain the more general result that *every set E can be expressed uniquely as the sum of two disjoint sets one of which is dense-in-itself and closed in E and the other scattered (where either of them may be empty).*

*In § **7** we pointed out that

$$R = E - N = \sum_{0 \leqslant \xi < \nu} (E_\xi - E_{\xi+1}), \tag{3}$$

where $\nu < \omega_{\mu+1}$ if $\bar{\bar{E}} = \aleph_\mu$, and $E_\xi - E_{\xi+1} \neq 0$ for $0 \leqslant \xi < \nu$. Here $\bar{\nu} \leqslant \bar{\bar{R}}$ and since R is scattered it is at most countable by Theorem 46. Consequently ν is a number of the first or second class.

It follows from the above that the nucleus of a set contained in a topological space with a countable basis, is obtained on removing from the set its isolated elements and the isolated elements of its derived sets of all orders in at most a countable number of steps.

We shall define by transfinite induction, for every set E and every ordinal number α, sets $E^{(\alpha)}$ as follows. $E^{(1)}$ denotes the derived set of E. Let α be a given ordinal number and assume that the sets $E^{(\xi)}$ have been defined for $\xi < \alpha$. If α is a number of the first kind, i.e., $\alpha = \xi + 1$ for some ξ, put $E^{(\alpha)} = (E^{(\xi)})'$. If α is a number of the second kind, put $E^{(\alpha)} = \prod_{\xi<\alpha} E^{(\xi)}$.

The set $E^{(\alpha)}$ so defined is called *the derived set of E of order α.* From the definition and Theorems 34 and 2, it follows by transfinite induction that *the* *derived set of order α of every set is a closed set* (for every ordinal number $\alpha > 0$).

If the set E is closed then the sets E_α defined by the relations (25) and (26) in § 7 are easily shown, by transfinite induction, to be identical with the sets $E^{(\alpha)}$ for every ordinal number $\alpha > 0$ (since $E_1 = E_0 . E'_0 = E . E' = E'$). Hence (3) becomes

$$E = E^{(\nu)} + \sum_{0 \leqslant \xi < \nu} (E^{(\xi)} - E^{(\xi+1)})$$

where $E^{(0)}$ is the set E and where ν is an ordinal number of the first or second class. $E^{(\nu)}$ is obviously the nucleus of the closed set E. Since $E^{(\nu)}$ is perfect, we have $E^{(\nu)} = E^{(\nu+1)}$ and so $E^{(\nu)} = E^{(\Omega)}$ $(\nu < \Omega)$. Hence $E^{(\Omega)}$ is the nucleus of a closed set. This gives immediately the result that E *is scattered if and only if* $E^{(\Omega)} = 0$.*

32. The Lindelöf and Borel-Lebesgue theorems.

THEOREM 48. *Every aggregate of disjoint open sets contained in a topological space with a countable basis is at most countable.*

Proof. Let M denote a given aggregate of disjoint open sets contained in a topological space with the countable basis (1). Let U be a set belonging to M. It is obvious from the definition of the sequence (1) that there exist natural indices n such that $U_n \subset U$; associate with U the smallest of these indices. Clearly, since the sets of M are disjoint, different sets will be associated with different indices. Ordering the sets of M according to increasing indices associated with them gives a sequence (finite or infinite) consisting of all the sets of M. This proves the theorem.

Concerning Theorem 48, we note that the condition that every aggregate of disjoint open sets contained in a topological space is at most countable is weaker than the condition of existence of a countable basis. (This condition is sometimes referred to as the Souslin condition[2].) In fact, there exists, as we know (§ 28), a countable topological space (hence necessarily satisfying the Souslin condition) which does not possess a countable basis.

Another example of a topological space without a countable basis but satisfying the Souslin condition is the non-countable space K of example 3, § 23. The open sets of this space are, apart from the null set and the set K itself, all subsets of K which differ from K only by a finite number of elements. Hence no two non-empty open sets are disjoint and so Souslin's condition is certainly satisfied. But K does not possess a countable basis. For suppose $U_1, U_2, \ldots$ is any sequence of non-empty open sets contained in K; then $K - U_n$ is finite and the set

$$K - U_1 . U_2 \ldots = \sum_{n=1}^{\infty} (K - U_n)$$

is at most countable. There exists, therefore, an element $p \in U_1 . U_2 \ldots$. The set $U = K - \{p\}$ is open and $p \in U_n - U$; this implies that $U_n - U \neq 0$

for $n = 1, 2, \ldots$ and so U cannot be a sum of sets of the sequence $U_1, U_2, \ldots$.

Since the space K (of example 3, § **23**) may have any cardinal $\geqslant \aleph_0$, we note that there exist topological spaces of any cardinal which satisfy the Souslin condition (and even a stronger condition, namely, that no two open sets are disjoint) whereas, by Theorem 41, a topological space with a countable basis must have cardinal $\leqslant \mathbf{c}$.

THEOREM 49. *If M is an aggregate of open sets contained in a topological space with a countable basis, then there exists a finite or infinite sequence of sets of M whose sum is identical with the sum of all the sets of the aggregate M.*

Proof. With each set U_n of the sequence (1) which is contained in at least one set of the aggregate M, associate one such set and denote it by V_n. These sets V_n of the aggregate M form a finite or infinite sequence S which satisfies the theorem.

For, on the one hand, $\sum V_n \subset \sum U$, where $U \in M$. On the other hand, let $a \in U \in M$. Since U is open, there exists a set U_m of the sequence (1) such that $a \in U_m \subset U$. But with U_m there is associated an open set $V_m \in M$ such that $U_m \subset V_m$; consequently $a \in V_m \in S$. Hence a is an element of at least one of the sets of S. This proves the theorem. The following is an immediate

COROLLARY (Lindelöf's Theorem). *If E is a set contained in a topological space with a countable basis, and M is an aggregate of open sets such that E is contained in their sum, then there exists a finite or countable sequence of sets of M whose sum contains E.*

Lindelöf's Theorem allows of an immediate generalization of Theorem 31 (Borel) as follows:

THEOREM 50 (Borel-Lebesgue). *If E is a closed and compact set contained in a topological space with a countable basis, and M is an aggregate of open sets whose sum contains E, then there exists a finite sequence of sets of M whose sum contains E.*

Note that the condition that Theorem 49 hold in a given space is weaker than the condition that the space be topological and possess a countable basis. In fact, Theorem 49 holds in every countable (V)space but, as we have seen (§ **28**), there exist countable topological spaces without a countable basis.

33. Transfinite descending sequences of closed sets.

THEOREM 51. *Every transfinite descending sequence*

$$E_0 \supset E_1 \supset \ldots \supset E_\omega \supset E_{\omega+1} \supset \ldots \supset E_\xi \supset E_{\xi+1} \supset \ldots$$

of different sets contained in a topological space with a countable basis, where $E_{\xi+1}$ is closed in E_ξ, is countable.

Proof. Let E_ξ denote any term of the sequence which has at least one successor. Since $E_\xi \supset E_{\xi+1}$ and $E_\xi \neq E_{\xi+1}$, there exists an element $a \in E_\xi - E_{\xi+1}$. Since $E_{\xi+1}$ is closed in E_ξ, we have $a \notin E'_{\xi+1}$. Hence there exists a neighbourhood U of a such that $a \in U$ and $U \cdot E_{\xi+1} = 0$. But a neighbourhood is an open set; hence there exists a rational set U_n such that $a \in U_n \subset U$ and so $U_n \cdot E_{\xi+1} = 0$. On the other hand, since $a \in U_n$ we have $U_n \cdot E_\xi \neq 0$. Thus for every set of the transfinite sequence other than the last (if there is one), there exists a rational set U_n such that $U_n \cdot E_\xi \neq 0$ but $U_n \cdot E_{\xi+1} = 0$. Associate with each set E_ξ the smallest index n for which $U_n \cdot E_\xi \neq 0$ while $U_n \cdot E_{\xi+1} = 0$.

We show next that to different sets correspond different indices. Suppose, on the contrary, that the sets E_ξ and E_η, $\xi < \eta$ are associated with the same index m; we then have $U_m \cdot E_\xi \neq 0$, $U_m \cdot E_{\xi+1} = 0$, $U_m \cdot E_\eta \neq 0$, $U_m \cdot E_{\eta+1} = 0$. This is impossible, since $\xi + 1 \leqslant \eta$. So $E_{\xi+1} \supset E_\eta$ and therefore $U_m \cdot E_{\xi+1} = 0$ gives $U_m \cdot E_\eta = 0$. If now we order all the sets of the transfinite sequence—except the last, if there is one—according to increasing indices, we obtain a countable sequence and this proves the theorem.

Analogously, this same theorem can be proved in connection with an ascending transfinite sequence in which each set is closed in the following one.[3]

Theorem 51 leads to the following

COROLLARY. *If in a topological space with a countable basis*

$$(4) \quad E_0 \supset E_1 \supset E_2 \supset \ldots \supset E_\omega \supset E_{\omega+1} \supset \ldots \supset E_\xi \supset E_{\xi+1} \supset \ldots \quad (\xi < \Omega)$$

is a descending transfinite sequence of type Ω of sets such that E_ξ is closed in the set E_η for all $\eta < \xi < \Omega$, then there exists an ordinal number $\alpha < \Omega$ such that $E_\xi = E_\alpha$ for all ξ such that $\alpha < \xi < \Omega$.

For if we retain in the above sequence only those sets which are different, the conditions of Theorem 51 will be satisfied; consequently the sequence (4) must be countable. Let its terms be

$$(5) \quad E_{\xi_1}, E_{\xi_2}, E_{\xi_3}, \ldots;$$

$\xi_1, \xi_2, \xi_3, \ldots$ are ordinal numbers $< \Omega$. Hence there exists an ordinal number $\alpha < \Omega$ such that $\alpha > \xi_n$ for $n = 1, 2, \ldots$. Let ξ be an ordinal number such that $\alpha < \xi < \Omega$. The set E_ξ must be identical with one of the sets, say E_{ξ_k}, of the sequence (5). Since $\xi_k < \alpha < \xi$ and, since the sequence (4) is descending, we have

$$E_{\xi_k} \supset E_\alpha \supset E_\xi$$

and so $E_\xi = E_\alpha$. This proves the corollary.

[3]As shown by Steckel, the condition that E_ξ be closed in E_η for all $\eta < \xi < \Omega$ cannot be replaced by the condition that the set $E_{\xi+1}$ be closed in the set E_ξ for all $\xi < \Omega$.

For let K denote a non-countable topological space (e.g., the set of all real numbers) and let

$$a_1, a_2, \ldots, a_\omega, a_{\omega+1}, \ldots, a_\xi, \ldots \qquad (\xi < \Omega) \tag{6}$$

be a given transfinite sequence of type Ω consisting of different elements of K. For $\alpha < \Omega$, denote by T_α the set of all terms a_ξ of (6) for which $\alpha \leqslant \xi < \Omega$. Corresponding to every ordinal number $\alpha < \Omega$ there exists, as we know, a unique ordinal number μ_α such that $\omega\mu_\alpha \leqslant \alpha < \omega(\mu_\alpha + 1)$. Put

$$E_\alpha = T_{\mu_\alpha}, \qquad \alpha < \Omega. \tag{7}$$

The transfinite sequence $\{E_\alpha\}$ is obviously descending and $E_{\xi+1} = E_\xi$ for all $\xi < \Omega$. The set $E_{\xi+1}$ is therefore closed in E_ξ for all $\xi < \Omega$. But there does not exist an ordinal number α such that $E_\xi = E_\alpha$ for all ξ such that $\alpha < \xi < \Omega$; otherwise (7) would give

$$T_{\mu_\xi} = T_{\mu_\alpha}$$

for all ξ in $\alpha < \xi < \Omega$ whereas, from the definition of the numbers μ_α, it follows that $\omega(\mu_\alpha + 1) \leqslant \alpha + \omega < \omega(\mu_{\alpha+\omega} + 1)$. Hence $\mu_\alpha + 1 < \mu_{\alpha+\omega} + 1$ or $\mu_\alpha < \mu_{\alpha+\omega}$ and this gives

$$T_{\mu_{\alpha+\omega}} \neq T_{\mu_\alpha}.^*$$

Similarly, it can be proved that if in a topological space with a countable basis

$$E_0 \subset E_1 \subset E_2 \subset \ldots \subset E_\omega \subset E_{\omega+1} \subset \ldots \subset E_\xi \subset \ldots \; (\xi < \Omega)$$

is an ascending transfinite sequence of type Ω such that each E_ξ is closed in E_η for all η in $\xi < \eta < \Omega$, then there exists an ordinal number $\alpha < \Omega$ such that $E_\xi = E_\alpha$ for all ξ such that $\alpha < \xi < \Omega$.

Problems

1. Show by an example that the expression *different* in Theorem 51 cannot be omitted.

Hint. Note the remark made by Steckel in connection with the corollary to Theorem 51.

2. A set contained in a topological space is said to be *open in a set* T if it is the intersection of the set T and an open set of the space considered.[4]

Prove that there exists, in every non-countable topological space, an ascending (descending) transfinite sequence of different sets $E_\xi (\xi < \Omega)$ such that the set E_ξ is open in $E_{\xi+1}$ ($E_{\xi+1}$ is open in E_ξ).

Proof. Since the given topological space is non-countable, there exists a transfinite sequence $\{p_\xi\}$, $\xi < \Omega$, of different elements of the given space. For $\alpha < \Omega$, put

$$E_\alpha = \underset{p_\xi}{\mathrm{E}}[\xi < \alpha] \text{ and } T_\alpha = \underset{p_\xi}{\mathrm{E}}[\alpha \leqslant \xi < \Omega].$$

It is obvious that in a topological space every set E is open in the set E increased by the addition of any element of the space. Since $E_{\xi+1} = E_\xi + \{p_\xi\}$ and $T_\xi = T_{\xi+1} + \{p_\xi\}$, the sequences $\{E_\xi\}$ and $\{T_\xi\}$ satisfy the required conditions.

34. Bicompact sets. A set E contained in a (V)space is called *bicompact* (Urysohn and Alexandroff) if every infinite subset $E_1 \subset E$ has a limit element of order equal to the cardinal of E_1. (A bicompact set is to be distinguished from a 2-compact set defined in § **16**.) Clearly a bicompact set is compact.

THEOREM 52. *In a topological space with a countable basis every compact set is bicompact.*

Proof. It is obviously sufficient to prove that if E is compact and non-countable then there exists an element a (belonging to E or not) such that every open set containing a contains a subset of E which has the same cardinal as E.

Denote by $\mathbf{m}$ the cardinal of E. We consider two cases:

(i) The cardinal number $\mathbf{m}$ is not the sum of a countable sequence of cardinal numbers smaller than $\mathbf{m}$. We shall show that in this case there exists an element a of E such that every open set containing a contains a subset of E of cardinal $\mathbf{m}$. For if not, then corresponding to every element $a \in E$ there exists an open set U containing a and such that the set $U . E$ has cardinal $<\mathbf{m}$. It may be assumed that U is a rational set (Lemma 1, § **27**). Let

$$U_{n_1}, U_{n_2}, \ldots$$

be those rational sets for which $U_{n_k}. E$ has cardinal $<\mathbf{m}$. Obviously

$$E \subset U_{n_1} + U_{n_2} + \ldots.$$

and so

$$E = U_{n_1} . E + U_{n_2} . E + \ldots.$$

Denoting the cardinal number of $U_{n_k}. E$ by $\mathbf{m}_k$ we obtain $\mathbf{m} \leqslant \mathbf{m}_1 + \mathbf{m}_2 + \ldots$, where $\mathbf{m}_k < \mathbf{m}$ for $k = 1, 2, \ldots$. But, since $\mathbf{m} > \aleph_0$, we have $\mathbf{m} = \mathbf{m} . \aleph_0 = \mathbf{m} + \mathbf{m} + \ldots \geqslant \mathbf{m}_1 + \mathbf{m}_2 + \ldots$; consequently $\mathbf{m} = \mathbf{m}_1 + \mathbf{m}_2 + \ldots$, contrary to assumption (i).

(ii) $\quad \mathbf{m} = \mathbf{m}_1 + \mathbf{m}_2 + \ldots$, where $\mathbf{m}_k < \mathbf{m}$ for $k = 1, 2, \ldots$.

Clearly, $\mathbf{s}_k = \mathbf{m}_1 + \mathbf{m}_2 + \ldots + \mathbf{m}_k < \mathbf{m}$, $k = 1, 2, \ldots$ (since a cardinal number $\geqslant \aleph_0$ cannot be the sum of a finite number of cardinal numbers each smaller than the given number). Let n be a given natural number. It will be shown that there exists an element $a_n \in E$ such that every neighbourhood U of a_n intersects E in a set of cardinal $\geqslant \mathbf{s}_n$. Assume the contrary. Then for every $a \in E$ there exists an open set V containing a and such that $V . E$

has cardinal $< \mathbf{s}_n$. As in (i) we may assume that V is a rational set and that

$$U_{n_1}, U_{n_2}, \ldots$$

are those rational sets for which $U_{n_k} . E$ has cardinal $< \mathbf{s}_n$. This would give

$$E = U_{n_1} . E + U_{n_2} . E + \ldots$$

and so

$$\mathbf{m} \leqslant \mathbf{s}_{n_1} + \mathbf{s}_{n_2} + \ldots = \aleph_0 \mathbf{s}_n;$$

this is impossible since $\mathbf{m} > \mathbf{s}_n$ and $\mathbf{m} > \aleph_0$. This establishes the existence of the element a_n for all n.

Consider next the sequence $a_1, a_2, \ldots, a_n, \ldots$ of elements of E, not all necessarily different. Since E is compact and, as a subset of a topological space, $\aleph_0$-compact, there exists an element $a \in K$ such that every neighbourhood of a contains infinitely many elements of the sequence $\{a_n\}$ (this is obviously true also in the case when only a finite number of the terms of the sequence are different).

Let U denote any open set containing a and n any natural number. It is evident from the properties of a that there exists an element $a_k \in U$ such that $k > n$ and so the cardinal of $U . E$ is $\geqslant \mathbf{s}_k \geqslant \mathbf{s}_n$. Consequently the cardinal of $U . E$ is $\geqslant \mathbf{s}_n$ for $n = 1, 2, \ldots$ and so $\geqslant \mathbf{m}$. Since $U . E \subset E$, the cardinal of $U . E \leqslant \mathbf{m}$; hence the cardinal of $U . E$ is $\mathbf{m}$ and Theorem 52 is proved.

Thus in a topological space with a countable basis bicompactness is equivalent to compactness. However, a (non-countable) topological space without a countable basis may be compact but not bicompact.

For example, the set of all ordinal numbers $< \Omega$ with the usual definition of neighbourhoods is, as is easily seen, a compact topological space; but it is not bicompact for although it is non-countable, it contains no element of condensation.

In § **18** we discussed topological limits of sequences of sets. It can be proved that in a topological space with a countable basis every infinite sequence of sets contains a topologically convergent infinite subsequence.[5]

Problems

1. Give an example of a non-countable topological space without a countable basis in which every compact set is bicompact.

Solution. The non-countable topological space K defined in example 3, § **23**. Every open set of that space contains all, except perhaps a finite number, of the elements of K.

2. Show that every topological space K may be considered as a subset of the bicompact topological space K_1 obtained by adjoining an element $a \notin K$ to K.

Proof. Let K be the given topological space, $\bar{E}$ the closure of $E \subset K$. Let $K_1 = K + \{a\}$, where $a \notin K$, and let the closure $f(E)$ of $E \subset K_1$ be defined as follows. If E is finite (or empty) let $f(E) = E$, and if $E \subset K_1$ is infinite put

$$f(E) = \overline{E - \{a\}} + \{a\}.$$

It is easy to see that this definition makes K_1 into a bicompact topological space (since every open set containing a contains all, except perhaps a finite number, of elements of K_1) and $\bar{E} = K \,.\, f(E)$ for all $E \subset K$.

CHAPTER IV

HAUSDORFF TOPOLOGICAL SPACES SATISFYING THE FIRST AXIOM OF COUNTABILITY

35. Hausdorff topological spaces. The limit of a sequence. Fréchet's (L) class. A topological space K is called a *Hausdorff* topological space,[1] or briefly an (HT)space, if it satisfies the following condition γ_1 (stronger than condition γ of § **19**):

γ_1. *For every pair a, b of different elements of the space K there exist neighbourhoods V_1 of a and V_2 of b such that $V_1 . V_2 = 0$.*

The space K where $\bar{E} = E$ for all $E \subset K$ (see example 2, § **23**) is an (HT)space (because sets consisting of single elements are open; consequently they are neighbourhoods of these elements). There exist, therefore, (HT) spaces of any cardinal.

On the other hand, the topological space given in example 3, § **23**, is not an (HT)space for any two non-empty open sets of K have infinitely many elements in common. There exist, therefore, topological spaces which are not Hausdorff topological spaces. This establishes the fact that condition γ_1 is indeed stronger than condition γ.

*Condition γ is called by some authors[2] the *first* (Fréchet) *separation axiom*, condition γ_1 the *second* (Hausdorff) separation axiom. They may be denoted by T_1 and T_2 respectively. A weaker axiom than T_1 has also been investigated, namely T_0 (Kolmogoroff): Given two different elements, there exists for at least one of them a neighbourhood which does not contain the other.

Moreover, a stronger condition than γ_1 is axiom T_3 (Vietoris): For every two disjoint closed sets, at least one of which consists of only one element, there exist two disjoint open sets each containing one of the closed sets. (Cf. the condition of regularity, § **42**.)

Finally, axiom T_4 (Tietze) is even stronger than axiom T_3: For every two disjoint closed sets there exist two disjoint open sets each containing one of the closed sets. (This is the so-called condition of normality, see § **42**.)*

We shall deduce in this chapter a number of theorems valid in Hausdorff topological spaces which satisfy the first axiom of countability (condition W, § **27**).

An infinite sequence $p_1, p_2, \ldots$ of elements of any (V)space is said to have the limit a (where a is an element of the same space) if for every neighbourhood V of a there exists a natural number μ such that

$$p_n \in V \qquad \text{for all } n > \mu.$$

In symbols,

$$\lim_{n\to\infty} p_n = a, \text{ or } p_n \to a \text{ as } n \to \infty.$$

In particular, for the set of real numbers with the usual definition of neighbourhoods, the above definition reduces to the definition of the limit of a sequence of real numbers as given in Analysis.

As a consequence of the above definition, we obtain the following properties of the limit of a sequence in an (HT)space.

Property 1. *If* $p_n = p$ *for* $n = 1, 2, \ldots,$ *then* $\lim\limits_{n\to\infty} p_n = p$.

In other words, an infinite sequence consisting of one element repeated has that element as limit. The proof is obvious.

Property 2. *If* $\lim\limits_{n\to\infty} p_n = a$ *and* $\lim\limits_{n\to\infty} p_n = b$, *then* $a = b$.

That is to say, an infinite sequence cannot have two different limits. Because, if $b \neq a$, there exists, by γ_1, a neighbourhood U of a and a neighbourhood V of b such that $U \cdot V = 0$. Thus, if $\lim p_n = a$, then $p_n \in U$ for $n > \mu$ and so $p_n \notin V$ for $n > \mu$; consequently b cannot be the limit of the sequence $\{p_n\}$.

Property 3. *If* $\lim\limits_{n\to\infty} p_n = a$ *and if* $n_1, n_2, \ldots$ *is any infinite sequence of increasing natural numbers, then*

$$\lim_{k\to\infty} p_{n_k} = a.$$

In other words, if a is the limit of an infinite sequence $p_1, p_2, \ldots,$ then a is also the limit of every infinite subsequence of the above sequence. The proof follows at once from the definition of the limit of a sequence.

Note that properties 1 and 3 hold in every (V)space but that property 2 is not necessarily true even in a topological space. It does not hold, for instance, in the space K defined in example 3, § **23**, where obviously each element is the limit of every infinite sequence of different elements of K.

Fréchet in his thesis[3] calls a set K consisting of any elements an (L)class provided there is given a definition of the limit whereby for a given element $a \in K$ and a given sequence $p_1, p_2, \ldots$ contained in K it is possible to say whether or not a is the limit of the sequence. This definition of the limit of a sequence may be quite arbitrary provided the limit has the properties 1, 2, and 3. In the thesis mentioned above Fréchet investigates the conclusions which follow from the above assumption (and appropriate definitions). Clearly every Hausdorff topological space is a Fréchet (L)class (but not conversely).

Indeed, a Fréchet (L)class need not even be a (V)space. For example, the set consisting of the two elements a and b will be an (L)class if we assign to the infinite sequence $\{a_n\}$, $a_n = a$ for all n, the element a as limit, to the sequence $\{b_n\}$, $b_n = b$ for all n, the element b as limit and no limit to any other infinite sequence obtained from the elements a and b. If the space under consideration were a (V)space, then the sequence $\{p_n\}$, where $p_1 = b$ and $p_n = a$ for $n \geqslant 2$, would have the limit a (in accordance with the definition of a limit given on page 72).

Of the other properties of limits (which follow readily from our postulates and definitions and which are true in all (V)spaces but not necessarily in Fréchet's (L)classes) we shall mention the following.

The limit of a sequence (or the existence of the limit) is independent of the order of the terms of the sequence.

The limit of a sequence remains unchanged if a finite number of terms be added to or removed from the sequence.

If $\lim_{n\to\infty} p_n = a$ *and* $\lim_{n\to\infty} q_n = a$ *then the infinite sequence* $p_1, q_1, p_2, q_2, p_3, q_3, \ldots$ *also has the limit* a.

If $\lim_{n\to\infty} p_n \neq a$, *there exists an infinite subsequence* $\{p_{n_k}\} \subset \{p_n\}$ *such that no subsequence of* $\{p_{n_k}\}$ *has the limit* a.

Problems

1. Show that a topological space satisfying the first axiom of countability is a Hausdorff topological space if and only if property 2 is satisfied.

Proof. The condition is obviously necessary since it is satisfied in every (HT)space. Suppose now that the topological space K satisfying the first axiom of countability is not an (HT)space. Consequently condition γ_1 is not satisfied in K and so there exist elements $a \in K$ and $b \in K$ such that for every neighbourhood U of a and every neighbourhood V of b we have $U \cdot V \neq 0$. Since the first axiom of countability is satisfied in K, there exists an infinite sequence $\{U_n\}$ of neighbourhoods of a such that for every neighbourhood U of a there exists a natural number μ such that $U_n \subset U$ for $n > \mu$. Similarly, there exists an infinite sequence $\{V_n\}$ of neighbourhoods of b such that for every neighbourhood V of b there exists a natural number ν such that $V_n \subset V$ for $n > \nu$. But for every natural number n we have $U_n \cdot V_n \neq 0$; there exists, therefore, an element $p_n \in U$ for $n > \mu$ and $p_n \in V$ for $n > \mu$. Since U and V are arbitrary neighbourhoods of a and b respectively, it follows that $\lim_{n\to\infty} p_n = a$ and $\lim_{n\to\infty} p_n = b$, a contradiction of property 2. Thus if property 2 holds in a topological space satisfying the first axiom of countability, the space must be a Hausdorff topological space.

2. Give an example of a non-countable topological space in which property 2 holds but which is not an (HT)space.

Solution. Let K be a non-countable set and $a \in K$. A subset $V \in K$ is a neighbourhood of a if and only if $a \in V$ and the cardinal of $K - V$ is $\leqslant \aleph_0$. K is obviously a topological space in which an element is the limit of an infinite sequence only when all except a finite number of the elements of the sequence are equal to a. Hence property 2 is satisfied in K. However, K is not an (HT)space because any two neighbourhoods have a non-countable set of elements in common and therefore condition γ_1 is not satisfied.

3. Give an example of a countable topological space satisfying the conditions of problem 2.

Solution. Such a space is provided by the set K of all natural numbers where a neighbourhood of any natural number k is defined to be the set V containing the number k and all those natural numbers for which the relation

$$\lim_{n\to\infty} \frac{N(n, V)}{n} = 1$$

is satisfied, where $N(n, V)$ denotes the number of all those numbers contained in V which are $\leqslant n$.

4. Does there exist a space satisfying the conditions of problem 2 which has a countable basis?

Answer. No, because in that case it would also satisfy the first axiom of countability and would therefore be an (HT)space; this is a contradiction.

5. Give an example of an (HT)space which does not satisfy the first axiom of countability.

Answer. The space of Appert and the space S defined in § **28**.

6. Prove that every infinite (HT)space contains an infinite isolated subset but that this need not be true in every infinite topological space.

Proof. Let K be an infinite (HT)space. If none of the elements of K is a limit element, then K is isolated. Suppose, therefore, that there exists an element a in K which is a limit element of K. Let a_1 be an element of K different from a. There exist, by γ_1, neighbourhoods U_1 of a and V_1 of a_1 such that $U_1 . V_1 = 0$. Since $a \in U_1$ and $a \in K'$ there exists an element $a_2 \in U_1$, $a_2 \neq a_1$, and, again by γ_1, there exist neighbourhoods U_2 of a and W_2 of a_2 such that $U_2 . W_2 = 0$. But the neighbourhoods U_1 and W_2 are open sets containing a_2; hence $U_1 . W_2$ is an open set and so there exists a neighbourhood V_2 of a_2 such that $V_2 \subset U_1 . W_2$. It follows that $V_2 . U_2 = 0$ and $V_1 . V_2 = 0$. Similarly, there exists an element $a_3 \in U_2$, $a_3 \neq a$, and neighbourhoods U_3 of a $(U_3 \subset U_2 \subset U_1)$ and V_3 of a_3 such that $U_3 . V_3 = 0$ and $V_1 . V_3 + V_2 . V_3 = 0$. Continuing this argument we obtain an infinite sequence $a_1, a_2, a_3, \ldots$ of elements of K which are contained in the disjoint neighbourhoods $V_1, V_2, V_3, \ldots$ respectively. This sequence is therefore an isolated set.

On the other hand, the topological space of example 3, § **23**, is infinite but none of its infinite subsets is isolated.

36. Properties of limit elements.

THEOREM 53. *An element a belonging to a Hausdorff topological space satisfying the first axiom of countability is a limit element of a set E contained in this space if and only if there exists an infinite sequence $p_1, p_2, \ldots$, such that*

$$a \neq p_n \in E, \qquad n = 1, 2, \ldots, \tag{1}$$

and

$$\lim_{n\to\infty} p_n = a. \tag{2}$$

Proof. The sufficiency of the conditions follows from the definition of a limit (§ **35**). It remains to prove the necessity of the condition. Suppose that a is a limit element of the set E contained in an (HT)space satisfying the first axiom of countability (condition W of § **27**). Let $V_1, V_2, \ldots$ denote an infinite sequence of open sets satisfying condition W in relation to the element a. Let n be a given natural number. Since $a \in V_n$ and $a \in E'$, there exists an element $p_n \in E$ such that $p_n \in V_n$ and $p_n \neq a$. It will be shown that the sequence $p_1, p_2, \ldots$ satisfies conditions (1) and (2). Condition (1) is obviously satisfied. Let V be any neighbourhood containing a and hence an open set containing a. It follows from the properties of the sequence $\{V_n\}$ that $V_n \subset V$ for $n > \mu$; hence $p_n \in V$ for $n > \mu$, and so (2) follows. Theorem 53 is therefore proved.

Remark. Condition (1) in Theorem 53 may be replaced by the condition that all elements of the sequence $\{p_n\}$ belong to E and be different. In fact, the sequence $\{p_n\}$ satisfying conditions (1) and (2) must contain infinitely many different terms; otherwise, one of the terms, say p_s, would be repeated infinitely many times. Condition (2) and properties 1, 2, and 3 of a limit would then give $p_s = a$, contrary to (1). If $p_{n_1}, p_{n_2}, \ldots$ is an infinite sequence obtained from the sequence $\{p_n\}$ by removing all those terms which are identical with any of the terms preceding them, then by (2) and property 3 of limits we obtain

$$\lim_{k\to\infty} p_{n_k} = a;$$

hence the remark is valid.

*Observe that Theorem 53 may not hold in (HT)spaces which do not satisfy the first axiom of countability. Consider, for example, the space K of all ordinal numbers $\leqslant \Omega$ with neighbourhoods defined as in the example on page 43. The number Ω is clearly a limit element (in fact, it is an element of condensation of the set K) but it is not the limit of any infinite sequence of elements different from Ω and contained in K.

On the other hand, there exist (HT)spaces which do not satisfy the first axiom of countability but in which Theorem 53 holds. Such a space, for example, is the space S of § **28**. This is evident from the fact that a set $E \subset S$ possesses a limit element (the element 1) if and only if there exists a natural number k such that for infinitely many natural numbers l the number $2^k(2l - 1) \in E$. For if $2^k(2l - 1) \notin E$ for $l < l_k$ $(k = 1, 2, \ldots)$, then there would exist a neighbourhood of the number 1 free of all elements of E contrary to the supposition that 1 is a limit element of E. If, however, $2^k(2l - 1) \in E$ for $l = m_1, m_2, \ldots$, where $m_1 < m_2 \ldots$, then $\lim_{i\to\infty} 2^k(2\,m_i - 1) = 1$.*

We recall that a set E is compact-in-itself if every infinite subset of E possesses at least one limit element which is contained in E. From Theorem 53 and the properties of a sequence we obtain the following

COROLLARY. *Every set compact-in-itself and contained in a Hausdorff topological space satisfying the first axiom of countability is closed.*

Proof. Let E denote a set which is compact-in-itself and contained in an (HT)space satisfying the first axiom of countability. Let a be a limit element of E. There exists then, by Theorem 53, an infinite sequence $\{p_n\}$ of different elements of E such that $\lim_{n\to\infty} p_n = a$. The set $T = \{p_1, p_2, \ldots\}$ is an infinite subset of E and, since E is compact-in-itself, T has a limit element $b \in E$. By Theorem 53, there exists an infinite sequence $n_1, n_2, \ldots$ of different natural numbers n_k such that

$$\lim_{k\to\infty} p_{n_k} = b$$

and, since the limit of a sequence is independent of the order of the terms, it may be assumed that $n_1 < n_2 < \ldots$. Since $\lim_{n\to\infty} p_n = a$, we have

$$\lim_{k\to\infty} p_n = a$$

and so, because of property 2 of limits, $a = b$. Since $b \in E$ we now have $a \in E$ and, since a is any limit element of E, E is closed.

It should be noted that in arbitrary (HT)spaces the Corollary is not in general true. It is not true, for example, in the space K consisting of all ordinal numbers $\leqslant \Omega$, discussed on page 43, because the set $K - \{\Omega\}$ is obviously compact-in-itself but is not closed.

On the other hand, *any compact and closed set contained in a (V)space is compact-in-itself.*

Problems

1. Prove that in any Hausdorff topological space a countable set which is compact-in-itself is necessarily closed.

Proof. Let $P = \{p_1, p_2, \ldots\}$ denote a countable set which is compact-in-itself and is contained in an (HT)space T and let $a \in P' - P$. For every natural number n there exist neighbourhoods U_n of p_n and V_n of a such that $U_n \cdot V_n = 0$. The intersection $V_1 \cdot V_2 \ldots V_n$ is open and so there exists a neighbourhood W_n of a such that $W_n \subset V_1 \cdot V_2 \ldots V_n$. Since $a \in P'$ there exists an element $p_{k_n} \in W_n$, where $p_{k_n} \in P$. For any k, $p_k \in U_k$ and therefore $p_k \notin V_k$; consequently $p_k \notin W_n$ for $k \leqslant n$. It follows that $k_n > n$ $(n = 1, 2, \ldots)$ since $p_{k_n} \in W_n$. The set

$$Q = \{p_{k_1}, p_{k_2}, \ldots\}$$

is therefore infinite. Since $Q \subset P$ and P is compact-in-itself, there exists an element $p_s \in Q'$. Now $p_s \in U_s$ and $p_{k_n} \in W_n \subset V_s$ for $n \geqslant s$; hence $p_{k_n} \notin U_s$ for $n \geqslant s$. There is therefore at most a finite number of elements of the set Q in the neighbourhood U_s of p_s, contrary to the fact that $p_s \in Q'$. Thus the assumption that $a \in P' - P$ is false, that is, $P' - P = 0$, so P is closed.

2. Give an example in a topological space of a countable set which is compact-in-itself but is not closed.

Solution. Let $T = \{p_1, p_2, \ldots\}$ be a countable topological space with closure defined as follows: $\bar{E} = E$ for E finite and $\bar{E} = E + \{p_1, p_2\}$ for E infinite. The set $Q = \bar{E} - \{p_1\}$, where E is infinite, is clearly compact-in-itself but is not closed (since $p_1 \in Q' - Q$).

37. Properties of functions continuous in a given set.

THEOREM 54. *A function f which is defined in a set E contained in a topological space satisfying the first axiom of countability and which takes on values in any Fréchet (V)space is continuous at an element p_0 of E if and only if for every infinite sequence $p_1, p_2, \ldots$ of elements of E for which*

$$\lim_{n\to\infty} p_n = p_0 \tag{3}$$

we have

$$\lim_{n\to\infty} f(p_n) = f(p_0). \tag{4}$$

Proof. Suppose that the function $f(p)$ defined in the set E is continuous at $p_0 \in E$ and let $p_n (n = 1, 2, \ldots)$ denote an infinite sequence of elements of E such that (3) holds. Further, let V denote any neighbourhood of $f(p_0)$. It follows from the definition of continuity of a function at a given element (§ **11**) that there exists a neighbourhood U of p_0 such that $p \in U \cdot E$ implies $f(p) \in V$. But, from (3) and the definition of a limit, we have $p_n \in U \cdot E$ for $n > \mu$; hence $f(p_n) \in V$ for $n > \mu$ and so, since V is an arbitrary neighbourhood of $f(p_0)$, (4) follows. The condition of the theorem is therefore necessary.

Assume now that the function $f(p)$ is not continuous in the set E at $p_0 \in E$. Hence there exists a neighbourhood V of $f(p_0)$ such that in every neighbourhood U of p_0 there is an element $p \in E$ for which $f(p) \notin V$.

Since, by hypothesis, the set E is contained in a topological space which satisfies the first axiom of countability, i.e., condition W of § **27**, there exists an infinite sequence $U_1, U_2, \ldots$ of open sets containing p_0 and such that for every neighbourhood U of p_0 we have $U_n \subset U$ for $n > \mu$ (where μ is a natural number dependent on U). Let n be a given natural number. Since $p_0 \in U_n$ and U_n is open, there exists a neighbourhood W_n of p_0 such that $W_n \subset U_n$. As stated above, there exists an element $p_n \in W_n$ such that $f(p_n) \notin V$.

Let U denote any neighbourhood of p_0. We have then $p_n \in W_n \subset U_n \subset U$ for $n > \mu$; consequently, (3) holds. But (4) is not true since $f(p_n) \notin V$ for $n = 1, 2, \ldots$; the condition of the theorem is therefore sufficient and Theorem 54 is proved.

We note that the axiom of choice is utilized in the proof of sufficiency since no rule is given for the selection of the elements p_n from W_n.

*Furthermore, the condition of Theorem 54 may not be sufficient for continuity in a Hausdorff topological space which does not satisfy the first axiom of countability. Thus, for example, let K be the set of all ordinal numbers $\leqslant \Omega$ (with neighbourhoods as previously defined) and hence an (HT)space which does not satisfy the first axiom of countability. Now define $f(\alpha) = 1$ for $\alpha < \Omega$ and $f(\Omega) = \Omega$ (or else, let $f(\alpha)$ be a real function defined in K by the conditions $f(\alpha) = 0$ for $\alpha < \Omega$ and $f(\Omega) = 1$). The condition of Theorem 54 is satisfied for the element Ω but $f(\alpha)$ is obviously not continuous in K at Ω.

Similarly, the function $f(1) = 1, f(n) = 2$ for $n \neq 1$ is not continuous in the space defined by Appert (§ **28**) at the element 1, although it satisfies the condition of Theorem 54 for this element.*

Problems

1. Give an example of a topological space which is dense-in-itself and in which every (1, 1) function taking on values in the same space is continuous.
Solution. The space defined in example 3, § **23**.

2. Prove that no (HT)space satisfies the condition of problem 1.

38. The power of the aggregate of functions continuous in a given set. Topological types. Let E denote a set contained in a topological space T, P a set contained in E and dense on E, and let $f(p)$ and $g(p)$ be two functions whose values are elements of a Hausdorff topological space H and which are defined and continuous in E with $f(p) = g(p)$ for $p \in P$. We shall show that $f(p) = g(p)$ for $p \in E$.

For suppose that for some element $p_0 \in E$ we have $f(p_0) \neq g(p_0)$. Since $f(p_0) \in H$ and $g(p_0) \in H$, there exists a neighbourhood V_1 of $f(p_0)$

and a neighbourhood V_2 of $g(p_0)$ such that $V_1 . V_2 = 0$. Moreover, since f and g are continuous in E, there exists a neighbourhood U_1 of p_0 such that

$$(5) \qquad f(p) \in V_1 \qquad \text{for } p \in U_1 . E$$

and a neighbourhood U_2 of p_0 such that

$$(6) \qquad g(p) \in V_2 \qquad \text{for } p \in U_2 . E.$$

Since T is a topological space there exists a neighbourhood U of p_0 such that $U \subset U_1 . U_2$. Consequently, (5) and (6) give

$$(7) \qquad f(p) \in V_1 \text{ and } g(p) \in V_2 \qquad \text{for } p \in U . E.$$

But the set P is dense on E and so, since $p_0 \in U$, $p_0 \notin P$, there exists in U an element $p_1 \in P$. Since $f(p) = g(p)$ for $p \in P$, we have

$$(8) \qquad f(p_1) = g(p_1)$$

and, since $p_1 \in U$, (7) gives

$$(9) \qquad f(p_1) \in V_1 \text{ and } g(p_1) \in V_2;$$

this is impossible since $V_1 . V_2 = 0$. It follows that $f(p) = g(p)$ for all $p \in E$. This result may be expressed in the following theorem:

A function which is continuous in a set E contained in a topological space and which takes on values in a Hausdorff topological space is defined in the whole set E whenever it is defined in a subset P of E which is dense on E.

*We note that, in the statement of the above theorem, the condition that the space be a Hausdorff space cannot be omitted. Indeed, let $T = \{1, 2, 3, \ldots\}$ and $a \in T$. Define a neighbourhood of a to be every subset of T containing a and all except perhaps a finite number of elements of T. It is obvious that T is a topological but not a Hausdorff topological space.

It is easily shown that every (1, 1) function defined in T and taking on values in T is continuous in T. In particular, the function $f(p) = p$ for $p \in T$ and the function $g(p)$ defined by the conditions that $g(2k - 1) = 2k - 1$ and $g(2k) = 2k + 2$ for $k = 1, 2, \ldots$ are continuous in T. But $f(p) \neq g(p)$ for p even, although $f(p) = g(p)$ in the set $P = \{1, 3, 5, \ldots\}$ which is obviously dense on T.*

It follows from Theorem 42 that every set E contained in a topological space with a countable basis contains a finite or countable subset which is dense on E. We may therefore take the set P which is dense on E to be finite or countable. It is possible, by Theorem 41, to define in a countable set at most **c** different functions which take on values in a given topological space with a countable basis. We thus obtain

THEOREM 55. *In every set E contained in a topological space with a countable basis we can define at most a continuum of continuous functions whose values belong to a given Hausdorff topological space with a countable basis.*

It is noteworthy that in the above theorem the condition that the space be a Hausdorff space may be omitted (in which case the proof of the theorem must be modified). We shall therefore prove

THEOREM 56. *In a topological space with a countable basis there can be defined at most a continuum of continuous functions whose values belong to a given topological space with a countable basis.*

Proof. Let $f(p)$ denote a function continuous in a topological space T with a countable basis B and whose values belong to a topological space T_1 with the countable basis $B_1 = \{V_1, V_2, \ldots\}$. The inverse images $f^{-1}(V_n)$ of the open sets $V_n \subset T_1$ are open sets in the space T because of the continuity of the function f (Theorem 24). But, by Theorem 43, there are at most $\mathbf{c}$ open sets in T. Consequently there is at most a continuum of different infinite sequences whose terms are the sets

$$f^{-1}(V_1), f^{-1}(V_2), f^{-1}(V_3), \ldots \tag{10}$$

where f runs through the set of all functions continuous in T.

To prove the theorem it is sufficient to show that the function f is defined whenever the sets of the sequence (10) are known. Hence let p_0 denote a given element of T; then $f(p_0)$ is the intersection $P(p_0)$ of all those sets V_n for which $p_0 \in f^{-1}(V_n)$.

For, on the one hand, $p_0 \in f^{-1}(V_n)$ implies $f(p_0) \in V_n$ and so $f(p_0) \in P(p_0)$. On the other hand, let q_1 be any element of T_1 such that $q_1 \neq f(p_0)$. There exists, therefore, a set $V_k \in B_1$ such that $f(p_0) \in V_k$ but $q_1 \notin V_k$. Hence $p_0 \in f^{-1}(V_k)$; but $q_1 \notin P(p_0)$, the intersection of all sets V_n such that $p_0 \in f^{-1}(V_n)$. Consequently $P(p_0)$ consists of the single element $f(p_0)$. This establishes Theorem 56.

It follows from Theorem 56 that *every set contained in a topological space with a countable basis possesses at most a continuum of different continuous images* (belonging to the same space) *and so at most a continuum of different homeomorphic images.*

Let now all the sets of such a space K be divided into so-called *topological types* by assigning two sets to the same topological type if and only if they are homeomorphic. Clearly each topological type contains at most a continuum of the subsets of the space considered. If K has cardinal $\mathbf{c}$ it possesses $2^{\mathbf{c}}$ different subsets and, since at most $\mathbf{c}$ of them belong to each type, it may be concluded that there are, in K, $2^{\mathbf{c}}$ different topological types.[4] Thus *in a topological space of cardinal* $\mathbf{c}$ *which possesses a countable basis there exist* $2^{\mathbf{c}}$ *different topological types.*

From this we deduce that there exist altogether $2^{\mathbf{c}}$ different topological types in all topological spaces with a countable basis. For a topological space K with a countable basis is determined by this basis (i.e., by a certain infinite sequence of subsets of K). Thus, given a set K of cardinal $\mathbf{c}$, the aggregate of all infinite sequences consisting of subsets of K has cardinal

$$(2^{\mathbf{c}})^{\aleph_0} = 2^{\mathbf{c}}.$$

Consequently there are at most $2^{\mathbf{c}}$ different topological spaces with a countable basis; each has cardinal $\mathbf{c}$. Since there are $2^{\mathbf{c}}$ different topological types in each such space there are obviously altogether $2^{\mathbf{c}}$ different topological types in all topological spaces with a countable basis.

It can be shown that, if we consider all topological spaces of cardinal $\mathbf{m} \geqslant \aleph_0$, we have altogether $2^{2^{\mathbf{m}}}$ different topological types.

39. Continuous images of compact closed sets. Continua.

THEOREM 57. *A continuous image of a closed and compact set contained in a Hausdorff topological space satisfying the first axiom of countability and itself contained in such a space, is closed and compact.*[5]

Proof. Let E be a given closed and compact set which is contained in an (HT)space K satisfying the first axiom of countability. Let $f(p)$ be a function defined and continuous in E whose values belong to an (HT)space K_1 satisfying the first axiom of countability (where we may have $K_1 = K$). Let T_1 denote any infinite subset of the set $T = f(E)$. There exists, therefore, an infinite sequence $\{q_n\}$ of different elements of T_1. Since

$$q_n \in T_1 \subset T = f(E), \qquad n = 1, 2, \ldots$$

there exists, for every natural number n, an element $p_n \in E$ such that $q_n = f(p_n)$ and all the terms of the sequence $\{p_n\}$ are different (since the terms of the sequence $\{q_n\}$ are all different). Denote by E_0 the set of all the terms of the sequence $\{p_n\}$. Then E_0 is an infinite subset of E and so has a non-empty derived set, since E is compact. Let a be an element of E'_0. Since $E_0 \subset E$ and E is closed, we have $a \in E$ and so $f(a) \in f(E) = T$. Let V denote any neighbourhood of $f(a)$ in K_1. Since f is continuous in E, there exists a neighbourhood U of a such that $f(p) \in V$ for $p \in U \,.\, E$. Since $a \in E'_0$, there exists, by Corollary 2 to Theorem 33 (§ **20**), two different elements p_m and p_n in the set $E_0 \,.\, U$. Hence $q_m = f(p_m) \in V$ and $q_n = f(p_n) \in V$. Since $m \neq n$ we have $q_m \neq q_n$ and so at least one of the elements, say q_m, is different from $f(a)$. Thus in every neighbourhood V of $f(a)$ there is at least one element of T_1 different from $f(a)$; consequently $f(a) \in T'_1$, that is $T'_1 \neq 0$. Since T_1 is any infinite subset of T, it follows that T is compact.[6]

Now let b denote any element of K_1 such that $b \in T'$. By Theorem 53 (and the remark to it), there exists an infinite sequence $\{q_n\}$ of different elements of the set T such that

$$\lim_{n\to\infty} q_n = b. \tag{11}$$

Since $q_n \in T = f(E)$ for $n = 1, 2, \ldots,$ there exists for every natural number n an element p_n such that

$$p_n \in E \text{ and } f(p_n) = q_n,$$

where the terms of the sequence $\{p_n\}$ are all different. Denote by E_0 the set of all terms of the sequence $\{p_n\}$. E_0 is an infinite set contained in the compact set E; hence $E'_0 \neq 0$. There exists, therefore, an element $a \in E'_0$. Hence, by Theorem 53, there exists an infinite sequence of elements of E_0, all different from a, whose limit is a. This sequence differs at most in the order of its terms from a subsequence of the sequence $\{p_n\}$ and, since the limit of a sequence is independent of the order of its terms (§ **35**), we may conclude that there exists an infinite subsequence $\{p_{n_k}\}$ of the sequence $\{p_n\}$ such that

$$\lim_{k\to\infty} p_{n_k} = a. \tag{12}$$

Since $a \in E'_0$ and E is closed, we know that $a \in E$; since f is continuous in E, it follows from (12) and Theorem 54 that

$$\lim_{k\to\infty} f(p_{n_k}) = f(a) \text{ i.e., } \lim_{k\to\infty} q_{n_k} = f(a).$$

As a consequence of properties 2 and 3 of a limit (§ **35**) we have $b = f(a)$ and so (since $a \in E$) $b \in f(E) = T$. Hence $b \in T'$ implies $b \in T$, i.e., T is closed. Theorem 57 is therefore proved.

We note that in an (HT)space with a countable basis (e.g., in the set of all real numbers) a continuous or even a homeomorphic image of a closed set need not be closed. Thus, for example, a finite open interval of real numbers (not a closed set) is a homeomorphic image of the set of all real numbers (a closed set). Similarly, a continuous or even a homeomorphic image of a compact set need not be compact; e.g., the set of all real numbers (not compact) is a homeomorphic image of a finite open interval of real numbers (a compact set).

Moreover, a homeomorphic image of a closed or a compact set contained in a topological space may be neither closed nor compact. For example, the set of all positive real numbers is neither closed nor compact yet it is a homeomorphic image of the closed set of all real numbers as well as a homeomorphic image of the compact set of all real numbers in the open interval (0, 1).

Furthermore, in an (HT)space which does not satisfy the first axiom of countability a homeomorphic image of a closed and compact set need not be closed.

For example, let K denote the set of all ordinal numbers $< \Omega + \Omega$. For $0 < \alpha < \Omega + \Omega$ we define a neighbourhood of α as a set of all ordinals ξ with $\beta < \xi < \gamma$, where $\beta < \alpha < \gamma$; for the ordinal number 0 we define as the only neighbourhood the set consisting of the single element 0. It is easily seen that K is an (HT)space. (Here $K - \{\Omega\}$ satisfies the first axiom of countability but K does not.) Denote by E the set of all ordinal numbers α such that $\Omega \leqslant \alpha < \Omega + \Omega$ and by T the set of all ordinal numbers $\alpha < \Omega$. The function $f(\alpha) = \Omega + \alpha$ establishes a homeomorphic mapping of the set T on E. The set E is clearly closed and compact in K but its homeomorphic image T is not closed in K since the number $\Omega \in T' - T$.

A closed, compact, and connected set containing more than one element is called a *continuum*. It follows from Theorems 57 and 25 that in an (HT)space satisfying the first axiom of countability *a continuous image of a continuum is a continuum.* Thus *the property that a subset of a Hausdorff topological space satisfying the first axiom of countability be a continuum is a topological property.* This need not be true in a general (HT)space.

For example, let K denote the set of all pairs (α, x) where α is an ordinal number $< \Omega + \Omega$, and x a real number such that $0 \leqslant x < 1$. We order K by the rule that of any two elements $p_1 = (\alpha_1, x_1)$, $p_2 = (\alpha_2, x_2)$, $p_1 \prec p_2$ if $\alpha_1 < \alpha_2$ and, in the case $\alpha_1 = \alpha_2$, if $x_1 < x_2$. We define an open interval (p_1, p_2), where $p_1 \in K$, $p_2 \in K$, to be the set of all elements $p \in K$ such that $p_1 \prec p \prec p_2$. If now $p_0 \neq (0, 0)$ be any element of K we define a neighbourhood of p_0 to be any open interval containing p_0. If $p_0 = (0, 0)$, we define a neighbourhood of p_0 to be a set consisting of p_0 and the elements of any open interval (p_0, p) for $p \in K$ and $p \neq p_0$. It is readily proved that with neighbourhoods so defined K is an (HT)space. Let E denote the set of all elements p of K such that $p \succ (\Omega, 0)$ and the element $(\Omega, 0)$, and let T be the set of all elements p of K for which $p \prec (\Omega, 0)$. The set E is obviously a continuum but T is not, since it is not closed (because it does not contain the limit element $(\Omega, 0)$). Nevertheless, the function $f(\alpha, x) = (\Omega + \alpha, x)$ establishes a homeomorphic mapping of T on the set E.

Problems

1. Prove that in an (HT)space a continuous image of a countable closed and compact set is closed.

Proof. Let $P = \{p_1, p_2, \ldots\}$ denote a countable closed and compact set contained in an (HT)space T and let $Q = f(P)$ denote a continuous image of P. Suppose $b \in Q' - Q$. Hence $b \neq q_n = f(p_n)$ for $n = 1, 2, \ldots$; so there exist, for $n = 1, 2, \ldots$, neighbourhoods U_n and V_n of the elements q_n and b respectively such that $U_n \,.\, V_n = 0$. Also, there exists for $n = 1, 2, \ldots$ a neighbourhood W_n of b such that $W_n \subset V_1 \,.\, V_2 \ldots V_n$. Since $b \in Q'$ there exists an element $q_{k_n} \in W_n$. But from $q_k \in U_k$ we have $q_k \notin W_n$ for all $k \leqslant n$; thus $q_{k_n} \in W_n$ for $k_n > n$. Consequently the set

$$\{q_{k_1}, q_{k_2}, \ldots\}$$

is infinite and so the set

$$E = \{p_{k_1}, p_{k_2}, \ldots\} \subset P$$

is infinite. Since P is compact we have $E' \neq 0$. There exists, therefore, an element $a \in E'$ and, since P is closed, $a \in P$, say $a = p_s$. Since f is continuous in P there exists, corresponding to the neighbourhood U_s of q_s, a neighbourhood U of p_s such that $f(p_k) \in U_s$ for $p_k \in U$. Because $p_s = a \in E'$, there exists an index $n \geqslant s$ such that $p_{k_n} \in U$. Thus

$$q_{k_n} = f(p_{k_n}) \in U_s.$$

This is impossible since $q_{k_n} \in W_n$ implies $q_{k_n} \in V_s$ for all $n \geqslant s$ and $U_s \,.\, V_s = 0$. We have, therefore, $Q' - Q = 0$, i.e., Q is closed.

2. Give an example of a countable topological space (not an (HT)space) in which a continuous image of a closed and compact set is not closed.

Solution. Let $P = \{p_1, p_2, \ldots\}$ and $Q = \{q_1, q_2, \ldots\}$ be two disjoint countable sets. Consider the space $T = P + Q$ in which closure is defined as follows: $\bar{X} = X$ for X finite, $\bar{X} = X + \{p_1, p_2\}$ if the set $X \,.\, P$ is infinite but $X \,.\, Q$ is finite, $\bar{X} = X + \{q_1\}$ if $X \,.\, P$ is finite but $X \,.\, Q$ infinite and, finally, $\bar{X} = X + \{p_1, p_2, q_1\}$ if both $X \,.\, P$ and $X \,.\, Q$ are infinite sets. It is easily seen that the set Q is closed and compact and that the function $f(q_n) = p_{n+1}$ $(n = 1, 2, \ldots)$ establishes a homeomorphism between the set Q and the set $P - \{p_1\}$; but the last set is not closed since it does not contain the limit element p_1.

40. The inverse of a function continuous in a compact closed set.

THEOREM 58. *If a function $f(p)$ defined in a closed and compact set E contained in a Hausdorff topological space satisfying the first axiom of countability and taking on values in a Hausdorff topological space satisfying the first axiom of countability is continuous and* (1, 1) *in E, then the inverse function of f is continuous in the set $T = f(E)$.*

Proof. Let $f(p)$ denote a function which is (1, 1) and continuous in the closed and compact set E and let $\phi(q)$ denote the inverse function of f; then ϕ is defined for $q \in T = f(E)$.

Suppose that $\phi(q)$ is not continuous at $q_0 \in T$. There exists, by Theorem 54, an infinite sequence $\{q_n\} \subset T$ such that

$$\lim_{n\to\infty} q_n = q_0 \tag{13}$$

but

$$\lim_{n\to\infty} \phi(q_n) \neq \phi(q_0).$$

Put

$$p_n = \phi(q_n), \qquad n = 0, 1, 2, \ldots. \tag{14}$$

Then

$$\lim_{n\to\infty} p_n \neq p_0.$$

Thus, by the definition of a limit, there exists an open set U containing p_0 and such that $p_n \notin U$ for infinitely many values of the index n. There exists, therefore, an increasing infinite sequence of indices n_k ($k = 1, 2, \ldots$) such that

$$p_{n_k} \notin U, \qquad k = 1, 2, \ldots. \tag{15}$$

Moreover, the set of all the different terms of the sequence $\{p_{n_k}\}$ is infinite. Otherwise, at least one element, say p_s, would be repeated infinitely many times and, since

$$f(p_s) = q_s, \quad f(p_0) = q_0,$$

we would have

$$f(p_s) = q_0 = f(p_0);$$

since f is (1, 1) in E it follows that $p_s = p_0 \in U$ contrary to (15).

The set of all different terms of the sequence $\{p_{n_k}\}$ is therefore infinite and, since it is a subset of the compact set E, it possesses a non-empty derived set. Hence there exists an infinite increasing sequence of indices k_r ($r = 1, 2, \ldots$) such that

$$\lim_{r\to\infty} p_{n_{k_r}} = p \tag{16}$$

where, since E is closed, $p \in E$. (For convenience in printing we replace the subscript (superscript) ${}^{n_k}{}_{r_s}$ by ${}_{n_{k_{r_s}}}$.) Since f is continuous in E, (16) gives

$$\lim_{r\to\infty} f(p_{n_{k_r}}) = f(p)$$

and so, since $f(p_n) = q_n$ for $n = 1, 2, \ldots,$

$$\lim q_{n_{k_r}} = f(p).$$

This gives $f(p) = q_0$. But $f(p_0) = q_0$ and f is (1, 1) in E; hence

$$p = p_0. \tag{17}$$

Now $p_0 \in U$ and U is open; hence from (16) and (17) we have, for sufficiently large r,

$$p_{n_{k_r}} \in U$$

contrary to (15). The assumption that $\phi(q)$ is not continuous in T leads to a contradiction. Theorem 58 is therefore proved.

An immediate consequence of Theorem 58 is the result that in an (HT)space satisfying the first axiom of countability *a continuous and* (1, 1) *image of a closed and compact set is a homeomorphic image of that set.*

*Note that in Theorem 58 the condition that the set E be closed and compact can not be weakened. Consider, for example, the set $E = \{1, 1/2, \ldots\}$ contained in the space K of all real numbers and put $f(1/n) = 1/n$ for $n = 2, 3, \ldots, f(1) = 0$. The function f is obviously continuous and (1, 1) in the compact set E but its inverse function is not continuous in the set $T = f(E) = \{0, 1/2, 1/3, \ldots\}$ at the element $0 \in T$.

Similarly, let $E = \{1, 2, \ldots\}$, and put $f(n) = 1/n$ for $n = 2, 3, \ldots, f(1) = 0$. Again f is continuous and (1, 1) in the closed set E but its inverse function is not continuous in $T = f(E)$ at the element $0 \in T$.

However, the set E of all positive real numbers is neither closed nor compact but if f is a continuous and (1, 1) function which maps E on any set of real numbers then the inverse function of f is continuous in $T = f(E)$. There exists, however, a function f which maps E continuously and in a (1, 1) manner on a plane set but the inverse function of f is not continuous in $T = f(E)$.

It can also be shown that the condition that the space satisfy the first axiom of countability can not be omitted from the statement of Theorem 58. For let K denote the set of all ordinal numbers $< \Omega + \Omega$ with neighbourhoods defined as in § **39**; K is then an (HT)space which does not satisfy the first axiom of countability. Let E denote the set of all ordinal numbers α such that $\Omega \leqslant \alpha < \Omega + \Omega$. The set E is obviously closed and compact. Put $f(\alpha) = \alpha - \Omega$ for $\Omega < \alpha < \Omega + \Omega$ and $f(\Omega) = \Omega$. It is easy to see that the function $f(\alpha)$ is continuous and (1, 1) in E but its inverse function (i.e. the function $\phi(\xi) = \Omega + \xi$ for $1 \leqslant \xi < \Omega$, $\phi(\Omega) = \Omega$) is not continuous in the set $T = f(E)$ of all ordinal numbers ξ, where $1 \leqslant \xi \leqslant \Omega$, at the element Ω.

This example proves at the same time that even the condition that the space to which the functional values belong satisfy the first axiom of countability cannot be omitted from the statement of Theorem 58. For the set E is an (HT)space compact-in-itself and satisfying the first axiom of countability but the set T does not satisfy the first axiom of countability. It follows that for (HT)spaces the property of satisfying the first axiom of countability is not an invariant under (1, 1) and continuous transformations. (It may be easily seen, however, that this property is a topological invariant.)

We note further that one could easily obtain a continuous and (1, 1) mapping f of a (V)space K consisting of two elements on a (V)space K_1 consisting of two elements such that the inverse of f is not continuous in K_1. In fact, let $K = \{a, b\}$ denote a (V)space in which the elements a and b have respectively the neighbourhoods $\{a\}$ and $\{b\}$ and let $K_1 = \{c, d\}$ denote a (V)space in which the elements c and d have each the neighbourhood $\{c, d\}$.

Put $f(a) = c, f(b) = d$. The function f is obviously continuous and (1, 1) in K but its inverse function is not continuous in $K_1 = f(K)$ at any of its elements.*

Problem

Prove that in Theorem 58 the condition that the space be a Hausdorff space can not be omitted.

Solution. Let T denote the countable topological space defined in the solution of problem 2, § **39**. It is easily proved that T satisfies the first axiom of countability and that the sets P and Q are closed and compact. Put $f(q_n) = p_n$ for $n = 1, 2, \ldots$. The function f is (1, 1) and continuous in Q but its inverse function $\phi(p_n) = q_n$, for $n = 1, 2, \ldots$, is not continuous in the set $P = f(Q)$ at the element p_2.

41. The power of an aggregate of open (closed) sets.

LEMMA. *An infinite Hausdorff topological space contains an infinite sequence of non-empty disjoint open sets.*

Proof. We consider two cases.

(i) The (HT)space K is an isolated set. Hence every subset of K is both open and closed. Consequently any infinite sequence of elements of K^7 is a sequence of non-empty disjoint open sets.

(ii) The element $a \in K$ is a limit element of K. Let p_1 be an element of K different from a. There exist, by condition γ_1, open sets U_1 and V_1 such that $p_1 \in U_1, a \in V_1$, and $U_1 . V_1 = 0$. Since $a \in V_1$ and $a \in K'$, there exists in V_1 an element of K, say p_2, where $p_2 \neq a$. Again, by condition γ_1, there exist open sets U_2 and V_2 such that $p_2 \in U_2, a \in V_2, U_2 . V_2 = 0$; it may be assumed that $U_2 \subset V_1, V_2 \subset V_1$ (for otherwise, it would be sufficient to consider the intersection with V_1 of U_2 and V_2 respectively).

Suppose we have already determined the sequence $p_1, p_2, \ldots, p_n$ of elements of K and the sequences $U_1, U_2, \ldots, U_n$ and $V_1, V_2, \ldots, V_n$ of open sets, where $p_n \in U_n \subset V_{n-1}, a \in V_n \subset V_{n-1}$, and $U_n . V_n = 0$. Since $a \in V_n$ and $a \in K'$, there exists in V_n an element $p_{n+1} \in K$ different from a; by condition γ_1, there exist open sets U_{n+1} and V_{n+1} such that $p_{n+1} \in U_{n+1}, a \in V_{n+1}$, $U_{n+1} . V_{n+1} = 0$. As previously, we may assume that $U_{n+1} \subset V_n$ and $V_{n+1} \subset V_n$.

The infinite sequences $U_1, U_2, \ldots$ and $V_1, V_2, \ldots$ of open sets are thus defined by induction and we have $V_1 \supset V_2 \supset \ldots, U_{n+1} \subset V_n, U_n . V_n = 0$, for $n = 1, 2, \ldots$; from this we conclude that the sets $U_1, U_2, \ldots$ are disjoint. This proves the lemma.

Note that this lemma need not be true in a topological space which does not satisfy condition γ_1. Thus, in example 3 of § **23**, no two non-empty different open sets are disjoint.

COROLLARY. *The aggregate of all open (closed) sets contained in an infinite Hausdorff topological space has cardinal* $\geqslant \mathbf{c}$.

For if K is an infinite (HT)space there exists in K, by the above lemma, an infinite sequence of non-empty disjoint open sets. The sum of any finite or infinite subsequence of these sets is an open set. But this sequence has $\mathbf{c}$ different subsequences; hence there are at least $\mathbf{c}$ different open sets in K. On taking complements we obtain at once the result for closed sets.

The corollary and Theorem 43 give

THEOREM 59. *The aggregate of all open (closed) sets contained in an infinite Hausdorff topological space with a countable basis has cardinal* $\mathbf{c}$.

CHAPTER V

NORMAL TOPOLOGICAL SPACES

42. Condition of normality. A topological space K is said to be ***normal*** (Urysohn) if it satisfies the following

Condition of normality: If P and Q are two disjoint closed sets contained in K there exist two disjoint open sets contained in K one of which contains P and the other Q.

It follows from the Corollary to Theorem 40 that *normality of a topological space is a topological invariant.*

Since in a topological space a set consisting of one element only is closed, it is clear that condition γ_1 of § **35** is satisfied. Hence a normal topological space is a Hausdorff topological space.

Certain Hausdorff topological spaces which are not normal have also been investigated, namely those that satisfy the so-called

Condition of regularity (Tychonoff): *If p is an element of K and U any open set of K containing p, there exists an open set V which contains p and whose closure is contained in U.*

The condition of regularity is equivalent to condition T_3 of § **35**. (The proof is left to the reader.)

We shall show that in a topological space the condition of normality implies the condition of regularity. Suppose that $p \in K$, K is a normal topological space, and U is an open set of K containing p. The sets $\{p\}$ and $K - U$ are obviously closed and disjoint. There exist, therefore, by the condition of normality, open sets V and W such that $p \in V$, $K - U \subset W$, and $V \cdot W = 0$. Hence $V \subset K - W$ and, since $K - W$ is closed,

$$\bar{V} \subset \overline{K - W} = K - W \subset U$$

(since from $K - U \subset W$ we have $K - W \subset U$). We conclude that $\bar{V} \subset U$.

Not every topological space which satisfies the condition of regularity is normal (but it can be shown that every regular topological space is a Hausdorff topological space).

*Niemytzki[1] has given the following example of a regular topological space K which is not normal. K is the set of all points (x, y) in the plane with $y \geqslant 0$. A neighbourhood of the point $p = (x, y)$, $y > 0$, is the interior of any circle with centre at p and radius $< y$. A neighbourhood of $p = (x, 0)$ is a set consisting of p and the interior of any circle tangent to the x-axis at p and having its centre above the x-axis. It can be shown (the proof is rather difficult) that the set P of all the points of the x-axis with rational abscissae and the set Q of points on the x-axis with irrational abscissae are closed disjoint sets of K but that they cannot be covered by two disjoint open sets U and V such that $U \supset P$ and $V \supset Q$.

On the other hand, it is easy to give an example of a Hausdorff topological space which is not normal. Such a space is, for example, the set of all real numbers x of the interval $0 \leqslant x \leqslant 1$, where a neighbourhood of the number x is any set consisting of x and all the rational numbers of the above interval which differ from x by less than $1/n$ for n any natural number.

Furthermore, it can be shown that a regular topological space with a countable basis is normal.[2*]

Problems

1. Let T denote the set of all real numbers. Define the closure $\bar{X}$ of a set $X \subset T$ to be the set obtained by adjoining to X all its elements of condensation (in the ordinary sense). Prove that T is a Hausdorff topological space which does not satisfy the first axiom of countability nor the condition of regularity.

2. Prove that if a topological space T is normal and P and Q are two closed sets contained respectively in two disjoint open sets U and V (contained in T), then there exist open sets U_1 and V_1 such that $P \subset U_1$, $\bar{U}_1 \subset U$, and $Q \subset V_1$, $\bar{V}_1 \subset V$.

3. Prove that if the sets P and Q contained in an (HT)space with a countable basis are closed, compact, and disjoint, then there exist open sets U and V such that $P \subset U$, $Q \subset V$, and $U \cdot V = 0$.

Proof. Since T is an (HT)space and so satisfies condition γ_1, there exist for every pair of different elements p and q open sets $U(p, q)$ and $V(p, q)$ such that $p \in U(p, q)$, $q \in V(p, q)$, and $U(p, q) \cdot V(p, q) = 0$. Let p be a given element of P and $R(p)$ the family of all open sets $V(p, q)$ for $q \in Q$. The closed and compact set Q is therefore covered by the sets of the family $R(p)$ and, since T possesses a countable basis, it follows from the Borel-Lebesgue Theorem (§ **32**) that there exists a finite set $q_1, q_2, \ldots, q_m$, of elements of Q such that $Q \subset V(p, q_1) + V(p, q_2) + \ldots + V(p, q_m) = V(p)$. The set $V(p)$ is obviously open and, since $U(p, q_i) \cdot V(p, q_i) = 0$ for $i = 1, 2, \ldots, m$, the open set $U(p) = U(p, q_1) \ldots U(p, q_m)$ contains p and has no elements in common with the set $V(p)$. The set P is clearly contained in the sum of all sets $U(p)$, where $p \in P$; consequently there exists, as in the case of the set Q, a finite sequence $p_1, p_2, \ldots, p_n$ of elements of the set P such that $P \subset U(p_1) + U(p_2) + \ldots + U(p_n) = U$. Since $U(p) \cdot V(p) = 0$ for $p \in P$, we have, for $V = V(p_1) \cdot V(p_2) \ldots V(p_n)$, $U \cdot V = 0$. The sets U and V are open and satisfy the required conditions.

4. Show that a compact Hausdorff topological space is normal. The proof follows from the result of problem 3 and the fact that compactness of a set is a hereditary property.

43. The powers of a perfect compact set and a closed compact set.

THEOREM 60. *Every non-empty, perfect, and compact set contained in a normal space has cardinal* $\geqslant \mathbf{c}$.

We shall first prove the following

LEMMA. *Corresponding to every non-empty perfect set P contained in a normal space, there exist two non-empty perfect sets P_0 and P_1 such that*

$$(1) \qquad P_0 \subset P, P_1 \subset P, \textit{and } P_0 \cdot P_1 = 0.$$

Proof. Let P denote a given non-empty perfect set contained in a normal space K, p_0 and p_1 two different elements of P (such elements exist since a non-empty perfect set cannot consist of only one element). Since K is normal, there exist open sets U_0 and U_1 such that

$$p_0 \in U_0,\ p_1 \in U_1, \text{ and } U_0 \cdot U_1 = 0.$$

But the condition of normality implies the condition of regularity (§ **42**); hence there exist open sets V_0 and V_1 such that $p_0 \in V_0$, $p_1 \in V_1$, $\bar{V}_0 \subset U_0$, $\bar{V}_1 \subset U_1$. Furthermore, since $U_0 \cdot U_1 = 0$, we have $\bar{V}_0 \cdot \bar{V}_1 = 0$. Since $p_0 \in V_0 \cdot P$ and P is perfect and hence dense-in-itself, it follows that $V_0 \cdot P$ is dense-in-itself (see the proof of Theorem 47). The set

$$P_0 = \overline{V_0 \cdot P} \neq 0$$

is dense-in-itself by Theorem 6; hence it is perfect. Moreover, $P_0 \subset \bar{P} = P$ and $P_0 \subset \bar{V}_0$ (since $\overline{V_0 \cdot P} \subset \bar{P}$ and $\overline{V_0 \cdot P} \subset \bar{V}_0$).

Similarly, we conclude the existence of a non-empty perfect set $P_1 \subset P$ and $P_1 \subset \bar{V}_1$. Since $\bar{V}_0 \cdot \bar{V}_1 = 0$, we have $P_0 \cdot P_1 = 0$. This proves the lemma.

Since U_0 may be considered to be a subset of an open set containing p_0, we have actually proved that *every open set containing an element of a perfect set P contains a non-empty perfect proper subset of P.*

To prove Theorem 60 let P denote a non-empty perfect and compact set contained in a normal space. By the above lemma there exist perfect sets P_0 and P_1 which satisfy condition (1).

Let k be a given natural number and suppose all the non-empty perfect sets

$$P_{\alpha_1\alpha_2\ldots\alpha_k}$$

already defined, where $\alpha_1\alpha_2\ldots\alpha_k$ is an arrangement of k numbers each of which is either 0 or 1. By our lemma, there exist non-empty perfect sets

$$P_{\alpha_1\alpha_2\ldots\alpha_k 0} \text{ and } P_{\alpha_1\alpha_2\ldots\alpha_k 1}$$

which satisfy the following conditions:

$$P_{\alpha_1\alpha_2\ldots\alpha_k 0} \subset P_{\alpha_1\alpha_2\ldots\alpha_k}, \quad P_{\alpha_1\alpha_2\ldots\alpha_k 1} \subset P_{\alpha_1\alpha_2\ldots\alpha_k}, \tag{2}$$

$$P_{\alpha_1\alpha_2\ldots\alpha_k 0} \cdot P_{\alpha_1\alpha_2\ldots\alpha_k 1} = 0. \tag{3}$$

The sets $P_{\alpha_1\alpha_2\ldots\alpha_k}$, where $\alpha_1\alpha_2\ldots\alpha_k$ is any finite arrangement of the numbers 0 and 1, are thus defined by induction. Now let

$$\alpha_1, \alpha_2, \alpha_3, \ldots \tag{4}$$

be an infinite sequence formed with the numbers 0 and 1. Consider the intersection

$$P_{\alpha_1} \cdot P_{\alpha_1\alpha_2} \cdot P_{\alpha_1\alpha_2\alpha_3} \cdots. \tag{5}$$

The sets

$$P_{\alpha_1\alpha_2\ldots\alpha_k} \qquad (k = 1, 2, \ldots)$$

form, by (2), an infinite descending sequence of non-empty sets. Furthermore, as subsets of the compact set P, they are compact and, since they are perfect, they are also closed. Consequently, by Theorem 30 (Cantor), the intersection (5) is not empty. Let $p(\alpha_1, \alpha_2, \ldots)$ denote an element of this intersection.

Thus to every infinite sequence (4) formed with the numbers 0 and 1 corresponds an element $p(\alpha_1, \alpha_2, \ldots)$ of the set P. There are $\mathbf{c}$ such sequences. Moreover, to different sequences (4) correspond different elements of P. For let

$$\beta_1, \beta_2, \ldots$$

denote an infinite sequence formed with the numbers 0 and 1 and different from the sequence (4); then there exist indices n such that $\alpha_n \neq \beta_n$. Let m denote the smallest of them. Hence

$$\alpha_i = \beta_i \qquad \text{for } i = 1, 2, \ldots, m-1, \tag{6}$$

but $\alpha_m \neq \beta_m$; assume

$$\alpha_m = 0, \ \beta_m = 1. \tag{7}$$

It follows from the definition of the elements $p(\alpha_1, \alpha_2, \ldots)$ and $p(\beta_1, \beta_2, \ldots)$ that

$$p(\alpha_1, \alpha_2, \ldots) \in P_{\alpha_1\alpha_2\ldots\alpha_m} \tag{8}$$

and

$$p(\beta_1, \beta_2 \ldots) \in P_{\beta_1\beta_2\ldots\beta_m};$$

but from (6) and (7)

$$P_{\alpha_1\alpha_2\ldots\alpha_m} = P_{\alpha_1\alpha_2\ldots\alpha_{m-1}0},$$

$$P_{\beta_1\beta_2\ldots\beta_m} = P_{\alpha_1\alpha_2\ldots\alpha_{m-1}1},$$

and

$$P_{\alpha_1\alpha_2\ldots\alpha_{m-1}0} \,.\, P_{\alpha_1\alpha_2\ldots\alpha_{m-1}1} = 0.$$

Hence, from (8), $p(\alpha_1, \alpha_2, \ldots) \neq p(\beta_1, \beta_2, \ldots)$.

The set consisting of all the elements $p(\alpha_1, \alpha_2, \ldots)$ which correspond to the infinite sequences $\alpha_1, \alpha_2, \ldots$ formed with the numbers 0 and 1, has cardinal **c**. But this set is a subset of P; consequently, the cardinal of the set P is $\geqslant$ **c**. This completes the proof.

Let p be any element of a perfect and compact set P and let U be an open set containing p. As shown in the preceding lemma, the intersection $P \,.\, U$ contains a non-empty perfect subset. Furthermore, this subset is compact; hence, by Theorem 60, it has cardinal $\geqslant$ **c**. Thus:

If p is an element of a perfect and compact set P contained in a normal space, then every open set containing p contains a subset of P of cardinal $\geqslant$ **c**.

This gives immediately

COROLLARY 1. *Every element of a perfect, compact set contained in a normal space is an element of condensation of the set.*

Theorems 41 and 60 lead to

COROLLARY 2. *In a normal space with a countable basis every non-empty perfect, compact set has cardinal* **c**.

From Corollary 2 and Theorem 47 we obtain

COROLLARY 3. *In a normal space with a countable basis every non-countable closed and compact set has the power of the continuum, i.e., cardinal* **c**.

We note that the condition of compactness cannot be omitted in Theorem 60 or in Corollaries 1, 2, and 3. In fact, the set of all rational numbers (with the usual definition of neighbourhoods) is clearly a normal and perfect but countable space.

Nor can the condition of normality be omitted from Theorem 60, as can be proved by the example of a countable topological space T in which the closure of every infinite set is the whole space T. The condition of normality may, however, be replaced by the condition of regularity, as is easily seen from the proof of Theorem 60.

Furthermore, in Corollaries 2 and 3 the condition that the space be normal may be replaced by the condition that the space be an (HT)space. This follows from the result of problem 4 and the fact that a subset of an (HT)space is an (HT)space.

Finally, the inequality in Theorem 60 cannot be replaced by the sign of equality. For, given any cardinal $\mathbf{m} \geqslant \mathbf{c}$, it is possible to give an example of a normal continuum (which is by definition a perfect, compact, and non-empty set) of cardinal $\mathbf{m}$.

Thus, for instance, let ϕ denote any ordinal number >0 and $\mathbf{m}$ its cardinal. Denote by M the set of all pairs (ξ, t) where ξ is an ordinal number satisfying the inequality $0 < \xi < \phi$ and t is a real number such that $0 \leqslant t < 1$, and the pair $(\phi, 0)$. The set M will be ordered if we assume that $(\xi_1, t_1) \prec (\xi_2, t_2)$ if $\xi_1 < \xi_2$ and, where $\xi_1 = \xi_2$, if $t_1 < t_2$. If $p_1 \in M$ and $p_2 \in M$ and $p_1 \prec p_2$, then the open interval (p_1, p_2) is the set of all elements $p \in M$ such that $p_1 \prec p \prec p_2$. We define a neighbourhood of any element $p \in M$ other than $(0, 0)$ and $(\phi, 0)$ to be any open interval containing p. In case $p = (0, 0)$ or $p = (\phi, 0)$, a neighbourhood of p is every set consisting of p and an open interval with p as the left or right end-point respectively. M thus becomes a topological space. It can be proved that M is normal, dense-in-itself (hence perfect), compact, and connected. It is, therefore, a continuum. The set M has obviously cardinal $\mathbf{m}.\mathbf{c}$ and therefore, if $\mathbf{m}>\mathbf{c}$, it has cardinal $\mathbf{m}$. Hence there exist normal spaces which are continua of cardinal $> \mathbf{c}$.

44. Urysohn's lemma.

URYSOHN'S LEMMA. *If P and Q are two disjoint closed sets contained in a normal space K, then there exists a real function $f(p)$ defined and continuous in K such that $0 \leqslant f(p) \leqslant 1$ for $p \in K$, $f(p) = 0$ for $p \in P$, and $f(p) = 1$ for $p \in Q$.*

Proof. Let P and Q be two disjoint closed sets contained in a normal space K. Corresponding to every number r of the form $k/2^n$ ($k = 0, 1, \ldots, 2^n$; $n = 1, 2, \ldots$) define an open set $G(r)$ as follows: put $G(1) = K - Q$; the set $G(1)$ is open and, since $P \,.\, Q = 0$, $P \subset G(1)$. Since K is normal there exist open sets U and V such that $P \subset U, Q \subset V$, and $U \,.\, V = 0$. Put $G(0) = U$; we thus have

$$P \subset G(0) \text{ and } \overline{G(0)} \subset G(1). \tag{9}$$

This is true because $U \,.\, V = 0$ implies that $U \subset K - V$ which is closed. Thus $\bar{U} \subset K - V \subset K - Q = G(1)$.

Let m be a given natural number and suppose that we have already defined all open sets $G(k/2^{m-1})$, $k = 0, 1, \ldots, 2^{m-1}$, and where

$$\bar{G}(k/2^{m-1}) \subset G((k+1)/2^{m-1}), \quad k = 0, 1, \ldots, 2^{m-1} - 1. \tag{10}$$

(This is certainly true for $m = 1$.)

Next, let k be a given number of the sequence $0, 1, 2, \ldots, 2^{m-1}-1$. Put $P_1 = \bar{G}(k/2^{m-1})$ and $Q_1 = K - G((k+1)/2^{m-1})$. The sets P_1 and Q_1 are obviously disjoint and closed and $P_1 \subset G((k+1)/2^{m-1})$, by (10); consequently, as in the case of $P \subset G(1)$, we conclude that there exists an open set—denote it by $G((2k+1)/2^m)$—such that

$$\bar{G}(k/2^{m-1}) \subset G((2k+1)/2^m)$$

and

$$\bar{G}((2k+1)/2^m) \subset G((k+1)/2^{m-1}).$$

The open sets $G(k/2^m)$, $k = 0, 1, \ldots, 2^m$; $m = 1, 2, \ldots$, are thus defined by induction and they all satisfy (10).

We now define a real function $f(p)$ for $p \in K$. Let $p \in K$; if $p \in Q$, put $f(p) = 1$. If $p \notin Q$, then $p \in K - Q = G(1)$ and so there exist numbers $k/2^n (n = 1, 2, \ldots; k = 0, 1, \ldots, 2^n)$ such that $p \in G(k/2^n)$ (e.g., $k/2^n = 1$). If t denotes the lower bound of such numbers $k/2^n$, put $f(p) = t$.

Clearly $f(p) = 0$ for $p \in P$ since, by (9), $P \subset G(0)$. It is also clear that $0 \leqslant f(p) \leqslant 1$ for all $p \in K$ (since $0 \leqslant k/2^n \leqslant 1$). It remains to show that $f(p)$ is continuous in K.

Now let p_0 denote a given element of K and ϵ an arbitrary positive number. We shall consider three cases.

Case 1. $f(p_0) = 0$. Let m be a natural number such that $1/2^m < \epsilon$ and put $U = G(1/2^m)$; then U is an open set which contains p_0. For since $f(p_0) = 0$, there exists a number $k/2^n < 1/2^m$ such that $p_0 \in G(k/2^n)$; consequently, by (10), $G(k/2^n) \subset G(1/2^m)$. Consider $p \in U$; then from the definition of the function f we have $f(p) \leqslant 1/2^m < \epsilon$ and so, since $f(p) \geqslant 0$, we have $0 \leqslant f(p) < \epsilon$. This establishes the continuity of the function f in K at the element p_0.

Case 2. $f(p_0) = 1$. Choose a natural number m to satisfy the inequality $1/2^m < \epsilon$ and put $U = K - \bar{G}((2^m - 1)/2^m)$; U is an open set which contains p_0 (for if

$$p_0 \in \bar{G}((2^m - 1)/2^m) \subset G((2^{m+1} - 1)/2^{m+1}),$$

we would have, from the definition of $f(p)$, $f(p_0) \leqslant (2^{m+1} - 1)/2^{m+1} < 1$, contrary to hypothesis). For $p \in U$, we have $p \notin \bar{G}((2^m - 1)/2^m)$; hence $p \notin G((2^m - 1)/2^m)$ and so $p \notin G(k/2^n)$ for $k/2^n < (2^m - 1)/2^m$. Consequently,

$$f(p) \geqslant (2^m - 1)/2^m = 1 - 1/2^m > 1 - \epsilon.$$

On the other hand, we always have $f(p) \leqslant 1$; hence $0 \leqslant f(p_0) - f(p) < \epsilon$. The function is therefore continuous in K at p_0.

Case 3. $0 < f(p_0) < 1$. In this case it is clear that there exist numbers n and k such that $1/2^{n-1} < \epsilon$ and $k/2^n < f(p_0) < (k+1)/2^n < 1$. Consequently $p_0 \notin G(k/2^n)$ and therefore $p_0 \notin \bar{G}((k-1)/2^n)$. However $p_0 \in G((k+1)/2^n)$. Put

$$U = G((k+1)/2^n) - \bar{G}((k-1)/2^n);$$

U is an open set containing p_0. Consider $p \in U$; then $p \in G((k+1)/2^n)$

but $p \notin \bar{G}((k-1)/2^n)$ and so, certainly, $p \notin G((k-1)/2^n)$. From the definition of the function f we therefore have

$$(k-1)/2^n \leqslant f(p) \leqslant (k+1)/2^n;$$

hence

$$|f(p) - f(p_0)| < \epsilon \qquad \text{for } p \in U.$$

This establishes the continuity of the function at every element of K.

45. The power of a connected set. S is a connected set consisting of more than one element and contained in a normal space K; let p_0 be a given element of S and U any open set containing p_0. Then U contains more than one element of S. Otherwise, the sets $\{p_0\}$ and $S - \{p_0\}$ would be separated, contrary to the assumption that S is connected. There exists, therefore, in U an element $p_1 \in S$, where $p_1 \neq p_0$. Put $P = \{p_0\}$ and $Q = \{p_1\} + \{K - U\}$; these are clearly two disjoint closed sets. Hence, by Urysohn's Lemma (§ **44**), there exists a real function $f(p)$ defined and continuous in the whole space K such that $0 \leqslant f(p) \leqslant 1$, $f(p_0) = 0$, and $f(p) = 1$ for $p \in Q$. Since $f(p)$ is continuous in K, it is certainly continuous in $S \subset K$ and, since $f(p_0) = 0$ and $f(p_1) = 1$ ($p_1 \in Q$), the function f, by the corollary to Theorem 25 (§ **13**), must take on at the elements of the connected set S every value between 0 and 1. But $K - U \subset Q$ and $f(p) = 1$ for $p \in Q$; consequently $f(p) \neq 1$ for $p \in U$. It follows that $f(p)$ takes on all values between 0 and 1 for $p \in S \,.\, U$. The set $S \,.\, U$ therefore has cardinal $\geqslant \mathbf{c}$. This gives

THEOREM 61. *If S is a connected set consisting of more than one element and contained in a normal space and if U is an open set such that $S \,.\, U \neq 0$, then the set $S \,.\, U$ has cardinal $\geqslant \mathbf{c}$.*

It is noteworthy that (under the hypothesis of Theorem 61) the set $S \,.\, U$ need not be connected and it need not even contain any connected subset consisting of more than one element.[3]

Theorems 61 and 41 give

COROLLARY 1. *Every connected set consisting of more than one element and contained in a normal space with a countable basis has the power of the continuum.*

There exist, however, as stated (without proof) by Urysohn,[4] countable connected topological spaces with countable bases. It is easy to see that the space K of example 3, § **23**, is such a space for $\mathbf{m} = \aleph_0$.

Since a continuum is a closed, connected, and compact set containing more than one element (§ **39**), we obtain, in conjunction with Theorems 61 and 41,

COROLLARY 2. *If a set C is a continuum contained in a normal space with a countable basis, then the intersection of every open set with C, if non-empty, has cardinal* $\mathbf{c}$.

This property justifies the term "continuum."

CHAPTER VI

METRIC SPACES

46. Metric spaces. A set M is a *metric space* (Hausdorff) if with every pair of elements a and b of M there is associated a real non-negative number[1] $\rho(a, b)$ called the *distance* between the elements a and b and subject to the following three conditions (so-called distance axioms):

1. $\rho(b, a) = \rho(a, b)$ (symmetry law);

2. $\rho(a, b) = 0$ if and only if $a = b$ (identity law);

3. $\rho(a, c) \leqslant \rho(a, b) + \rho(b, c)$ for every three elements a, b, c of M (the triangle law).[2]

A subset of a metric space is evidently a metric space. Elements of a metric space are frequently called points.

*Menger[3] calls a space M *semi-metric* if with every pair of elements of M there is associated a non-negative number subject to conditions 1 and 2 (but not necessarily 3). Semi-metric spaces have also been investigated by Chittenden.[4] There exist semi-metric spaces which are not topological and there are topological spaces whose homeomorphic images are not semi-metric spaces.[5]

Even more general spaces have been investigated in which the distance function satisfies neither the triangle law nor the symmetry law.[6] Alexandroff and Hopf[7] call a space in which there is defined a distance function satisfying no specific conditions an abstract metric space. Birkhoff[8] has investigated spaces in which condition 2 is replaced by the condition $a = b$ implies $\rho(a, b) = 0$. Spaces with a so-called *weak metric* have been studied by Ribeiro.[9] Here the distance function satisfies condition 3 and the condition that $a = b$ implies $\rho(a, b) = 0$ but does not necessarily satisfy conditions 1 and 2 (for example, the distance in a mountainous countryside might be measured by the time required to traverse it).

Furthermore, spaces with the conditions 1, 2, and a weaker form of 3 have also been studied as, for example, the *almost-metric* spaces of Menger.[10] Menger[11] has investigated spaces in which the distance is not necessarily a real number.

Finally, Bieberbach[12] calls a (V)space *locally metric* if a distance function is defined in it in such a manner that each of the elements possesses a neighbourhood which is a metric space.*

Examples of metric spaces

1. Let M be a given set and $\rho(x, y)$ an arbitrary symmetric function of the two variables x and y defined for $x \in M$, $y \in M$ in such a manner that $\rho(x, y) = 0$ for $x = y$, and $1 \leqslant \rho(x, y) \leqslant 2$ for $x \neq y$. M is evidently a metric space since, for $x \neq y \neq z$, we have

$$\rho(x, y) + \rho(y, z) \geqslant 1 + 1 = 2 \geqslant \rho(x, z).$$

In particular, we may assume $\rho(x, y) = 1$ for $x \neq y$. Thus every set becomes a metric space if we assume that the distance between any two different elements is equal to 1. *There exist, therefore, metric spaces of arbitrary power.*

2. The set P of all points in the plane becomes a metric space if the distance $\rho(p_1, p_2)$ between the points $p_1 = (x_1, y_1)$ and $p_2 = (x_2, y_2)$ is defined to be the number

$$\rho(p_1, p_2) = ((x_1 - x_2)^2 + (y_1 - y_2)^2)^{\frac{1}{2}},$$

where the radical is to be taken with the positive sign.

But the set P will also become a metric space if the distance between the points $p_1 = (x_1, y_1)$ and $p_2 = (x_2, y_2)$ is defined to be the number

$$\rho_1(p_1, p_2) = |x_1 - x_2| + |y_1 - y_2|$$

(since the function ρ_1, as well as the function ρ, satisfies the three distance axioms).

It is easily seen that we do not have $\rho(p_1, p_2) = \rho_1(p_1, p_2)$ for every pair p_1, p_2 of elements of P; there are, therefore, *two different metrics* ρ and ρ_1 defined in the set P.

*There arises now the question—*how many different metrics can be defined in a given set?*

Consider first a set consisting of two elements. A metric in this set is defined as soon as the distance between the two elements is given. Since this distance may be any real positive number, it is clear that $\mathbf{c}$ different metrics can be defined in a set consisting of two elements. The same result follows readily for every finite set containing more than one element. We shall now prove

THEOREM 62. *The set of all different metrics that can be established in an infinite set of cardinal* $\mathbf{m}$ *has cardinal* $2^{\mathbf{m}}$.

Proof. A metric in a set M of cardinal $\mathbf{m} \geqslant \aleph_0$ is defined by a real function of two variables which are elements of the set M and, since there are $\mathbf{c}^{\mathbf{m}^2} = 2^{\mathbf{m}}$ such functions, there are at most $2^{\mathbf{m}}$ different metrics in a set of cardinal $\mathbf{m}$.

Now let E denote any subset of M and put

$$\rho_E(a, b) = 1 \qquad \text{for } a \in E \text{ and } b \in E,$$

$$\rho_E(a, b) = 0 \qquad \text{if } a = b,$$

$$\rho_E(a, b) = 2$$

otherwise. The function $\rho_E(a, b)$ obviously satisfies the distance axioms (see example 1). If now E_1 and E_2 be any two different subsets of M, each consisting of more than one element, and if $a \in E_1 - E_2$, $b \in E_1$, or $a \in E_2 - E_1$, $b \in E_2$, then $\rho_{E_1}(a, b) \neq \rho_{E_2}(a, b)$. Hence to different subsets of M, consisting of more than one element, correspond different metrics; since the set of these subsets has cardinal $2^{\mathfrak{m}}$, the set of all different metrics in the set M has cardinal $\geqslant 2^{\mathfrak{m}}$. But, as proved above, there can be at most $2^{\mathfrak{m}}$ different metrics in the set M; hence the cardinal of the set of all different metrics is $2^{\mathfrak{m}}$. This proves the theorem.*

47. Congruence of sets. Equivalence by division. Let E and E_1 be subsets of metric spaces M and M_1 respectively with metrics ρ in M and ρ_1 in M_1. (In particular, we may have $\rho = \rho_1$ and even $M = M_1$.) If it is possible to establish a (1,1) correspondence f between the elements of E and E_1 in such a manner that

$$\rho_1(f(p), f(q)) = \rho(p, q), \qquad p \in E \text{ and } q \in E \tag{1}$$

then the set E is said to be *congruent* to the set E_1. The set E_1 is then congruent to the set E since (1) is equivalent to the relation

$$\rho(f^{-1}(a), f^{-1}(b)) = \rho_1(a, b), \qquad a \in E_1 \text{ and } b \in E_1.$$

The sets E and E_1 are then said to be congruent to each other or *isometric;* in symbols $E \simeq E_1$. The relation of congruence of sets is therefore symmetric. Furthermore, it is transitive and reflexive.

A function f satisfying relation (1) is said to establish an isometric mapping of the set E on the set $E_1 = f(E)$. It follows readily from (1) that f is (1,1) in E. For if $p \neq q$, $p \in E$ and $q \in E$, we have $\rho(p, q) \neq 0$, by condition 2. Hence, by (1), $\rho_1(f(p), f(q)) \neq 0$ and so, since ρ_1 satisfies condition 2, $f(p) \neq f(q)$.

Relation (1) may be satisfied by two (or more) different functions f defined in E. Thus two sets may be congruent in different ways (for example, two line segments of equal length are congruent either by translation or by rotation). In some cases, however, the congruence of two sets may be unique (e.g., the two half rays $x \leqslant 0$ and $x \geqslant 0$).

Two metric spaces consisting of the same elements but with different metrics may be isometric. For example, the space consisting of the three elements p_1, p_2, and p_3 with the metric ρ, where $\rho(p_1, p_2) = 3$, $\rho(p_1, p_3) = 4$,

$\rho(p_2, p_3) = 5$, and the space consisting of the same elements with metric ρ_1, where $\rho_1(p_1, p_2) = 5$, $\rho_1(p_1, p_3) = 3$ and $\rho_1(p_2, p_3) = 4$. For, putting $f(p_1) = p_3$, $f(p_2) = p_1$, and $f(p_3) = p_2$, we obtain an isometric mapping f of the first space on the second.

*Certain theorems of an apparently simple nature concerning the congruence of sets are actually very difficult to establish, even if we restrict ourselves to the case of linear point sets, and have been proved only on the assumption of the axiom of choice. One such theorem is the theorem of Kuratowski[13] which states: If a set E contained in a metric space be divided in two ways into the sum of two disjoint and congruent sets, say $E = M_1 + M_2 = N_1 + N_2$, where $M_1 \simeq M_2$ and $N_1 \simeq N_2$, then the sets M_1 and N_1 are each sums of four disjoint sets correspondingly congruent, that is $M_1 = P_1 + P_2 + P_3 + P_4$, $N_1 = Q_1 + Q_2 + Q_3 + Q_4$, where $P_i \simeq Q_i$ for $i = 1, 2, 3, 4$.

The generalization of this theorem by König and Valko[14] is even more difficult. This states: If a set E be divided in two ways into a finite number n of disjoint and congruent subsets, say $E = M_1 + M_2 + \ldots + M_n = N_1 + N_2 + \ldots + N_n$, where $M_i \simeq M_k$, $N_i \simeq N_k$ for $i = 1, 2, \ldots, n$, $k = 1, 2, \ldots, n$, then the sets M_1 and N_1 are each sums of n^2 disjoint sets: $M_1 = P_1 + P_2 + \ldots + P_{n^2}$, $N_1 = Q_1 + Q_2 + \ldots + Q_{n^2}$, where $P_i \simeq Q_i$ for $i = 1, 2, \ldots, n^2$. The proof is very difficult even for $n = 3$.

A set may be congruent to a proper subset of itself; for example the half-line $x \geqslant 0$ and the half-line $x \geqslant 1$. A set which is not congruent to any of its proper subsets is called monomorphic by Lindenbaum.[15]

It can be shown that every bounded linear set is monomorphic. It is easy, however, to construct a bounded set of points in the plane which is not monomorphic. For example, starting from a fixed point on the circumference of a circle, measure off along the circumference an infinite number of times an arc of fixed length which is not a rational multiple of the circumference, proceeding always in the same direction. The end-points of these arcs form the required set.

One can also construct *an unbounded plane set which is the sum of two disjoint sets each of which is congruent to the original set.*[16] This set is obtained as follows:

Let ϕ denote a translation of the plane through unit distance in the direction of the x-axis and ψ a rotation of the plane through one radian about the point (0, 0). Let E be the set obtained by the application of the two transformations ϕ and ψ to the point (0, 0) a finite number of times in any order.[17] Put $A = \phi(E)$ and $B = \psi(E)$. It is obvious that $A \simeq E$, $B \simeq E$, $A \subset E$, $B \subset E$, $E = A + B$. Further, it can be shown that $A \cdot B = 0$ (the proof rests on the transcendental nature of the radian).

Furthermore,[18] for every cardinal number $\mathbf{m} \leqslant \mathbf{c}$ there exists a plane set P which is the sum of $\mathbf{m}$ disjoint sets each congruent to P. It can be shown, however, that no linear set can be expressed as the sum of two disjoint sets each congruent to the original set. Nor can a bounded plane set be so expressed. However, such sets exist in R_3 on the surface of a sphere.

With the aid of the axiom of choice it can be proved that every segment of a straight line is the sum of a countable aggregate of disjoint sets all congruent to each other;[19] and that, for every cardinal $\mathbf{m} \leqslant \mathbf{c}$, the straight line is the sum of $\mathbf{m}$ disjoint sets all congruent to each other.[20]

It is readily deduced that if a metric space is congruent to a proper subset of itself it is itself a proper subset of a metric space to which it is congruent and conversely.*

If a metric space E is isometric with a subset of a metric space M, then we say that E can be *embedded* (metrically) in the space M. Clearly every metric space consisting of two elements can be embedded in a linear space and one consisting of three elements can be embedded in a plane (with the usual definition of distance). There exist, however, metric spaces consisting of four elements which cannot be embedded in a three-dimensional space (and not even in an n-dimensional space, as we shall show in § **57**).

*If Φ is any family of metric spaces, then there exists a metric space K in which every metric space M of the family Φ can be embedded.

For let K denote the set of all pairs (p, M), where $p \in M \in \Phi$. Let p_M be a fixed element of the space M for each $M \in \Phi$. Define the distance $\rho(q_1, q_2)$ of two elements of K as follows:

If $q_1 = (p_1, M_1)$, $q_2 = (p_2, M_2)$ and $M_1 = M_2 = M$, put

$$\rho(q_1, q_2) = \rho_M(p_1, p_2),$$

where ρ_M is the distance in the space M; if $M_1 \neq M_2$, put

$$\rho(q_1, q_2) = \rho_{M_1}(p_1, p_{M_1}) + \rho_{M_2}(p_2, p_{M_2}) + 1.$$

It is easy to see that the function ρ so defined satisfies the three distance axioms and that the space M $(M \in \Phi)$ is congruent to the subset of K consisting of all the elements (p, M), where $p \in M$. Thus K has the required properties.

We note further that from the so-called Cantor hypothesis about the alephs it follows that, for every cardinal number $\mathbf{m} > \aleph_0$, there exists a metric space $U_{\mathbf{m}}$ of cardinal $\mathbf{m}$ in which every metric space of cardinal $\mathbf{m}$ can be embedded.[21] In particular, the continuum hypothesis is equivalent to the existence of such a space $U_{\aleph_1}$.

Two metric spaces are said to be *locally congruent* if there exists a (1, 1) correspondence between their points under which some neighbourhood[22] of every point in one space is isometric with some neighbourhood of the corresponding point in the other.[23] Two metric spaces may be locally isometric without being isometric. For example, the linear space R_1 and the space P consisting of all real numbers in which the distance $\rho(x, y) = \min(|x-y|, 1)$.*

Examples

1. Give an example of two linear sets which are not congruent although each is congruent to a certain subset of the other.

Let A consist of all real numbers $\geqslant 1$ and $B = A + \{0\}$. The set B is congruent to a subset of A obtained from B by a translation through unit length. But the set A is not congruent to the set B.

Another example is supplied by the countable linear sets $A_1 = \{2, 3, 4, \ldots\}$ and $B_1 = A_1 + \{0\}$.

2. A set T contained in a metric space is called *k times* larger than a set E if there exists a (1, 1) mapping of E on T under which the distance between any two points in T is k times the distance between the corresponding points in E.

(a) Give an example of a linear set congruent to a set twice as large.

The set of all rationals, the set of all reals or the set of all numbers 2^k, where k is an integer, are such sets.

(b) Construct a linear set T which is the sum of two disjoint sets each of which is twice as large as T.

T may be the set of all natural numbers and the two disjoint sets may be the set of all odd numbers and the set of all even numbers.

(c) Construct a linear set T which is the sum of a countable aggregate of disjoint sets each of which is at least twice as large as T.

Put $T = \{1, 2, 3, \ldots\}$ and $T_n = \{2^{n-1}.1, 2^{n-1}.3, 2^{n-1}.5, \ldots\}$. Hence $T = T_1 + T_2 + \ldots$, $T_k . T_l = 0$ for $1 \leqslant k < l$, and T_n is 2^n times larger than T.

If each of two given sets A and B contained in metric spaces can be divided into n disjoint subsets, say $A = A_1 + A_2 + \ldots + A_n$, $B = B_1 + B_2 + \ldots + B_n$, correspondingly congruent (i.e., $A_1 \simeq B_1$, $A_2 \simeq B_2, \ldots, A_n \simeq B_n$), then the sets A and B are said to be *equivalent by division* (into n parts). In symbols, $A \underset{n}{=} B$.

Two sets, each of which is congruent to a subset of the other, need not be congruent to each other (see example 1). Such sets are, however, congruent by division into two subsets. (The proof of this fact rests on the so-called Banach theorem in the general theory of sets). It can be proved, more generally, that, if P is equivalent by division to a subset of Q and Q to a subset of P, then the sets P and Q are equivalent by division.[24] Furthermore,[25] if $A \supset E \supset B$ and $A \underset{n}{=} B$, then $A \underset{n+1}{=} E$.

Two sets which are equivalent by finite division to a third are equivalent by finite division.[26] If $A_1 \subset A$, $B_1 \subset B$, $A \simeq B$, $A_1 \simeq B_1$, then the sets $A - A_1$ and $B - B_1$ need not be congruent and they may not even be equivalent by division.

Among the more curious results obtained in this theory we may mention the following: In space of three dimensions, two spheres of different radii are equivalent by division; but in a plane two circles of different radii are not equivalent by division.[27] It is not known whether a circle and a square of the same area are equivalent by division, although a sphere and a cube (even of different volumes) are equivalent by division.

We note here that the equivalence of polygons (or polyhedra) studied in elementary geometry is not the same as equivalence by division of these configurations. Two polygons (polyhedra) are said to be equivalent (in the elementary sense) if they can be divided into the same finite number of non-overlapping polygons (polyhedra) correspondingly congruent (by superposition).[28] Thus, for example, we can prove that an isosceles triangle is equivalent (in the elementary sense) to a rectangle by dividing both into two right-angled triangles by cutting the first along its altitude and the second along a diagonal.

*The following theorem, usually assumed without proof in elementary geometry and sometimes called the *axiom of de Zolt*,[28] plays a fundamental part in the theory of equivalence of polygons:

If a polygon P is a part of a polygon G, then the two polygons are not equivalent.

Assuming axioms of elementary geometry, Hilbert[29] has shown that the above theorem can be proved; since the proof is difficult, it is usually omitted from textbooks in elementary geometry on the high school level.

Assuming de Zolt's axiom, the following theorem in the theory of measure, which gives at the same time a necessary and sufficient condition for the equivalence of two polygons, can be proved:

Two polygons P and G are equivalent if and only if they have the same area.

Tarski[28] has proved that both these theorems remain valid when the equivalence of polygons be understood in the set—theoretic sense (i.e., as equivalence by division). It is noteworthy that the proofs of these theorems, the first of which may even appear obvious, are very involved. They rest on results of Banach[30] which require the complete apparatus of modern set theory, including the axiom of choice.

We remark further that statements analogous to the above two theorems, but concerning polyhedra instead of polygons, are not true in the case of equivalence by division. For we can derive the result: *Any two solids are equivalent by division.* This implies, in particular, that *any cube can be divided into a finite number of disjoint subsets which may be put together to form another cube with edge twice as large.*

On the other hand, as proved by Dehn,[31] two polyhedra even of the same volume need not be equivalent in the sense of elementary geometry. In particular, a regular tetrahedron is not equivalent (in the elementary sense) to the sum of two regular tetrahedra.[32*]

48. Open spheres. Let M denote a given metric space with metric ρ, p a point of M, and r a given positive number. The set of all points q of M which satisfy the condition

$$\rho(p, q) < r$$

is called an open sphere (or the interior of a sphere) with centre at p and radius r and is denoted by $S(p, r)$.

We shall now show that if the neighbourhoods of a point p of M are taken to be all open spheres with centre p, then M becomes a normal topological space satisfying the first axiom of countability.

With this definition of neighbourhoods, conditions α and β of § **19** are immediately satisfied. Condition γ is also satisfied for, if b is an element of M different from a, the open sphere $S(a, r)$, where $0 < r < \rho(a, b)$, is a neighbourhood of a which does not contain b.

Let $b \in S(a, r)$; hence $\rho(a, b) < r$. Put $r_1 = r - \rho(a, b)$, i.e., $r_1 > 0$. If $q \in S(b, r_1)$, then $\rho(q, b) < r$ and so, from properties 3 and 1 of distance, we obtain

$$\rho(q, a) \leqslant \rho(q, b) + \rho(b, a) < r_1 + \rho(b, a) = r.$$

Hence $\rho(q, a) < r$ which proves that $q \in S(a, r)$. Consequently $S(b, r_1) \subset S(a, r)$ and so condition δ of § **19** is satisfied.

The metric space M is therefore a topological space (with neighbourhoods as defined above). We shall next show that M possesses a property which implies normality (§ **42**).

THEOREM 63. *If P and Q are two separated sets contained in a metric space M, then there exist open sets U and V such that $P \subset U$, $Q \subset V$, and $U \,.\, V = 0$.*

Proof. P and Q are two separated sets contained in a metric space M. Hence $P \,.\, Q = 0$ and $P \,.\, Q' = 0$. Thus, if $p \in P$, then $p \notin Q + Q'$; there exists, therefore, a neighbourhood of p free of elements of Q and so, from the definition of neighbourhoods in M, there exists a positive number $r = \phi(p)$ such that $S(p, r) \,.\, Q = 0$. Similarly, there exists for every element $q \in Q$ a positive number[33] $r = \phi(q)$ such that $S(q, r) \,.\, P = 0$. Put

$$U = \sum_{p \in P} S(p, \tfrac{1}{2}\phi(p)), \quad V = \sum_{q \in Q} S(q, \tfrac{1}{2}\phi(q)). \tag{2}$$

The sets U and V are open and $P \subset U$, $Q \subset V$. It remains to show that $U \cdot V = 0$.

Assume that $U \cdot V \neq 0$; then there exists an element a such that $a \in U$ and $a \in V$. Hence there are elements $p \in P$ and $q \in Q$ such that $a \in S(p, \frac{1}{2}\phi(p))$ and $a \in S(q, \frac{1}{2}\phi(q))$. This gives

$$\rho(p, a) < \tfrac{1}{2}\phi(p), \ \rho(q, a) < \tfrac{1}{2}\phi(q)$$

from which

$$\rho(p, q) \leqslant \rho(p, a) + \rho(a, q) < \tfrac{1}{2}\phi(p) + \tfrac{1}{2}\phi(q).$$

If $\phi(p) \geqslant \phi(q)$, then $\rho(p, q) < \phi(p)$; hence $q \in S(p, \phi(p))$ contrary to the definition of the number $\phi(p)$. Similarly, we arrive at a contradiction if we assume that $\phi(p) \leqslant \phi(q)$. Hence $U \cdot V = 0$ and Theorem 63 is thus proved.

Since two non-empty disjoint closed sets are separated, it follows from Theorem 63 that a metric space is normal.

Furthermore, let a be a given element of a metric space M. Put $V_n = S(a, 1/n)$ for $n = 1, 2, \ldots$ and let V be any open set containing a. Then there exists a neighbourhood U of a which is contained in V, where $U = S(a, r)$ is an open sphere and where r is a certain positive number. But, for $n > \mu \geqslant 1/r$, we have $V_n \subset U$ and so, since $a \in V_n$ for $n = 1, 2, \ldots$, the space M satisfies the first axiom of countability (§ **27**).

We have thus proved that *every metric space* (with neighbourhoods as defined above) *is a normal topological space satisfying the first axiom of countability*. Hence all theorems proved in chapters I, II, IV, and V hold in every metric space. It can be shown, however, that of all theorems of chapter III only Theorem 52 holds in every metric space.

We know that if a set E contained in a topological space K is the derived set of some subset of K then $E' \subset E \subset K'$ (§ **20**). It can be shown that in a metric space the converse also is true. Consequently:

A set E contained in a metric space M is the derived set of a subset of M if and only if

$$E' \subset E \subset M'.$$

49. Continuity of the distance function. From the definition of continuity in (V)spaces (§ **11**) and the definition of neighbourhoods in metric spaces it follows that *a function f defined in a set E contained in a metric space with distance function ρ and taking on values in a metric space with distance function ρ_1 is continuous at $p_0 \in E$ if and only if for every number $\epsilon > 0$ there exists a number $\delta > 0$ such that the conditions*

$$p \in E, \ \rho(p, p_0) < \delta$$

imply the inequality

$$\rho_1(f(p), f(p_0)) < \epsilon;$$

or, what amounts to the same thing, *if and only if the conditions*

$$p_n \in E, \lim_{n\to\infty} \rho(p_n, p_0) = 0$$

imply

$$\lim_{n\to\infty} \rho_1(f(p_n), f(p_0)) = 0.$$

We have seen (§ **35**) that if $\lim_{n\to\infty} p_n = p_0$, then for every $\epsilon > 0$ there exists a natural number μ such that $p_n \in S(p_0, \epsilon)$ for $n > \mu$, i.e., $\rho(p_n, p_0) < \epsilon$ for $n > \mu$. Consequently

$$\lim_{n\to\infty} \rho(p_n, p_0) = 0.$$

On the other hand, if U is an arbitrary neighbourhood of p_0 then, from the definition of neighbourhoods in a metric space, we have, for some $r > 0$, $U = S(p_0, r)$. If now $\lim_{n\to\infty} \rho(p_n, p_0) = 0$, then there exists a natural number μ (which depends on r and therefore on U) such that $\rho(p_n, p_0) < r$ for $n > \mu$. Hence $p_n \in S(p_0, r) = U$ and

$$\lim_{n\to\infty} p_n = p_0.$$

Thus in a metric space the relations

$$\lim_{n\to\infty} p_n = p_0 \text{ and } \lim_{n\to\infty} \rho(p_n, p_0) = 0$$

are equivalent.

THEOREM 64. *The distance $\rho(p, q)$ between two elements p and q of a metric space is a continuous function of the two variables p and q.*

Proof. Let ρ be the distance function in the metric space M, p_0 and q_0 two given points of M, and ϵ a given positive number. For

$$p \in M, \ q \in M, \ \rho(p, p_0) < \epsilon/2, \ \rho(q, q_0) < \epsilon/2,$$

we have

$$\rho(p, q) \leqslant \rho(p, p_0) + \rho(p_0, q_0) + \rho(q_0, q) < \rho(p_0, q_0) + \epsilon$$

and

$$\rho(p_0, q_0) \leqslant \rho(p_0, p) + \rho(p, q) + \rho(q, q_0) < \rho(p, q) + \epsilon;$$

hence

$$-\epsilon < \rho(p, q) - \rho(p_0, q_0) < \epsilon.$$

This establishes the continuity of the function ρ in M.

50. Separable metric spaces.

THEOREM 65. *A separable metric space possesses a countable basis.*

Proof. Let M denote a separable metric space. Then there exists a finite or countable set $P \subset M$ such that $M = \bar{P}$. Let $P = \{p_1, p_2, \ldots\}$. The aggregate B of all sets $S(p_m, 1/n)$, where m and n are natural numbers, is obviously countable. We shall show that every open set contained in M is the sum of sets belonging to B.

Let U be an open set $\subset M$ and $p \in U$. Then there exists a number $r > 0$ such that $S(p, r) \subset U$. Choose a natural number $n > 2/r$. Since $M \subset \bar{P}$ there exists an element $p_m \in P$ such that $\rho(p, p_m) < 1/n$. If $q \in M$ and $\rho(q, p_m) < 1/n$, we have

$$\rho(p, q) \leqslant \rho(p, p_m) + \rho(p_m, q) < 2/n < r.$$

This implies that $q \in S(p, r) \subset U$ from which we conclude that $S(p_m, 1/n) \subset U$. Since $\rho(p, p_m) < 1/n$, $p \in S(p_m, 1/n)$. Thus every element p of U is contained in at least one set of B which, in turn, is contained in U. Consequently U is the sum of all those sets of B which are contained in U. Therefore the aggregate B of open sets constitutes a countable basis in the space M.

It follows from Theorems 65 and 42 that *in a metric space the properties of being separable and possessing a countable basis are equivalent.* We have furthermore

THEOREM 66. *Every separable metric space is hereditarily separable.*

Theorem 66 does not hold in every topological space (see example 5, § **23**). It can be shown, however, that it holds in all topological spaces satisfying the first axiom of countability.

It follows further from Theorem 65 that all theorems of chapter III hold in every separable metric space.

*A metric space M satisfies *the generalized Bolzano-Weierstrass theorem* if every infinite sequence of sets contained in M contains a topologically convergent subsequence (§ **18**). It has been shown[34] that every separable metric space satisfies the generalized Bolzano-Weierstrass theorem and, with the aid of the continuum hypothesis, that every metric space satisfying the generalized Bolzano-Weierstrass theorem is separable.[35]

Local separability. A metric space M is *separable at a point p* if for some number $r > 0$ the set $M \,.\, S\,(p, r)$ is separable. Urysohn[36] has given an example of a metric space which is not separable at any of its points; another example is given at the end of § **63**.

A metric space M is said to be *locally separable* if it is separable at each of its points. It has been proved[37] that *a non-separable metric space is locally separable if and only if it is the sum of two disjoint separable open sets.* It is clear that *a connected metric space which is locally separable is separable.*[38]

Note that a metric space which is not locally separable may become locally separable by the removal of a single point. Consider, for example, the connected space M consisting of the pair (0, 0) and all pairs (x, y) of positive real numbers, where $\rho((x_1, y_1), (x_2, y_2)) = |y_1 - y_2|$ for $x_1 = x_2$ and is equal to $y_1 + y_2$ for $x_1 \neq x_2$. This space is not locally separable, but becomes locally separable when the point (0, 0) is removed.*

51. Properties of compact sets.

THEOREM 67. *If E is a compact subset of a metric space then, corresponding to every number $\epsilon > 0$, there exists a finite sequence $p_1, p_2, \ldots, p_m$ of elements of E such that every element p of E is at a distance less than ϵ from at least one element of the sequence.*

Some authors (e.g., Kuratowski) call sets for which Theorem 67 holds *totally bounded.*

Proof. Let E be a compact set contained in a metric space M with distance function ρ, and let ϵ be a given positive number. Let p_1 be any element of E. If there are elements p of E such that $\rho(p, p_1) \geqslant \epsilon$, let p_2 be one of them. Let n be a natural number and suppose we have already defined the elements $p_1, p_2, \ldots, p_n$ of the set E. If there are elements $p \in E$ such that $\rho(p, p_k) \geqslant \epsilon$ for $k = 1, 2, \ldots, n$, let p_{n+1} be one of them.

The set $p_1, p_2, \ldots$ defined by induction cannot be infinite. For, suppose it were; then the set E_1, consisting of all the elements of the sequence, would have, as an infinite subset of a compact set, at least one limit element, say a. The sphere $S(a, \epsilon/2)$ would contain infinitely many elements of E_1. Hence there would exist indices k and $l > k$ such that

$$\rho(a, p_k) < \epsilon/2 \text{ and } \rho(a, p_l) < \epsilon/2$$

and so, by the triangle law,

$$\rho(p_k, p_l) \leqslant \rho(p_k, a) + \rho(a, p_l) < \epsilon$$

contrary to the definition of the sequence $\{p_n\}$ where, for $l > k$, $\rho(p_l, p_k) \geqslant \epsilon$. This proves the theorem.

In the above proof use was made of the axiom of choice in the definition of the infinite sequence $p_1, p_2, \ldots$ since no rule was given for the selection of these elements. In fact, we can give an example of a compact linear set E from which, even for $\epsilon = 1$, no finite set of points $p_1, p_2, \ldots, p_n$ could be selected to satisfy Theorem 67. It is sufficient here to take a certain non-empty subset of the interval (0,1) from which we could not select an element.[39]

Theorem 67 may be restated as follows:

Every compact subset of a metric space can be covered by a finite number of open spheres of arbitrarily small radii. Or else: *Every compact set contained in a metric space is totally bounded.*

The converse of the last statement need not be true; the set of all rational numbers in the interval (0, 1) is totally bounded but not compact. Total boundedness of a set is clearly a hereditary property but it is not a topological one.

In Theorem 67, assume $\epsilon = 1/n$, where n is a given natural number, and denote corresponding sequence by $p_1^n, p_2^n, \ldots, p_m^n$. Further, let P denote the set of all different terms of the infinite sequence $p_1^1, p_2^1, \ldots p_{m_1}^1$, $p_1^2, p_2^2, \ldots, p_{m_2}^2, p_1^3, p_2^3, \ldots, p_{m_3}^3, p_1^4, \ldots$.

For every element $p \in E$ and every natural number n there exists an index $k \leqslant m_n$ (dependent on p and n) such that

$$\rho(p, p_k^n) < 1/n;$$

hence $p_k^n \in S(p, 1/n)$ and so $P \,.\, S(p, 1/n) \neq 0$. This being true for all n it follows that p is either an element or a limit element of P; consequently $E \subset \bar{P}$. Since P is either a finite or a countable subset of E, this gives

THEOREM 68. *Every compact set contained in a metric space is separable.*

COROLLARY. *A closed and compact set contained in a metric space is the closure of some finite or countable set.*

52. The diameter of a set and its properties. If E is a subset of a metric space then the upper bound of the distances of all pairs of points of E (i.e., $\sup \rho(p, q)$ for $p \in E$ and $q \in E$) is called *the diameter of* E and is denoted by $\delta(E)$. Thus the diameter of every non-empty set contained in a metric space is a uniquely defined real non-negative number, finite or infinite. A set is said to be bounded if it has a finite diameter.

A set of diameter r can obviously be covered by a closed sphere of radius r. The diameter of a sphere $S(p, r)$ is always $\leqslant 2\,r$, but it may be $< 2\,r$. (If, for instance, in a given metric space there are no points at a distance $< r$ from p, then the diameter of $S(p, r)$ is 0.) We note further that the diameter of a set E is not in general the lower bound of the diameters of all spheres containing E.

Diameters of non-empty sets have the following properties:

$\delta(E) = 0$ *if and only if* E *consists of a single element. If* $E_1 \subset E$, *then* $\delta(E_1) \leqslant \delta(E)$. (This follows from the properties of the upper bound of a set of real numbers.)

If $E_1 \,.\, E_2 \neq 0$, *then* $\delta(E_1 + E_2) \leqslant \delta(E_1) + \delta(E_2)$. (The proof is left to the reader.)

$\delta(\bar{E}) = \delta(E)$ *for every set* E. (This follows from the continuity of the distance function and the definition of a diameter; the proof is left to the reader.)

Theorem 67 may be restated as follows:

Every compact set contained in a metric space is the sum of a finite number of sets of arbitrarily small diameters.

Problems

1. Show that for a set D to be the set of all distances between points of some metric space M it is necessary and sufficient that D be a set of real non-negative numbers including the number zero.

Proof. The condition is obviously necessary. Let D denote a set of real non-negative numbers including the number 0. For $x \in D$ and $y \in D$, let $\rho(x, y) = 0$ for $x = y$ and $\rho(x, y) = \max(x, y)$ for $x \neq y$. D is obviously a metric space and, since $0 \in D$ and $\rho(0, x) = x$, the set of all distances between the points of D is identical with D.

2. Show that if M is a bounded metric space there exists a metric space $M_1 = M + \{q\}$ such that the point q is equidistant from all points of M. In other words, every bounded metric space lies on the surface of some sphere.

Proof. Let M be a bounded metric space with the distance function ρ, a a positive number, and $\rho(x, y) \leqslant a$ for $x \in M$, $y \in M$. Let q be a point not contained in M; put $\rho_1(x, y) = \rho(x, y)$ for

$$x \in M, \quad y \in M, \quad \rho_1(q, q) = 0, \ \rho_1(q, x) = \rho_1(x, q) = a$$

for $x \in M$ and let $M_1 = M + \{q\}$. M_1 is readily seen to be a space satisfying the required conditions.

3. Show that the intersection of all sets of a family Φ of non-empty closed sets, at least one of them compact, is non-empty if and only if every finite number of sets of Φ have a non-empty intersection.

Proof. The necessity of the condition is obvious. Assume the condition to be satisfied. Let E_0 be a compact set of the family Φ. There exists, by Theorem 67, a natural number n_1 such that E_0 is the sum of n_1 sets of diameters $\leqslant 1$. Taking the closure of these sets, we obtain n_1 closed sets $E_1, E_2, \ldots, E_{n_1}$ of diameters $\leqslant 1$ such that $E_0 = E_1 + E_2 + \ldots + E_{n_1}$. It will be shown that there exists a natural number $k_1 \leqslant n_1$ such that every finite aggregate of sets of Φ has at least one element in common with the set E_{k_1}. Suppose, on the contrary, that such a number does not exist. Then for every $i = 1, 2, \ldots, n_1$ there exist sets $X_{i,1}, X_{i,2}, \ldots, X_{i,m_i}$ of the family Φ such that $E_i \,.\, X_{i,1} \ldots X_{i,m_i} = 0$. Since $E_0 = E_1 + E_2 + \ldots + E_{n_1}$, it follows that

$$E_0 \,.\, \prod_{i=1}^{n_1} X_{i,1} \,.\, X_{i,2} \ldots X_{i,m_i} = 0$$

and so the intersection of a finite number $(m_1 + m_2 + \ldots + m_{n_1} + 1)$ of sets of Φ is empty, contrary to hypothesis. Consequently the number k_1 exists.

Since the set $T_1 = E_{k_1}$ is compact (as a subset of a compact set), we conclude similarly that it is the sum of a finite number of closed sets of diameters $\leqslant 1/2$ and, furthermore, that there exists a closed set $T_2 \subset T_1$ of diameter $\leqslant 1/2$ and such that every finite aggregate of sets of Φ has at least one point in common with T_2. Proceeding in this manner we obtain an infinite sequence $T_1 \supset T_2 \supset T_3 \supset \ldots$ of sets, where T_n is a closed and compact set of diameter $\leqslant 1/n$ which meets every finite aggregate of sets of Φ in at least one point. It follows from Cantor's theorem that there exists a point $p \in T_1 \,.\, T_2 \,.\, T_3 \ldots$. Let E be any set of Φ and n any natural number. Since $E \,.\, T_n \neq 0$, $p \in T_n$, and $\delta(T_n) \leqslant 1/n$, there exists a point $q_n \in E \,.\, T_n$ such that $\rho(p, q_n) \leqslant 1/n$. Hence $\lim_{n\to\infty} q_n = p$ and so, since E is closed, $p \in E$. Since E is any set of Φ, p is contained in every set of Φ. The condition is therefore sufficient.

4. Prove that there are only three different topological types of separable (countable) metric spaces which possess only one limit element.

Hint. Let $E_1 = \{0, 1, 1/2, 1/3, \ldots\}$,

$$E_2 = E_1 + \{2, 3, 4, \ldots\},$$

$$E_3 = \{0\} + \{\ldots, 2^{-m-n-2} + 7 \,.\, 2^{-m-3}, \ldots\},$$

where m and n run through all natural numbers. It can be first shown that each of the sets E_1, E_2, E_3 has only one limit element; secondly, that no two of the sets E_1, E_2, E_3 are homeomorphic; and thirdly, that every separable metric space which has only one limit element, is homeomorphic with one of the sets E_1, E_2, E_3. (Note that E_1 is compact, E_2 is locally compact, and E_3 is neither.)

5. Prove that the following metric ρ can be introduced into the set of all complex numbers:

$$\rho(z_1, z_2) = (|z_1 - z_2|)^{\frac{1}{2}},$$

where z_1 and z_2 are complex numbers.

6. Determine why neither of the distance functions

$$\rho(x, y) = (x - y)^2 \text{ and } \rho(x, y) = |x^2 - y^2|$$

establishes a metric in the set of all real numbers.

Hint. The first does not satisfy the triangle law, nor the second the identity law, since $\rho(1, -1) = 0$.

7. Given two metrics ρ_1 and ρ_2 in a space M, prove that there exists a metric ρ in M such that $\rho(p, q) \geqslant \rho_1(p, q)$ and $\rho(p, q) \geqslant \rho_2(p, q)$ for $p \in M$, $q \in M$. Show by an example (in a suitably chosen space M with metrics ρ_1 and ρ_2) that there does not always exist a metric ρ' such that $\rho'(p, q) \leqslant \rho_1(p, q)$, $\rho'(p, q) \leqslant \rho_2(p, q)$ for $p \in M$ and $q \in M$.

Hint. To obtain ρ it is sufficient to put $\rho(p, q) = \rho_1(p, q) + \rho_2(p, q)$. To prove the second part, let M be the straight line with the usual metric ρ_1. Put $f(x) = x$ for all real numbers other than 0 and 1, $f(0) = 1, f(1) = 0$, and put $\rho_2(p, q) = \rho_1(f(p), f(q))$.

8. Show that the metric space M consisting of all pairs (x, y) of real numbers with the metric $\rho_1((x, y), (x_1, y_1)) = |x - x_1| + |y - y_1|$ is not isometric with R_2 (i.e., the Euclidean plane with the usual metric).

Proof. The points (1/2, 1/2) and (0, 1) of M are at a distance 1 from each of the points (0, 0) and (1, 1) while in R_2 there is only one point bisecting the distance between two given points. It follows readily from this proof that there exist metric spaces of four elements which cannot be embedded in a three-dimensional space.

9. Prove that the metric space M consisting of all pairs of real numbers and having the metric

$$\rho((x, y), (x_1, y_1)) = (a(x - x_1)^2 + b(y - y_1)^2)^{\frac{1}{2}},$$

a and b two given real positive numbers, is isometric with R_2.

Proof. The function $f(x, y) = ((a)^{\frac{1}{2}}x, (b)^{\frac{1}{2}}y)$ is obviously an isometric mapping of M on R_2.

10. Show that the metric space M consisting of all real numbers with the distance $\rho_1(x, y) = |x^3 - y^3|$ is isometric with the straight line.

Proof. The function $f(x) = x^3$ maps M isometrically on R_1.

11. Prove that the metric space M consisting of all real numbers with the metric $\rho_1 = |\, x|x| - y|y| \,|$ is isometric with R_1.

Proof. The function $f(x) = x|x|$ is an isometric mapping of M on R_1.

12. The function $f(x)$ of a real variable is defined to equal x if x is rational, and to equal $2x$ if x is irrational. Show that the metric space M of all real numbers with the distance $\rho_1 = |f(x) - f(y)|$ is isometric with R_1.

Proof. It is easy to see that the function $f(x)$ is an isometric mapping of M on R_1.

13. Prove that the plane is not isometric with the straight line.

The proof follows readily from the fact that there are three points in the plane at a distance 1 from each other but there are no such three points on the straight line.

14. Show that in the set M of all real numbers the distance can be so defined that M becomes isometric with the plane.

The proof follows easily from Cantor's theorem on the (1, 1) correspondence between the set of all real numbers and the set of all points in the plane.

15. Show that the metric space M consisting of all real numbers x, where $-1 < x < 1$, with the metric

$$\rho_1(x, y) = |x/(1 - |x|) - y/(1 - |y|)|$$

is isometric with the straight line.

The proof follows from the fact that the function $f(x) = x/(1 - |x|)$ establishes an isometric mapping of M on the straight line.

16. Define a metric ρ_1 in the set of all real positive numbers in such a manner that the metric space thus obtained is isometric with the straight line.

Solution. Let $f(x) = x - 2$ for $x \geqslant 1$, $f(x) = -1/x$ for $0 < x < 1$. We now define, for $a > 0$, $\rho_1(x, y) = a\,|f(x) - f(y)|$.

17. Show that a straight line is equivalent by division into two parts to a straight line from which any bounded set is removed.

Proof. Let P denote the x-axis, E a given bounded subset of P of diameter $\delta(E) = d$. Let E_k $(k = 0, 1, 2, \ldots)$ denote the set obtained from E by a translation through the length $2kd$ in the positive direction of the x-axis. The sets $E = E_0, E_1, E_2, \ldots$ are obviously disjoint and congruent to each other. Put $A = E_0 + E_1 + \ldots$, $B = P - A$, $P_1 = P - E$, $A_1 = A - E_0 = E_1 + E_2 + \ldots$. Then $P = A + B$, $P_1 = A_1 + B$, $A \,.\, B = A_1 \,.\, B = 0$, $A \simeq A_1$ (since $E_{k-1} \simeq E_k$, $k = 1, 2, \ldots$) and so $P \underset{2}{=} P_1$.

18. Show that a straight line is equivalent by division into two parts to the set of all irrational numbers.

Proof. Let P be the set of all real numbers, E the set of all rational numbers, and $Q = P - E$. Denote by E_k, $k = 0, 1, 2, \ldots$, the set of all numbers of the form $r + k2^{\frac{1}{2}}$, where $r \in E$. The sets $E = E_0, E_1, E_2, \ldots$ are clearly disjoint and congruent to each other. Put $A = E_0 + E_1 + \ldots, B = P - A$, $A_1 = A - E_0$. Then $P = A + B$, $Q = A_1 + B$, $A \,.\, B = A_1 \,.\, B = 0$, $A \simeq A_1$, and so $P \underset{2}{=} Q$.

19. Prove that the straight line is equivalent by division into two parts to the set of all transcendental numbers.

The solution is similar to that of example 18. We replace the set of rational numbers by the set of algebraic numbers and the number $2^{\frac{1}{2}}$ by any transcendental number (e.g. π).

20. Show that the set of all irrational numbers is equivalent by division into two parts to the set of all real transcendental numbers.

The proof follows easily from the fact that, if Q is the set of all irrational numbers and T the set of all transcendental real numbers, then Q contains a countable aggregate of disjoint sets which are congruent to the set $Q - T$ (for instance, consecutive translations of the set $Q - T$ through multiples of a given transcendental number).

21. Two subsets B and B_1 of a linear set A are equivalent by division into a finite number of parts. Show that $A - B$ and $A - B_1$ may not be equivalent by finite division.

Proof. Let A be the straight line, B and B_1 any finite subsets with a different number of elements.

Remark. It can be shown that the set of all irrational numbers is not equivalent by finite division to the set of all algebraic numbers (the proof is much more difficult).

Furthermore, according to a theorem of Tarski,[40] a *non-empty linear set is not the sum of two disjoint sets each equivalent to it by finite division.*

22. Prove that the set of all rational numbers $< 2^{\frac{1}{2}}$ is not congruent to the set of all rational numbers $> 2^{\frac{1}{2}}$ nor to the set of all rational numbers $< 3^{\frac{1}{2}}$.

23. Show that, for a and b real, the set of all rational (or irrational) numbers $< a$ is congruent to the set of all rational (or irrational) numbers $< b$ if and only if $b - a$ is rational but that the two sets are always equivalent by division into two parts.

24. Let M be a given metric space, a and b two different points of M. Define a continuous real function in M which takes on the value α at a and the value β at b, where α and β are two given real numbers.

Answer. The required function is

$$f(p) = \frac{\alpha\rho(p, b) + \beta\rho(p, a)}{\rho(p, a) + \rho(p, b)},$$

where ρ is the metric in M.

25. A (1, 1) mapping of a set E, contained in a metric space M with the distance function ρ, on a set $f(E)$, contained in a metric space with distance function ρ_1 is called a *contraction* of the set E if

$$\rho_1(f(p), f(q)) < \rho(p, q), \qquad p \in E, q \in E.$$

Show that a contraction of a bounded plane set may have the same diameter as the set itself.

Proof.[41] Let E be the set consisting of the points

$$(-1 + 1/2n, 1 - 1/2n) \text{ and } (1 - 1/3n, 1 - 1/3n), n = 1, 2, \ldots.$$

Let $f(p)$ be the projection of the point p on the x-axis. The set $f(E)$ is obviously a contraction of the set E and each set has diameter $= 2$.

26. Show that a contraction of a set is a continuous mapping on itself.[42]

27. Prove that no bounded metric space can be a contraction of itself.

53. Properties equivalent to separability. A set E contained in a metric space is said to possess the Lindelöf property if in every aggregate of open sets whose sum contains E there exists a finite or countable aggregate of these sets whose sum contains E. (Cf. Theorem 49, § **32**.)

THEOREM 69. *In a metric space the following four properties of a set are equivalent:*

A. *A set E is separable.*

B. *A set E possesses the Lindelöf property.*

C. *Every non-countable subset of E has at least one element of condensation which is contained in E.*

D. *Every non-countable subset of E possesses at least one limit element (which may or may not be contained in E).*

It is obviously sufficient to show that A $\to$ B $\to$ C $\to$ D $\to$ A.

Proof. Assume that the set E contained in a given metric space M is separable. There exists, therefore, a finite or countable subset $P = \{p_1, p_2, \ldots\}$ of E such that $E \subset \bar{P}$. Further let Φ denote a family of open sets contained in M whose sum contains E. If now p be a given element of E and U any open set of the family Φ containing p, there exists a real number $r > 0$ such that $S(p, r) \subset U$. Let n be a positive integer such that $1/n < r/2$; since $p \in \bar{P}$ there exists an index k such that $\rho(p, p_k) < 1/n$. It is easily seen that $p \in S(p_k, 1/n) \subset S(p, r) \subset U$. Thus each element of E is associated with a pair of indices k, n such that $p \in S(p_k, 1/n) \subset U$. Denote the open set U containing the sphere $S(p_k, 1/n)$ by $U_{k,n}$. Clearly the set of all such sets $U_{k,n}$ is at most countable and the sum of these sets contains E. This proves that E has the Lindelöf property; hence A implies B.

Next assume that E possesses property B. If E does not possess property C, then there exists a non-countable subset N of E without an element of condensation in E. Consequently, there exists for each element $p \in E$ a sphere $S(p, r)$, where $r > 0$, which contains at most a countable subset of N. By property B, there exists at most a countable aggregate of these spheres whose sum S contains E. We thus have $N \subset E \subset S$; this is impossible, since S contains a set of elements of N which is at most countable. Hence B implies C. It is obvious that C implies D.

To prove that D implies A, we shall first prove the following

LEMMA. *If a set E contained in a metric space M with distance function ρ is not separable, then there exists a positive number d and a non-countable subset N of E such that $\rho(p, q) \geqslant d$ for $p \in N$, $q \in N$, $p \neq q$.*

Proof. Suppose the set E is not separable. If for every positive integer n there existed an at-most-countable set $P_n \subset E$ such that every element $p \in E$ were at a distance $< 1/n$ from some element of P_n, then E would be contained in the closure of the at-most-countable set $P_1 + P_2 + \ldots$, contrary to the assumption that E is not separable. There exists, therefore, an integer n for which the corresponding subset P_n does not exist. Let p_1 be any element of E. Let α be an ordinal number, $1 < \alpha < \Omega$, and suppose that all elements

$p_\xi \in E$, where $\xi < \alpha$, are already defined. The sequence $\{p_\xi\}$ is at most countable since $\alpha < \Omega$. Since we concluded that P_n does not exist, there exists an element $p_\alpha \in E$ such that $\rho(p_\alpha, p_\xi) \geqslant 1/n$ for all $\xi < \alpha$. The elements p_α are thus defined by induction for all $\alpha < \Omega$. Denote the sequence $\{p_\alpha\}$ by N. N is obviously a non-countable subset of E with $\rho(p, q) \geqslant 1/n$ for $p \in N$, $q \in N$, $p \neq q$. The lemma is therefore proved.

Next suppose that N is a non-countable subset of E satisfying the condition of the above lemma. It is obvious that no element of the space M is a limit element of N. For if a, an element of M, were a limit element of N, the sphere $S(a, d/2)$ would contain two different elements p and q of N; this would imply that $\rho(p, q) \leqslant \rho(p, a) + \rho(a, q) < d$, contrary to the definition of the set N. Thus a set which is not separable does not possess property D. Consequently, D implies A and Theorem 69 is proved.

Example. Show that *a metric space M is separable if and only if every aggregate of disjoint open sets contained in M is at most countable.*

Proof. The necessity of the condition follows from Theorem 48, § **32**, and the fact that a separable metric space is a topological space with a countable basis. To prove the sufficiency assume M is not separable. Then there exists, by the above lemma, a positive number d and a non-countable subset $N \subset M$ such that $\rho(p, q) \geqslant d$ for $p \in N$, $q \in N$, $p \neq q$. Consequently $S(p, d/2) \,.\, S(q, d/2) = 0$ for $p \in M$, $q \in M$, $p \neq q$. The family of spheres $S(p, d/2)$, where $p \in N$, is obviously a non-countable aggregate of non-empty open disjoint sets. Thus, if every family of non-empty open and disjoint sets contained in a metric space M is at most countable, then M must be separable.

54. Properties equivalent to closedness and compactness. A set E contained in a metric space is said to possess *the Borel property* if for every infinite sequence $U_1, U_2, \ldots$ of open sets such that $E \subset U_1 + U_2 + \ldots$ there exists an integer n such that $E \subset U_1 + U_2 + \ldots + U_n$.

A set E is said to possess *the Borel-Lebesgue property* if for every aggregate of open sets whose sum contains E there exists a finite class of the sets of the aggregate whose sum contains E.

It follows from the definition of the Lindelöf property (§ **53**) that a set possessing the Lindelöf and Borel properties possesses also the Borel-Lebesgue property and conversely.

THEOREM 70. *In a metric space the following three properties of a set E are equivalent:*

I. *A set E is compact-in-itself.*

II. *A set E possesses the Borel-Lebesgue property.*

III. *A set E possesses the Borel property.*

To prove the theorem it is sufficient to show that I $\rightarrow$ II $\rightarrow$ III $\rightarrow$ I. Suppose that the set E contained in a metric space M is compact-in-itself; it is therefore closed and compact and so, by Theorem 31, possesses the Borel property. Furthermore, it is separable, by Theorem 68, and so by Theorem 69, it possesses the Lindelöf property. Since it also possesses the Borel property, it must possess the Borel-Lebesgue property. We have thus proved that I implies II. Obviously II implies III.

Suppose now that a set E possesses the Borel property and assume E is not compact. Hence E contains an infinite subset E_1 such that $E'_1 = 0$. Let $p_1, p_2, \ldots$ denote an infinite sequence of different elements of E_1. Denoting by P_n the set of all elements of the sequence $p_n, p_{n+1}, p_{n+2}, \ldots$, we have $P_n \subset E_1$. Since $P'_n \subset E'_1$, we have $P'_n = 0$. Hence the sets $P_n (n = 1, 2, \ldots)$ are closed and so the sets $U_n = M - P_n (n = 1, 2, \ldots)$ are open. Moreover, $E \subset U_1 + U_2 + \ldots$. For, if $p \in E - P_1$, then $p \in U_1$, and if $p \in P_1$, e.g. $p = p_k$, then $p \in M - P_{k+1} = U_{k+1}$. On the other hand, whatever the integer n, we cannot have $E \subset U_1 + U_2 + \ldots + U_n$ because $p_n \in E$ and $p_n \in P_k$ for $k \leqslant n$ and so $p_n \notin U_k = M - P_k$ for $k = 1, 2, \ldots n$. Hence E does not possess the Borel property contrary to assumption. The set E is therefore compact.

We next prove that E is closed. Suppose, on the contrary, that $a \in E' - E$. For every integer n, denote by U_n the set of all elements $p \in M$ for which $\rho(p, a) > 1/n$; the sets $U_n (n = 1, 2, \ldots)$ are open and, as is easily seen, $M - \{a\} = U_1 + U_2 + \ldots$. Since $a \notin E$ we have $E \subset U_1 + U_2 + \ldots$. But $a \in E'$ and so there exists for every natural number n an element $p_n \in E$ such that $\rho(p_n, a) < 1/n$; consequently $p_n \notin U_n$ and, since $U_1 + U_2 + \ldots + U_n = U_n$, $p_n \notin U_1 + U_2 + \ldots + U_n$. Hence the relation $E \subset U_1 + U_2 + \ldots + U_n$ is impossible for any n contrary to the assumption that E possesses the Borel property. The set E is therefore closed and, since it is compact, it is compact-in-itself. It follows that III implies I and this proves Theorem 70.

Problems

1. Show that a set E contained in a metric space is compact if and only if the intersection of every infinite sequence $E_1 \supset E_2 \supset \ldots$, where $E_1 \subset E$, of closed non-empty sets is non-empty.

Proof. The necessity of the condition follows at once from Cantor's theorem. If, on the other hand, the set E were not compact, it would contain a countable subset $E_1 = \{p_1, p_2, \ldots\}$ such that $E'_1 = 0$. Put $E_n = \{p_n, p_{n+1}, \ldots\}$ for $n = 1, 2, \ldots$. The sets E_n form a descending sequence of non-empty closed sets whose intersection is the null set. The condition is therefore sufficient.

2. Prove that a set compact-in-itself is not congruent to a proper subset of itself.[43]

3. Prove that *a metric space is compact-in-itself if and only if every real and continuous function defined in the space is bounded.*

The necessity follows from Theorems 57 and 67.

Suppose that the metric space M with distance function ρ is not compact. There exists, therefore, an infinite sequence $p_1, p_2, \ldots$ of different elements of M such that the set $P = \{p_1, p_2, \ldots\}$ has an empty derived set. Hence for every positive integer n there exists a number r_n such that $0 < r_n < 1/n$ and such that the sphere $S(p_n, r_n)$ contains no element of P other than p_n. For a given integer n and $x \in M$, put $f_n(x) = n(1 - 2(\rho(x, p_n))/r_n)$, if $\rho(x, p_n) \leqslant r_n/2$, and $f(x) = 0$, if $\rho(x, p_n) > r_n/2$. Finally, let

$$f(x) = f_1(x) + f_2(x) + \ldots.$$

The function $f(x)$ is obviously continuous in M but it is not bounded, since $f_n(p_n) = n$, for $n = 1, 2, \ldots$. The condition is therefore sufficient.

4. Give an example of a function discontinuous in R_1 which maps every subset of R_1 into a compact-in-itself subset of the same space.

Solution. The function $f(x)$ equal to 1 for all rational x and to 0 for all irrational x. This function maps every linear set on a linear set compact-in-itself (consisting of two points at most) but it is not continuous at any point of the line.

55. The derived set of a compact set.

THEOREM 71. *The derived set of a compact set contained in a metric space is compact.*

Proof. Let E denote a compact set contained in a metric space M and T an infinite subset of the derived set E'. Then there exists an infinite sequence $q_1, q_2, \ldots$ of elements of T which are all different. Since $q_1 \in E'$, there exists an element $p_1 \in E$ such that $\rho(q_1, p_1) < 1$. Let n denote a natural number > 1 and suppose that the elements $p_1, p_2, \ldots, p_{n-1}$ of the set E are already defined. Since $q_n \in E'$ there exists an element $p_n \in E$, different from $p_1, p_2, \ldots, p_{n-1}$, and such that $\rho(q_n, p_n) < 1/n$. We have thus defined by induction an infinite sequence $p_1, p_2, \ldots$ of different elements of E. Denoting this sequence by E_1, we conclude that, since it is an infinite subset of E, $E'_1 \neq 0$. Let $a \in E'_1$. The sphere $S(a, \epsilon/2)$, for a given $\epsilon > 0$, contains an infinity of different elements of E_1. On the other hand, since all elements of the sequence $q_1, q_2, \ldots$ are different, there exists a natural number μ such that $q_n \neq a$ for all $n > \mu$. But there exists a natural number $n > 2/\epsilon + \mu$ such that $p_n \in S(a, \epsilon/2)$. Hence, since $q_n \neq a$, $\rho(p_n, a) < \epsilon/2$, $\rho(q_n, p_n) < 1/n < \epsilon/2$, we have, by the triangle law, $\rho(q_n, a) < \epsilon$. Thus, for every positive ϵ, there exists an element q_n of the set T, different from a, and such that $\rho(q_n, a) < \epsilon$. Consequently $a \in T'$ and so $T' \neq 0$. It follows that E' is compact and Theorem 71 is proved.

Since the sum of two compact sets is compact (§ **16**), Theorem 71 gives immediately the following

COROLLARY. *The closure of a compact set is compact.*

Note that this corollary (and hence Theorem 71) does not hold in general Hausdorff topological spaces. It is not true, for example, in the space K consisting of all numbers x of the interval $0 \leqslant x \leqslant 1$, where a neighbourhood of x is every set consisting of x and all rational numbers of K which differ from x by less than $1/n$ for n any natural number. That K is a Hausdorff topological space is obvious. It is also readily seen that the set E of all rational numbers is compact. However, its derived set $E' = K$ is not compact, since F, the set of all irrational numbers contained in K, does not possess a single limit element in K.

56. Condition for connectedness. ϵ-chains. Let ϵ denote a given positive number. Two elements p and q of a given set E contained in a metric space M with distance function ρ are said to be joined by an ϵ-chain in E if there exists a finite sequence $p_0, p_1, \ldots, p_n$ of elements of E such that

$$p_0 = p, \quad p_n = q, \quad \text{and } \rho(p_{k-1}, p_k) < \epsilon \quad \text{for } k = 1, 2, \ldots, n.$$

THEOREM 72. *Any two elements of a connected set E contained in a metric space can be joined in E by an ϵ-chain for every $\epsilon > 0$.*

Proof. Suppose that the elements a and b of a set E contained in a metric space M with distance function ρ cannot be joined in E by an η-chain for some $\eta > 0$. Let A denote the set of all elements of E (a included) which can be joined to a by an η-chain and put $B = E - A$. Then $a \in A, b \in B$, and therefore $A \neq 0$, $B \neq 0$. It follows that $A \, . \, B' = A' \, . \, B = A \, . \, B = 0$. For if, on the contrary, $p \in A \, . \, B'$, then $p \in B'$; so there exists an element $q \in B$ such that $\rho(q, p) < \eta$. But $p \in A$ and can therefore be joined to a by an η-chain; hence q can be joined to a by an η-chain; this is impossible since $q \in B = E - A$. Similarly, if $p \in A' \, . \, B$, there exists an element $q \in A$ such that $\rho(q, p) < \eta$. We conclude, as before, that p can be joined to a by an η-chain; this contradicts the fact that $p \in B$. Thus E is the sum of two separated sets and so is not connected. The theorem is therefore proved. It is easy to see that the converse is not true. For example, the set of all rational numbers is totally disconnected, although any two rational numbers can be joined in the set by an ϵ-chain for every $\epsilon > 0$. We have, however,

THEOREM 73. *A set E which is compact-in-itself and contained in a metric space is connected if and only if, for every $\epsilon > 0$, any two elements of E can be joined by an ϵ-chain in E.*

Proof. The condition is necessary by Theorem 72. To prove the sufficiency let E denote a set which is compact-in-itself, contained in a metric space M

with distance function ρ, and not connected. Since E is closed, it is, by Theorem 14, the sum of two non-empty closed and disjoint sets A and B. Hence there exist elements a and b such that $a \in A$ and $b \in B$. Suppose that a and b can be joined by an ϵ-chain in E for every $\epsilon > 0$. Since a, the first element of the chain, is in A and b, the last element of the chain, is in B, there must be two consecutive elements of the chain one of which is in A and the other of which is in B. Hence, for every natural number n, there exist elements p_n and q_n such that $p_n \in A$, $q_n \in B$ and $\rho(p_n, q_n) < 1/n$. If the sequence $p_n (n = 1, 2, \ldots)$ contains only a finite number of different elements, then one of them, say p, must be repeated infinitely many times and so, for infinitely many values of n, $\rho(p, q_n) < 1/n$. Since $q_n \in B$ for $n = 1, 2, \ldots,$ it follows that $p \in B'$; this is impossible since $A \,.\, B' = 0$. The set of all different elements of the sequence $p_1, p_2, \ldots$ is therefore infinite and, since it is a subset of the compact set E, it must posses a limit element p_0. Since $p_n \in A$ for $n = 1, 2, \ldots,$ and A is closed, we have $p_0 \in A$.

Now for every $\eta > 0$ there exists a natural number $n > 1/\eta$ such that $\rho(p_0, p_n) < \eta$ and, at the same time, $\rho(p_n, q_n) < 1/n < \eta$; hence $\rho(p_0, q_n) < 2\eta$. Since $q_n \in B$ for $n = 1, 2, \ldots,$ and since B is closed, it follows that $p_0 \in B$. But $p_0 \in A$ and $A \,.\, B = 0$; hence a contradiction arises from the assumption that in a compact-in-itself but disconnected set a pair of elements can be joined by an ϵ-chain for every $\epsilon > 0$. The condition is therefore sufficient and Theorem 73 is proved.

Theorem 73 may not be true for a set that is closed but not compact. For example, the set consisting of the points of an hyperbola and its asymptotes is not connected although it is closed and any two of its points can be joined in it by an ϵ-chain for every $\epsilon > 0$.

Problems

1. Show that every metric space is a subset of a connected metric space.

Hint. If M is a given metric space with distance function ρ and if p_0 is a given point of M, let S denote the space consisting of the point p_0 and all ordered pairs (p, x) where $p \in M$, $p \neq p_0$, and $0 < x < \rho(p, p_0)$. Let the distance ρ_1 in S be defined by the conditions:

$$\rho_1(p_0, (p, x)) = \rho(p, p_0) - x \text{ for } p \in M, p \neq p_0, 0 \leqslant x < \rho(p, p_0),$$

$$\rho_1((p, x), (p, y)) = |x - y|,$$

and

$$\rho_1((p, x), (q, y)) = \min\,(\rho(p, q) + x + y, \rho(p, p_0) + \rho(q, p_0) - x - y)$$

$$\text{for } p \neq q.$$

Each point of S which is different from p_0 can be joined to p_0 by a set which is isometric with a line segment (that is, with a connected set). M will be contained in S if we put $(p, 0) = p$ for $p \in M$ and $p \neq p_0$.

2. Give an example of a plane connected set which is the sum of an infinite sequence of sets, each separated from each of the others.

57. Hilbert space and its properties. Let H denote the set of all infinite sequences $x_1, x_2, \ldots$ of real numbers such that the series

$$x_1^2 + x_2^2 + \ldots$$

is convergent. If $p = (x_1, x_2, \ldots)$ and $q = (y_1, y_2, \ldots)$ are two elements of H, then the series $\sum_1^\infty x_n^2$ and $\sum_1^\infty y_n^2$ are convergent. Hence the series $\sum_1^\infty (x_n - y_n)^2$ is convergent. Put

$$\rho(p, q) = ((x_1 - y_1)^2 + (x_2 - y_2)^2 + \ldots)^{\frac{1}{2}};$$

this is a real non-negative number determined by p and q. It is sufficient to show that the function ρ satisfies the triangle law to prove that H is a metric space with distance ρ.

Let $a_1, a_2, \ldots, a_n$ and $b_1, b_2, \ldots, b_n$ be any two given finite sequences of real numbers. We have, for $k = 1, 2, \ldots, n, i = 1, 2, \ldots, n,$

$$(a_k b_i - a_i b_k)^2 \geqslant 0;$$

so

$$a_k^2 b_i^2 + a_i^2 b_k^2 \geqslant 2 a_k a_i b_k b_i.$$

Hence

$$2 \sum_{k=1}^n a_k^2 \sum_{i=1}^n b_i^2 = \sum_{k=1}^n \sum_{i=1}^n (a_k^2 b_i^2 + a_i^2 b_k^2) \geqslant 2 \sum_{k=1}^n \sum_{i=1}^n a_k a_i b_k b_i$$

$$= 2 \sum_{k=1}^n a_k b_k \sum_{i=1}^n a_i b_i = 2\left(\sum_{k=1}^n a_k b_k\right)^2;$$

so

$$\left(\sum_{k=1}^n a_k b_k\right)^2 \leqslant \sum_{k=1}^n a_k^2 \sum_{i=1}^n b_i^2,$$

or

$$\left|\sum_{k=1}^n a_k b_k\right| \leqslant \left(\sum_{k=1}^n a_k^2 \sum_{i=1}^n b_i^2\right)^{\frac{1}{2}}.$$

Therefore,

$$\sum_{k=1}^n (a_k + b_k)^2 \leqslant \sum_1^n a_k^2 + \sum_1^n b_k^2 + 2\left(\sum_1^n a_k^2 \sum_1^n b_k^2\right)^{\frac{1}{2}}$$

$$= \left(\left(\sum_1^n a_k^2\right)^{\frac{1}{2}} + \left(\sum_1^n b_k^2\right)^{\frac{1}{2}}\right)^2.$$

So

$$(3) \qquad \left(\sum_1^n (a_k + b_k)^2\right)^{\frac{1}{2}} \leqslant \left(\sum_1^n a_k^2\right)^{\frac{1}{2}} + \left(\sum_1^n b_k^2\right)^{\frac{1}{2}}.$$

In particular, let $a_k = x_k - y_k$, $b_k = y_k - z_k$ for $k = 1, 2, \ldots, n$; then (3) gives

$$(4) \qquad \left(\sum_1^n (x_k - z_k)^2\right)^{\frac{1}{2}} \leqslant \left(\sum_1^n (x_k - y_k)^2\right)^{\frac{1}{2}} + \left(\sum_1^n (y_k - z_k)^2\right)^{\frac{1}{2}}$$

for all n. Letting $n \to \infty$ we obtain for $p = (x_1, x_2, \ldots)$, $q = (y_1, y_2, \ldots,)$ and $r = (z_1, z_2, \ldots)$, the inequality

$$\rho(p, r) \leqslant \rho(p, q) + \rho(q, r).$$

Hence ρ satisfies the triangle law and the set H is a metric space. It is called *Hilbert space.*

Euclidean m-dimensional space R_m (m a given natural number) is the set of all finite sequences $(x_1, x_2, \ldots, x_m)$ of real numbers in which the distance of two elements $p = (x_1, x_2, \ldots, x_m)$ and $q = (y_1, y_2, \ldots, y_m)$ is defined to be the number

$$\rho(p, q) = ((x_1 - y_1)^2 + (x_2 - y_2)^2 + \ldots + (x_m - y_m)^2)^{\frac{1}{2}}.$$

We see from (4) that this function ρ satisfies the triangle law; R_m is therefore a metric space for all m. It is clear that R_m is congruent to a subset of H, namely, that one which is composed of all infinite sequences $x_1, x_2, \ldots$ such that $x_k = 0$ for $k > m$. Hence

A Euclidean space R_n (n any natural number) *can be embedded in Hilbert space.*

There exists, however, a metric space consisting of four elements which cannot be embedded in Hilbert space.

For let $E = \{p, q, r, s\}$ where

$$(5) \quad \rho(p, s) = \rho(q, r) = 2 \text{ and } \rho(p, q) = \rho(p, r) = \rho(q, s) = \rho(r, s) = 1.$$

It can be shown that E is such a space. To prove this result, we shall first establish the following property of Hilbert space:

Every pair a, b of points of Hilbert space determines a unique mid-point, that is, a point $x \in H$ such that

$$\rho(a, x) = \rho(x, b) = \tfrac{1}{2}\rho(a, b).$$

Suppose $a \in H$, $b \in H$, $x \in H$, $a = (a_1, a_2, \ldots)$, $b = (b_1, b_2, \ldots)$, $x = (x_1, x_2, \ldots)$, and $\rho(a, x) = \rho(x, b) = \frac{1}{2}\rho(a, b)$. Then

$$4\sum_1^\infty (a_k - x_k)^2 = 4\sum_1^\infty (b_k - x_k)^2 = \sum_1^\infty (a_k - b_k)^2.$$

Employing the identity

$$(\alpha + \beta - 2\xi)^2 = 2(\alpha - \xi)^2 + 2(\beta - \xi)^2 - (\alpha - \beta)^2,$$

we obtain

$$\sum_1^\infty (a_k + b_k - 2x_k)^2 = 2\sum_1^\infty (a_k - x_k)^2 + 2\sum_1^\infty (b_k - x_k)^2 - \sum_1^\infty (a_k - b_k)^2 = 0;$$

so $a_k + b_k - 2x_k = 0$ for $k = 1, 2, \ldots$. This gives

$$x = (\tfrac{1}{2}(a_1 + b_1), \tfrac{1}{2}(a_2 + b_2), \ldots). \tag{6}$$

The point x which satisfies the required conditions is obviously unique. Conversely, a point x given by (6) satisfies the required conditions.

There exists, therefore, only one mid-point between any two points of Hilbert space. But in the space E defined above and satisfying (5) the elements q and r are each at the same distance from the elements p and s. We therefore conclude that E cannot be embedded in Hilbert space.

We remark further that every three points of E lie on a straight line (i.e., they form a set which can be embedded in R_1). Hence, if every three points of a set lie on a straight line, it does not follow that the whole set lies on a straight line. It can be shown, however, that if in a metric (or semi-metric) space every four points lie on a straight line, the space itself lies on a straight line. This is a special case of a more general theorem by Menger[44] that if in a semi-metric space every $n + 3$ points can be embedded in R_n, the space itself can be embedded in R_n. If, however, a metric (or semi-metric) space consists of more than four points and every three of them lie on a straight line, then the whole space lies on a straight line.[45]

The set E consisting of the four elements considered above can clearly be mapped isometrically on the set of points dividing the circumference of the circle of radius $2/\pi$ into four equal parts provided the distance between any two points is taken to be the length of the shorter arc joining the points.

Furthermore, it can be shown that every set consisting of four elements and contained in Hilbert space can be embedded in a three-dimensional Euclidean space (but not necessarily in a plane) and every finite subset of Hilbert space consisting of n elements ($n > 1$) can be embedded in an $(n - 1)$-dimensional Euclidean space. There exist, however, countable subsets of Hilbert space which cannot be embedded in any Euclidean space of a finite number of dimensions (for example, the set consisting of the elements $p_1, p_2, \ldots$, where, for $n = 1, 2, \ldots, p_n$ is an infinite sequence whose nth term is 1 while all others are zero).

No sphere $S(p, r)$ of Hilbert space is compact. For let $p = (x_1, x_2, \ldots)$ denote a point of Hilbert space and put $p_n = (x_1, x_2, \ldots, x_{n-1}, x_n + r/2,$

$x_{n+1}, \ldots)$ for $n = 1, 2, \ldots$. Hence $p_n \in S(p, r)$ for $n = 1, 2, \ldots$ and the distance between any two different points of the set $P = \{p_1, p_2, \ldots\}$ is $r/2^{\frac{1}{2}}$; consequently, the set $S(p, r) \supset P$ is not compact (in any metric space containing it). Hilbert space is, therefore, not locally compact at any of its points. It will, however, be shown later that Hilbert space is homeomorphic with a certain compact subset of itself.

We prove next that *Hilbert space is isometric with a certain nowhere-dense subset of itself.*

To this end let $f(p) = (0, x_1, x_2, \ldots)$ for $p = (x_1, x_2, \ldots) \in H$. Clearly H and its subset $E = f(H)$ are isometric. Moreover, the set E is nowhere-dense on H. For let $q = (y_1, y_2, \ldots)$ denote any point of H and ϵ an arbitrary positive number. Let $a \neq 0$ be a number such that $y_1 < a < y_1 + \epsilon/2$ and put $q_0 = (a, y_2, y_3, \ldots)$. Then $\rho(q_0, q) = a - y_1 < \epsilon/2$. Put $r = \min(|a|, \epsilon/2)$. Since $\rho(q_0, q) < \epsilon/2$ and $r \leqslant \epsilon/2$, we have $S(q_0, r) \subset S(q_0, \epsilon/2) \subset S(q, \epsilon)$.

Suppose that $p = (x_1, x_2, \ldots) \in S(q_0, r)$; then $\rho(p, q_0) < r$ and so $|x_1 - a| < r$. Since $|a| \geqslant r$, $x_1 \neq 0$. It follows from the definition of the set E that $p \notin E$; this proves that $S(q_0, r) \,.\, E = 0$.

Thus every open sphere $S(q, \epsilon)$ contained in H contains an open sphere which does not meet E. This result implies that E is nowhere-dense on H. The set H is open in H but the set E, since it is nowhere-dense on H, is clearly not open in H. Hence the result:

In Hilbert space an isometric image of an open set need not be open.

This is not the case in Euclidean spaces where it can be shown that openness of a set is a topological invariant. The proof, however, is not easy, even for plane sets.

We next show that *Hilbert space is separable.*

The aggregate of all finite sequences of rational numbers is countable. Hence the set P of all points $p = (x_1, x_2, \ldots) \in H$ with rational coordinates, only a finite number of which are different from zero is countable (the so-called rational points of the space H). It will be shown that $H = \bar{P}$.

Let $p = (x_1, x_2, \ldots)$ denote an arbitrary point of H. Put

$$p_n = \left(\frac{[nx_1]}{n}, \frac{[nx_2]}{n}, \ldots \frac{[nx_n]}{n}, 0, 0, \ldots\right) \text{ for } n = 1, 2, \ldots,$$

(where $[x]$ is the largest integer contained in $|x|$ and has the sign of x). Obviously $p_n \in P$ for all n. Also, $\lim_{n\to\infty} p_n = p$. For let $\epsilon > 0$ be a given number. Since $p \in H$ the series $x_1^2 + x_2^2 + \ldots$ is convergent; there exists, therefore, a natural number μ such that $1/\mu < \epsilon^2$ and $x_{n+1}^2 + x_{n+2}^2 + \ldots < \epsilon^2$ for $n > \mu$. But

$$\left|x_k - \frac{[nx_k]}{n}\right| < 1/n, \qquad k = 1, 2, \ldots, ;$$

hence, for $n > \mu$,

$$\rho(p, p_n) = \left(\sum_{k=1}^{n}\left(x_k - \frac{[nx_k]}{n}\right)^2 + \sum_{k=n+1}^{\infty} x_k^2\right)^{\frac{1}{2}} < (1/n + \epsilon^2)^{\frac{1}{2}} < \epsilon 2^{\frac{1}{2}}.$$

Since ϵ is arbitrary, this gives $\lim_{n\to\infty} p_n = p$. Hence $p \in \bar{P}$, from which we conclude that, since $\bar{P} \subset H$, we have $H = \bar{P}$.

There exist in Hilbert space closed and bounded sets which are not totally bounded, that is, they do not satisfy Theorem 67. For example, the set of all points of the set H with just one coordinate equal to 1 and all the others equal 0 is a countable set with the distance between any two points equal to $\sqrt{2}$; it is therefore closed and bounded but not totally bounded.

Let P and Q be two metric spaces in which the distance functions are denoted by ρ_1 and ρ_2 respectively. Fréchet denotes by $[[P, Q]]$ the space composed of all ordered pairs (p, q), where $p \in P$, $q \in P$ and where the distance ρ between two pairs (p_1, q_1) and (p_2, q_2) is given by

$$\rho((p_1, q_1), (p_2, q_2)) = (((\rho_1(p_1, p_2))^2 + (\rho_2(q_1, q_2))^2)^{\frac{1}{2}}.$$

It is easily proved that the distance function ρ satisfies the three distance axioms.

Some authors[46] call a metric space $[[P, Q]]$ the *metric product* $P \times Q$ of the two spaces P and Q.

Another metric in the space $P \times Q$ is obtained by putting

$$\rho((p_1, q_1), (p_2, q_2)) = \max\,(\rho_1(p_1, p_2), \rho_2(q_1, q_2)).$$

It is easily seen that if H is a Hilbert space then the space $H \times H$ i.e. the metric square of H) is isometric with H. For it is sufficient to map the point $x = (x_1, x_2, \ldots) \in H$ on the pair $((x_1, x_3, \ldots), (x_2, x_4, \ldots)) \in H \times H$.

Problems

1. Show that if, in Hilbert space, a point d is mid-way between the points a and b as well as mid-way between the points a and c, then $b = c$.

Proof. Let $a = (a_1, a_2, \ldots)$, $b = (b_1, b_2, \ldots)$, $c = (c_1, c_2, \ldots)$ and $d = (d_1, d_2, \ldots)$. Since $\rho(a, d) = \rho(b, d) = \rho(c, d)$, we have, for $n = 1, 2, \ldots$, $a_n + b_n - 2d_n = 0$ and $a_n + c_n - 2d_n = 0$; hence $b_n = c_n$ and so $b = c$.

2. Employ problem 1 to show that the metric space $\{a, b, c, d\}$ with distance function ρ, where $\rho(a, b) = \rho(a, c) = \rho(b, c) = 2$, $\rho(a, d) = \rho(b, d) = \rho(c, d) = 1$, cannot be embedded metrically in Hilbert space.

Proof. The point d is the mid-point of a and b and of a and c; this gives $b = c$ (problem 1) in Hilbert space, contrary to the fact that $\rho(b, c) = 2$.

3. Let M be a metric space consisting of the three elements a, b, and c with distance function ρ.

What necessary and sufficient conditions must be satisfied by the numbers $\rho(a, b)$, $\rho(a, c)$, and $\rho(b, c)$ in order that M may be embedded metrically in R_1?

Answer. One of the numbers must be the sum of the remaining two.

4. Give an example of a metric space that cannot be embedded in the plane, but which has the property that every subset consisting of four points can be so embedded.

Answer. The space $M = \{a, b, c, d, e\}$ with distance function ρ satisfying the following conditions:

$$\rho(a, b) = \rho(a, c) = \rho(a, d) = \rho(a, e) = 1,$$

$$\rho(b, e) = \rho(c, e) = \rho(d, e) = 1,$$

$$\rho(b, c) = \rho(c, d) = \rho(d, b) = 3^{\frac{1}{2}}.$$

If M could be embedded in a plane, then the points a and e which are at a distance 1 from each other would each be the centre of the triangle b, c, d; this is impossible.

5. Give an example of a metric space consisting of four points which cannot be embedded in a three-dimensional Euclidean space, and which is such that no three of its points lie on a straight line.

Answer. The space $\{a, b, c, d\}$ where $\rho(a, b) = \rho(a, c) = \rho(b, c) = 7$, and $\rho(a, d) = \rho(b, d) = \rho(e, d) = 4$.

6. M is the set of all infinite sequences $z = (z_1, z_2, \ldots)$ of complex numbers such that the series $\sum_{n=1}^{\infty} |z_n|^2$ converges; the distance between two points $u = (u_1, u_2, \ldots)$ and $v = (v_1, v_2, \ldots)$ is given by $\rho(u, v) = (|u_1 - v_1|^2 + |u_2 - v_2|^2 + \ldots)^{\frac{1}{2}}$.

Show that M is a metric space isometric with Hilbert space.

Hint. To obtain an isometric mapping of M on H let the image of the point $(x_1 + i\, y_1, x_2 + i\, y_2, \ldots) \in M$ be the point $(x_1, y_1, x_2, y_2, \ldots) \in H$.

7. $M_1, M_2, \ldots$ is an infinite sequence of metric spaces with distance function ρ_n in M_n for $n = 1, 2, \ldots$; M is the set consisting of the infinite sequence $0 = (0, 0, \ldots)$ and all infinite sequences $p = (p_1, p_2, \ldots p_n, \ldots)$, where $p_n \in M_n$ and $0 \in M_n$ for $n = 1, 2, \ldots$ and where $\sum_{n=1}^{\infty} [\rho_n(p_n, 0)]^2$ is convergent. Setting, for $p \in M$ and $q \in M$,

$$\rho(p, q) = \left(\sum_{n=1}^{\infty} [\rho_n(p_n, q_n)]^2\right)^{\frac{1}{2}},$$

show that M is a metric space with distance function ρ.

Hint. To establish the triangle law for the function ρ employ relation (3) (obtained for real a_k and b_k) letting $a_k = \rho_k(p_k, q_k)$, $b_k = \rho_k(q_k, r_k)$. This gives

$$\rho_k(p_k, r_k) \leqslant \rho_k(p_k, q_k) + \rho_k(q_k, r_k),$$

from which we conclude that

$$\sum_{k=1}^{n} [\rho_k(p_k, r_k)]^2 \leqslant \sum_{k=1}^{n} [\rho_k(p_k, q_k) + \rho_k(q_k, r_k)]^2.$$

8. Show that if, in example 7, $M_n = H$ for $n = 1, 2, \ldots$, then the space M is isometric with Hilbert space H.

Hint. To establish an isometric mapping f of the space H on M put, for $x = (x_1, x_2, \ldots) \in H$, $f(x) = (p_1, p_2, \ldots) \in M$, where $p_k = (x_{1.2^{k-1}}, x_{3.2^{k-1}}, x_{5.2^{k-1}}, \ldots)$ for $k = 1, 2, \ldots$.

9. Give an example of an infinite subset of Hilbert space such that all points are at a distance 1 from each other.

Answer. The set of points $(1/2^{\frac{1}{2}}, 0, 0, \ldots)$, $(0, 1/2^{\frac{1}{2}}, 0, \ldots)$, $(0, 0, 1/2^{\frac{1}{2}}, 0, \ldots)$, ...

58. Urysohn's theorem. Dimensional types.

THEOREM 74 (Urysohn). *Every normal topological space with a countable basis is homeomorphic with a certain subset of Hilbert space.*

Proof. Let K denote a normal (§ **42**) topological space with the countable basis $U_1, U_2, \ldots$ (§ **27**). Corresponding to every index l there exist indices k such that $\bar{U}_k \subset U_l$. For if $p \in U_l$, there exists an open set U such that $p \in U$ and $\bar{U} \subset U_l$ (§ **42**) and so, by virtue of the properties of the sequence $U_1, U_2, \ldots$, there exists an index k such that $p \in U_k \subset U$; hence $\bar{U}_k \subset \bar{U} \subset U_l$.

Consider all pairs (U_k, U_l) such that $\bar{U}_k \subset U_l$ and let

$$(U_{k_1}, U_{l_1}), (U_{k_2}, U_{l_2}), \ldots \tag{7}$$

be the infinite sequence consisting of all these pairs.

Let n denote a given natural number. The sets $\bar{U}_{k_n}$ and $K - U_{l_n}$ are closed and disjoint. Hence there exists, by Urysohn's Lemma (§ **44**), a real function $f_n(p)$ defined and continuous in K such that $f_n(p) = 0$ for $p \in \bar{U}_{k_n}$, $f_n(p) = 1$ for $p \in K - U_{l_n}$, and $0 \leqslant f_n(p) \leqslant 1$ throughout K. For every element $p \in K$, let $\phi(p)$ denote the infinite sequence

$$(f_1(p), 2^{-1}f_2(p), 2^{-2}f_3(p), \ldots, 2^{-n+1}f_n(p), \ldots). \tag{8}$$

The sequence (8) is obviously an element of Hilbert space since the series of the squares of its terms is convergent. The set $\phi(K)$ is therefore a subset of H. We first show that the function ϕ is (1, 1) in K.

Suppose that $p \in K$, $q \in K$, and $p \neq q$. It follows from condition γ (§ **19**) and Lemma 1 (§ **27**) that there exists an index l such that $p \in U_l$ and $q \notin U_l$; hence there exists an index k such that $p \in U_k$ and $\bar{U}_k \subset U_l$. The pair (U_k, U_l) is therefore a member of the sequence (7). Thus $U_k = U_{k_n}$ and $U_l = U_{l_n}$ for some n and $p \in U_{k_n}$ while $q \notin U_{l_n}$, that is, $q \in K - U_{l_n}$. It follows from the definition of the function f_n that $f_n(p) = 0$, $f_n(q) = 1$ and so $\phi(p) \neq \phi(q)$ (since the sequences $\phi(p)$ and $\phi(q)$ differ in their nth terms). The function $\phi(p)$ is therefore (1, 1) in K.

Next, let p_0 be a given element of K and ϵ a given positive number. Let m be a natural number such that

$$2^{-2m+2} < \epsilon^2. \tag{9}$$

Since the real functions $f_n(p)$ are continuous in K there exists for every n an open set $V_n \subset K$ such that $p_0 \in V_n$ and

$$|f_n(p) - f_n(p_0)| < \epsilon/2m \qquad \text{for } p \in V_n. \tag{10}$$

Put $V = V_1 . V_2 \ldots V_m$. Then V is open; it contains p_0 and from (10) we have

$$|f_n(p) - f_n(p_0)| < \epsilon/2m \text{ for } p \in V \text{ and } n = 1, 2, \ldots, m. \tag{11}$$

Therefore,

$$\rho(\phi(p), \phi(p_0)) = \left(\sum_{n=1}^{\infty} 2^{-2n+2}[f_n(p) - f_n(p_0)]^2\right)^{\frac{1}{2}}. \tag{12}$$

Since $|f_n(p) - f_n(p_0)| \leqslant 1$ for $n = 1, 2, \ldots$, (11) and (9) give, for $p \in V$,

$$\sum_{n=1}^{\infty} 2^{-2n+2}[f_n(p) - f_n(p_0)]^2 \leqslant \sum_{n=1}^{m} 2^{-2n+2}[f_n(p) - f_n(p_0)]^2$$
$$+ \sum_{n=m+1}^{\infty} 2^{-2n+2} < m\epsilon^2/4m^2 + 2^{-2m+1} \leqslant \epsilon^2/4 + \epsilon^2/2 < \epsilon^2.$$

Thus for $p \in V$ we have from (12)

$$\rho(\phi(p), \phi(p_0)) < \epsilon. \tag{13}$$

Hence $\phi(p)$ is continuous in K. To prove that the inverse function ϕ^{-1} is continuous in the set $\phi(K) \subset H$, let q_0 be a given element of $\phi(K)$ and U an arbitrary open set (in K) containing the element $p_0 = \phi^{-1}(q_0)$. Since $p_0 \in U$, there exist indices l and k such that $p_0 \in U_k$ and $\bar{U}_k \subset U_l$. The pair $(U_k\ U_l)$ is therefore a member of the sequence (7), say $U_k = U_{k_m}$, and $U_l = U_{l_m}$. Thus

$$p_0 \in U_{k_m} \text{ and } \bar{U}_{k_m} \subset U_{l_m}. \tag{14}$$

Now let q be an element of $\phi(K)$ such that

$$\rho(q, q_0) < 2^{-m+1}; \tag{15}$$

then

$$\phi^{-1}(q) \in U. \tag{16}$$

For if $p = \phi^{-1}(q) \notin U$, then for any $U_l = U_{l_m} \subset U$, we would have $p \notin U_{l_m}$, i.e. $p \in K - U_{l_m}$ and therefore $f_m(p) = 1$. But $f_m(p_0) = 0$ since

$p_0 \in U_{k_m} \subset \bar{U}_{k_m}$. Consequently $f_m(p) - f_m(p_0) = 1$ and so, taking into consideration the definition of the function ϕ, we obtain from (12) $\rho(q, q_0) = \rho(\phi(p), \phi(p_0)) \geqslant 2^{-m+1}$; this contradicts (15). Thus (15) implies (16) and the continuity of the inverse function ϕ^{-1} in $\phi(K)$ is established. This proves that the function ϕ maps K topologically on a subset $\phi(K)$ of Hilbert space. Theorem 74 is therefore established.

It should be noted that Theorem 74 may not be true in a Hausdorff topological space with a countable basis.

For example, let K be the set of all real numbers x of the interval $0 \leqslant x \leqslant 1$ with the usual definition of neighbourhoods, with one exception: all numbers of the form $1/n$, $n = 1, 2, \ldots$, are deleted from every neighbourhood of zero. It is easily seen that K is a Hausdorff topological space with a countable basis but it is not regular and therefore not normal. The set $V = K - \{1, \frac{1}{2}, \ldots\}$ is open (since it is a neighbourhood of zero) but the closure of any open set containing 0 and contained in V cannot itself be contained in V since it contains an infinity of elements of the form $1/n$. Hence K is not homeomorphic with any metric space.[47]

If a set P is homeomorphic with a subset of a set Q and Q is homeomorphic with a subset of P, then P and Q are said to have the same *dimensional type* (Fréchet); in symbols $dP = dQ$. Clearly if $dP = dQ$ then $dQ = dP$ and if $dP = dQ$, $dQ = dR$, then $dP = dR$. Homeomorphic sets have obviously the same dimensional type but the converse need not be true. For example, a closed linear segment and an open one are of the same dimensional type but they are not homeomorphic. If P is homeomorphic with a subset of Q but Q is not homeomorphic with any subset of P, then P is said to have a smaller dimensional type than Q and we write $dP < dQ$ (or $dQ > dP$). Clearly if $dP < dQ$ and $dQ \leqslant dR$ then $dP < dR$.

It has been proved by Banach[48] that if two sets P and Q have the same dimensional type there exist decompositions $P = P_1 + P_2$ and $Q = Q_1 + Q_2$, where $P_1 . P_2 = Q_1 . Q_2 = 0$, such that $P_1 \, h \, Q_1$ and $P_2 \, h \, Q_2$.

The proof that two sets are not of the same dimensional type is generally very difficult (as, for example, in the case of Euclidean spaces of different dimensions). Furthermore, not all sets can be compared as to dimensional type.

There exist in a sense two smallest dimensional types which are not commensurable. Let E be the set consisting of the number 0 and all numbers of the form $2^{-m} + 2^{-m-n}$, m and n natural numbers, and T the set consisting of the numbers 0, n^{-1} and $1 + n^{-1}$, $n = 1, 2, \ldots$. It can be shown[49] that the dimensional types of E and T are incommensurable and that if E_1 and T_1 are any two metric spaces whose dimensional types are incommensurable then either $dE \leqslant dE_1$ and $dT \leqslant dT_1$, or $dE \leqslant dT_1$ and $dT \leqslant dE_1$ (or both).

It has been proved by Kuratowski[50] that there exists a family of $2^{\mathbf{c}}$ linear sets whose topological types are pairwise incommensurable. On the other hand, an example can easily be given of a family of $2^{\mathbf{c}}$ different linear sets all of the same dimensional type. (For instance, the family of all linear sets containing the interval $(0, 1)$.) It is easily seen that for every linear set E there exist at least $\mathbf{c}$ different linear sets of the same dimensional type as that of E (if E is finite or countable, there are precisely $\mathbf{c}$).

If a metric space N_1 is homeomorphic with a subset M_1 of a metric space M, it does not follow that there exists a metric space N containing N_1 and homeomorphic with M. Thus, for example, the space N_1 of all natural numbers is homeomorphic with the set $M_1 = \{1, 1/2, 1/3, \ldots\}$ contained in the space $M = \{0, 1, 1/2, \ldots\}$ (with the usual metric); there exists, however, no metric space $N \supset N_1$ which is homeomorphic with M. This follows because M is compact-in-itself, whereas N_1, as an unbounded set, cannot be a subset of a space N compact-in-itself. (Consequently, by Theorem 55, N cannot be homeomorphic with M.) If, however, as has been proved by Hausdorff,[51] a metric space N_1 is homeomorphic with a closed subset M_1 of a metric space M, then there exists a metric space N containing N_1 and homeomorphic with M.

*If $dE < dT$ and if there exists no set Q such that $dE < dQ < dT$ then the topological type of T is said to *follow immediately* after the topological type of E (or to be *next* to the topological type of E, or else, to be the *immediate successor* of the topological type of E). For example, the topological type of the circle is next to that of the straight line. It can be shown that certain linear sets (for instance, the set of all rational numbers) have a dimensional type without an immediate predecessor. This result, however, cannot be proved without the continuum hypothesis.

It is noteworthy that the addition of a single element to a set E may give rise to a set E_1 for which there exist two sets Q_1 and Q_2 such that $dE < dQ_1 < dQ_2 < dE_1$. (For example,[52] where E is the set of all numbers of the form $2^{-m} + 2^{-m-n}$, m and n natural numbers, $E_1 = E + \{0\}$, $Q_1 = \{0, 1, 1/2, 1/3, \ldots\}$, $Q_2 = Q_1 + \{2 + 1, 2 + 1/2, 2 + 1/3, \ldots\}$.)

It can also be shown that if T_1 denotes a plane set consisting of finite segments radiating from the point 0 and T is the set $T_1 + \{0\}$, then there exists an infinite sequence of sets $R_1, R_2, \ldots$ such that $dT_1 < dR_1 < dR_2 < \ldots < dT$. Thus the addition of a single element to a set may change its dimensional type considerably.

It can be proved that of all countable metric spaces the set of rational numbers has the greatest dimensional type.

We note further that a dimensional type may have two different immediate successors (incommensurable with each other). For example, let $E_2 = Q_2 + \{2\}$ (where E_1 and Q_2 are the sets defined above); then dE_1 and dE_2 are each next to dQ_2 but $dE_1 \neq dE_2$.

To express the fact that dH follows immediately after dE Ruziewicz[53] writes $dE\ p\ dH$ (dE precedes dH); he gives an example of five linear sets E_1, E_2, E_3, H_1, and H_2, such that, $dE_1\ p\ dE_2\ p\ dE_3$ and $dE_1\ p\ dH_1\ p\ dH_2\ p\ dE_3$. Consequently, if a dimensional type dE_2 is next to dE_1 and precedes dE_3 it does not follow that it is the unique dimensional type between dE_1 and dE_3. As in this case, there may even be two consecutive dimensional types between dE_1 and dE_2.

It can be shown that the dimensional type of the set N of all rational numbers is followed immediately by the dimensional type of the set X of all real numbers but dN is not preceded immediately by any other dimensional type. Moreover, if E is any linear set with $dE > dN$ then $dE \geqslant dX$. However, if we consider sets other than linear, this result does not hold; for there exist plane sets P such that $dN < dP$ but dX is neither less than nor equal to dP (e.g. the plane continuum P of Janiszewski which does not contain any simple arc).

The Euclidean m-dimensional space R_m can be embedded metrically, for every natural m, in the Hilbert space H (since it is obviously isometric with the set of all points of H whose coordinates, beginning with the $(m+1)$th are all equal to zero). As shown by Kunugui,[54] there exists a subset H_1 of H with $dH_1 < dH$, and such that every Euclidean space R_n, $n = 1, 2, \ldots$, can be embedded metrically in H_1. This set H_1 consists of all those points of H which have a finite number of coordinates different from zero. (The proof that $dH_1 < dH$ is not easy.) It can be shown, however, that there exists a set H_2 with $dH_2 < dH_1$ and such that every Euclidean space R_n ($n = 1, 2, \ldots$) can be embedded metrically in it and that no set of dimensional type smaller than that of H_2 has this property.[55*]

Since Hilbert space, as a separable metric space, is a topological and normal space with a countable basis, Theorem 74 gives the following

COROLLARY 1. *Hilbert space has the greatest dimensional type of all topological normal spaces with a countable basis.*

Theorem 74 leads also immediately to

COROLLARY 2. *In every normal topological space E with a countable basis, a metric can be established so that the space becomes a metric space homeomorphic with E.*

There exist, however, countable Hausdorff topological spaces into which a metric cannot be introduced (for example, the set constructed by Appert (§ **28**), which possesses no countable basis, although it is separable (Theorem 65).

COROLLARY 3. *A metric space is separable if and only if it is homeomorphic with a subset of Hilbert space.*

The necessity of the condition follows from Theorem 74 since a separable metric space is a topological normal space with a countable basis—the sufficiency from the fact that Hilbert space is separable (§ 57), from Theorem 66, and from the fact that, by Theorem 40, separability is a topological property.

Corollaries 1 or 3 permit us to state that *Hilbert space has the greatest dimensional type of all separable metric spaces.* Therefore, the topology of separable metric spaces reduces to the topology of Hilbert space and its subsets.

Let R denote the set of all points of Hilbert space which have only a finite number of coordinates different from zero. It is easily seen that a Euclidean space of any dimension may be embedded metrically in R. It has been proved by Kunugui[56] that *the space R has a smaller dimensional type than Hilbert space.*

Furthermore, a metric space of dimensional type smaller than that of R can be constructed in which a Euclidean space of any dimension may be embedded.[55] This is the set consisting of all finite sequences of real numbers where the distance between two sequences $x = (x_1, x_2, \ldots, x_m)$ and $y = (y_1, y_2, \ldots, y_n)$ for $m \leqslant n$, is given by

$$\rho(x, y) = n - m + ((x_1 - y_1)^2 + (x_2 - y_2)^2 + \ldots + (x_m - y_m)^2 + y_{m+1}^2 + \ldots + y_n^2)^{\frac{1}{2}}.$$

Kunugui[57] has proved that this set has the smallest dimensional type of all metric spaces with the above property.

From Theorems 74 and 63 we easily deduce

COROLLARY 4. *A metric can be introduced into a topological space with a countable basis if and only if the space is normal.*

COROLLARY 5. *A metric can be introduced into a compact topological space if and only if the space is normal and possesses a countable basis.*[58]

Proof. The necessity of the condition follows immediately from the fact that a metric space is normal and from Theorems 68 and 65. The sufficiency follows from Corollary 2.

59. Fréchet's space E_ω and its properties. Hilbert space is a natural generalization of Euclidean m-dimensional space (because of the definition of distance), but it has a somewhat artificial limitation on the coordinates, namely the condition of convergency for the sum of their squares (a condition necessary to assure that the distance between two points is always finite). Fréchet raised the following question: let E be a set whose elements are infinite sequences of real numbers

$$x_1, x_2, x_3, \ldots.$$

Is it possible to introduce a metric in E so that in the resulting metric space the element

(17) $$p = (x_1, x_2, \ldots)$$

will be the limit of the infinite sequence $\{p_1, p_2, \ldots\} \subset E$, where

(18) $$p_n = (x_1^{(n)}, x_2^{(n)}, \ldots),$$

if and only if

(19) $$\lim_{n\to\infty} x_k^{(n)} = x_k \qquad \text{for } k = 1, 2, \ldots\,?$$

Fréchet answered the question positively by choosing the distance between two elements

$$p = (x_1, x_2, \ldots) \text{ and } q = (y_1, y_2, \ldots)$$

to be the number

(20) $$\rho(p, q) = \sum_{n=1}^{\infty} \frac{1}{n!} \frac{|x_n - y_n|}{1 + |x_n - y_n|}.$$

First, it is necessary to show that the function (20) satisfies the distance axioms. It is evident that it satisfies the first two; it remains to prove the triangle law.

Since for any two numbers a and b we have

$$\frac{|a+b|}{1+|a+b|} \leqslant \frac{|a|}{1+|a|} + \frac{|b|}{1+|b|},$$

then for $a = x_n - y_n$, $b = y_n - z_n$,

(21) $$\frac{|x_n - z_n|}{1 + |x_n - z_n|} \leqslant \frac{|x_n - y_n|}{1 + |x_n - y_n|} + \frac{|y_n - z_n|}{1 + |y_n - z_n|}.$$

Thus, if $r = (z_1, z_2, \ldots)$, (20) and (21) give the inequality

$$\rho(p, r) \leqslant \rho(p, q) + \rho(q, r).$$

The set of all infinite sequences of real numbers with the metric given by (20) becomes a metric space which Fréchet denotes by E_ω.

Let $p_n (n = 1, 2, \ldots)$ denote an infinite sequence of elements of E_ω, such that $\lim_{n\to\infty} p_n = p$, where p_n and p are the sequences (18) and (17) respectively. Then,

(22) $$\lim_{n\to\infty} \rho(p, p_n) = 0.$$

But for $k = 1, 2, 3 \ldots$,

$$\frac{1}{k!} \frac{|x_k - x_k^{(n)}|}{1 + |x_k - x_k^{(n)}|} \leqslant \rho(p, p_n);$$

therefore, $\lim_{n\to\infty} |x_k - x_k^{(n)}| = 0$, which gives (19).

On the other hand, let $\{p_n\}$ be a sequence of elements of E_ω, for which (19) holds. Let ϵ be any positive number. Choose a natural number m sufficiently large so that

$$\sum_{m+1}^{\infty} 1/k! < \epsilon/2 \tag{23}$$

(this is possible, since the series $\sum_1^\infty 1/k!$ is convergent). Then

$$\sum_{m+1}^{\infty} \frac{1}{k!} \frac{|x_k - x_k^{(n)}|}{1 + |x_k - x_k^{(n)}|} < \epsilon/2, \qquad n = 1, 2, \ldots. \tag{24}$$

There exists, by (19), an index μ such that

$$|x_k - x_k^{(n)}| < \epsilon/2m \qquad \text{for } n > \mu, \text{ and } k = 1\ 2, \ldots, m.$$

Therefore

$$\sum_{1}^{m} \frac{|x_k - x_k^{(n)}|}{k!(1 + |x_k - x_k^{(n)}|)} < m\epsilon/2m = \epsilon/2 \qquad \text{for } n > \mu;$$

this and (24) give

$$\rho(p, p_n) < \epsilon \qquad \text{for } n > \mu;$$

consequently $\lim\limits_{n\to\infty} p_n = p$.

Hence, in order that an element p of E_ω be the limit of an infinite sequence of elements of E_ω, it is necessary and sufficient that, for every index k, the kth coordinates of the terms of the sequence approach the kth coordinate of the element p. In Hilbert space this condition is necessary (as can be proved similarly as in the case of E_ω) but it is not sufficient. For example, the sequence $p_n (n = 1, 2, \ldots)$, where $p_n \in H$ and has all coordinates equal to zero except the nth which is equal to 1, does not approach $p = (0, 0, \ldots)$ since $\rho(p, p_n) = 1$ although $x_k^{(n)} \to 0$ as $n \to \infty$ for all k. It follows that the function $f(p) = p$ maps H on a subset of E_ω continuously but not topologically.

Let the infinite sequences of rational numbers, with all but a finite number of terms in each sequence equal zero, be called the rational elements of E_ω. The set P of these rational elements is obviously countable. From the property of E_ω deduced above, it follows that P is dense on E_ω. For let (17) be any element of E_ω. Put, for $n = 1, 2, \ldots,$

$$x_k^{(n)} = \frac{[nx_k]}{n} \text{ for } k \leqslant n; \quad x_k^{(n)} = 0 \text{ for } k > n. \tag{25}$$

It is easily seen that relations (19) are satisfied; therefore, for the sequence (18) we have $\lim\limits_{n\to\infty} p_n = p$. But (18) and (25) give $p_n \in P$ for $n = 1, 2, \ldots$; hence $p \in \bar{P}$ and, since $p \in E_\omega$, we obtain $E_\omega \subset \bar{P}$. Consequently, *E_ω is separable.*

THEOREM 75. *Every separable metric space is homeomorphic with some subset of E_ω.*

Proof. Let M denote a separable metric space. Then there exists a set $P = \{p_1, p_2, \ldots\}$ which is at most countable, such that $M = \bar{P}$. Denoting the distance in M by ρ_1, put, for $p \in M$,

$$f(p) = \{\rho_1(p, p_1), \rho_1(p, p_2), \ldots\}; \tag{26}$$

then $f(p) \in E_\omega$. It will be shown that the function f establishes a homeomorphism between M and the set $f(M) \subset E_\omega$.

We first prove that the function f is biuniform in M. Suppose that p and p' are two different elements of M; then $\rho_1(p, p') > 0$, and so, since $p \in \bar{P}$ there exists an index k such that

$$\rho_1(p, p_k) < \tfrac{1}{2}\rho_1(p, p').$$

This gives

$$2\rho_1(p, p_k) < \rho_1(p, p') \leqslant \rho_1(p, p_k) + \rho_1(p_k, p'),$$

that is,

$$\rho_1(p, p_k) < \rho_1(p', p_k).$$

Hence $f(p) \neq f(p')$ since they differ in their kth coordinates. Let $p^{(n)} \in M$ for $n = 1, 2, \ldots$ and $\lim_{n\to\infty} p^{(n)} = p \in M$. Since ρ_1 is a continuous function in M of the two variables (Theorem 64), we have

$$\lim_{n\to\infty} \rho_1(p^{(n)}, p_k) = \rho_1(p, p_k) \qquad \text{for } k = 1, 2, \ldots.$$

Hence for all k, the kth coordinate of $f(p^{(n)})$ tends to the kth coordinate of $f(p)$ and so (by the property of E_ω)

$$\lim_{n\to\infty} f(p^{(n)}) = f(p).$$

Thus the relations

$$p^{(n)} \in M \qquad \text{for } n = 1, 2, \ldots,$$

and

$$\lim_{n\to\infty} p^{(n)} = p \in M$$

imply that

$$\lim_{n\to\infty} f(p^{(n)}) = f(p).$$

Next, suppose that $p^{(n)} (n = 1, 2, \ldots)$ is an infinite sequence of elements of M such that $\lim_{n\to\infty} f(p^{(n)}) = f(p)$. From (10) and the property of E_ω, we obtain

$$\lim_{n\to\infty} \rho_1(p^{(n)}, p_k) = \rho_1(p, p_k) \qquad \text{for } k = 1, 2, \ldots.$$

Let ϵ be an arbitrary positive number. Since $p \in M = \bar{P}$, there exists an index k such that $\rho_1(p, p_k) < \epsilon$ and, since $\lim_{n\to\infty} \rho_1(p^{(n)}, p_k) = \rho_1(p, p_k)$, there exists a natural number μ such that $\rho_1(p^{(n)}, p_k) < 2\epsilon$ for $n > \mu$. Consequently,

$$\rho_1(p^{(n)}, p) \leqslant \rho_1(p^{(n)}, p_k) + \rho_1(p_k, p) < 3\epsilon \text{ for } n > \mu;$$

hence

$$\lim_{n\to\infty} p_n = p.$$

It therefore follows that for $p^{(n)} \in M$ $(n = 1, 2, \ldots)$ and $p \in M$ the relations

$$\lim_{n\to\infty} p^{(n)} = p \text{ and } \lim_{n\to\infty} f(p^{(n)}) = f(p)$$

are equivalent. This proves that the function f maps the set M topologically on the set $f(M) \subset E_\omega$.

In particular, we obtain at once from Theorem 75 the result that Hilbert space is homeomorphic with a certain subset of E_ω and since, by Theorem 74, E_ω is homeomorphic with a certain subset of Hilbert space, we have the

COROLLARY. *Hilbert space and the space E_ω have the same dimensional type.*

However, the question whether Hilbert space and the space E_ω are homeomorphic is not as yet settled.

The space E_ω is not compact (although it is bounded); we shall show that E_ω is not even semi-compact, that is, it is not the sum of a countable aggregate of compact sets. For suppose that $E_\omega = E_1 + E_2 + \ldots$, where E_n $(n = 1, 2, \ldots)$ is compact. Then, for every natural number k, the kth coordinates of the elements of E_n form a bounded set. There exists, therefore, a real number $a_k^{(n)}$ such that the kth coordinates of the elements of E_n are numerically $\leqslant a_k^{(n)}$. Hence the element $p = (a_1^{(1)} + 1, a_2^{(2)} + 1, \ldots) \in E_\omega$ is not contained in any of the sets $E_n (n = 1, 2, \ldots)$; this is impossible.

It will be shown, however, that E_ω *is homeomorphic with a certain compact subset of itself.*

Let T denote the set of all those elements $p = (x_1, x_2, \ldots)$ of E_ω for which $|x_k| < 1$, $k = 1, 2, \ldots$ It is easily seen that the function

$$f(p) = \left(\frac{x_1}{1 - |x_1|}, \frac{x_2}{1 - |x_2|}, \ldots\right)$$

establishes a homeomorphism between T and E_ω.

It remains to prove that T is compact in E_ω. Let $p^{(n)} = (x_1^{(n)}, x_2^{(n)}, \ldots)$, $n = 1, 2, \ldots$, be an infinite sequence of different elements of T. We therefore have $|x_1^{(n)}| < 1$ for $n = 1, 2, \ldots$; so there exists an increasing sequence of natural numbers $1 < a_1 < a_2 < \ldots$ such that $\lim_{n\to\infty} x_1^{(a_n)} = x_1$ exists. Similarly, since $|x_2^{(a_n)}| < 1$ for $n = 1, 2, \ldots$, there exists an infinite sequence of indices $1 < b_1 < b_2 < \ldots$, such that

$$\lim_{n\to\infty} x_2^{(a_{b_n})} = x_2$$

(for notation see p. 86). We conclude similarly the existence of a sequence of indices $1 < c_1 < c_2 < \ldots$, such that

$$\lim_{n\to\infty} x_3^{(a_{b_{c_n}})} = x_3$$

exists, and so on. Put $p = (x_1, x_2, \ldots)$ and $m_1 = 1$, $m_2 = a_1$, $m_3 = a_{b_1}$, $m_4 = a_{b_{c_1}}$ and so forth. Hence $m_1 < m_2 < \ldots < m_n < \ldots$, and

$$\lim_{n\to\infty} x_k^{(m_n)} = x_k \qquad \text{for } k = 1, 2, \ldots;$$

therefore

$$\lim_{n\to\infty} p^{(m_n)} = p.$$

The elements of the sequence $p^{(m_n)}$ $(n = 1, 2, \ldots)$ are obviously all different elements of T; consequently $p \in T'$. Since the sequence $\{p^{(n)}\}$ is any sequence of different elements of T, this proves that T is compact in E_ω.

Thus E_ω is homeomorphic with a certain compact subset of itself. However, since E_ω is not compact, it is not homeomorphic with any of its subsets which are compact-in-themselves. It also follows from Theorem 75 that every separable metric space is homeomorphic with a certain compact subset of E_ω. On the other hand, every compact subset of a metric space is separable (Theorem 68) and separability is a hereditary property as well as a topological invariant. We may therefore state that *a metric space is a homeomorphic (or continuous) image of a subset of a compact metric space if and only if it is separable.* Since every compact subset of a metric space is totally bounded, it follows that every separable metric space is homeomorphic with a totally bounded metric space. But the converse is also true; for, as we know, a totally bounded metric space is separable and separability is a topological invariant. Hence *a metric space is separable if and only if it is homeomorphic with a totally bounded space.*

Example. Show that E_ω is homeomorphic with the "fundamental cube" of the Hilbert space H, that is, with the set Q of all the points $x = (x_1, x_2, \ldots)$ of H such that $|x_n| < 1/n$, for $n = 1, 2, \ldots$.

Proof. For $p \in E_\omega$, put

$$(27) \qquad f(p) = \left(\frac{x_1}{1(1 + |x_1|)}, \frac{x_2}{2(1 + |x_2|)}, \ldots, \frac{x_k}{k(1 + |x_k|)}, \ldots\right);$$

hence $f(p) \in Q$. Obviously $f(E_\omega) = Q$. Let $p^{(n)} = (x_1^{(n)}, x_2^{(n)}, \ldots)$ for $n = 1, 2, \ldots$ denote an infinite sequence of different elements of E_ω such that $\lim_{n\to\infty} p^{(n)} = p \in E_\omega$, where $p = (x_1, x_2, \ldots)$. Hence $\lim_{n\to\infty} x_k^{(n)} = x_k$ for $k = 1, 2, \ldots$ and therefore,

$$(28) \qquad \lim_{n\to\infty} \frac{x_k^{(n)}}{1 + |x_k^{(n)}|} = \frac{x_k}{1 + |x_k|}, \qquad k = 1, 2, \ldots.$$

Let ϵ be an arbitrary positive number. Choose a natural number $j > 1/\epsilon^2$. Since

$$\sum_{m=1}^{\infty} 1/(j+m)^2 < \sum_{m=1}^{\infty} 1/(j+m-1)(j+m) = 1/j,$$

we have

$$\sum_{j+1}^{\infty} 1/k^2 < \epsilon^2.$$

There exists, by (28), a natural number μ such that

$$\left|\frac{x_k^{(n)}}{1+|x_k^{(n)}|} - \frac{x_k}{1+|x_k|}\right| < \epsilon/j \text{ for } n > \mu, \text{ and } k = 1, 2, \ldots, j. \tag{29}$$

Now $t/(1+|t|) < 1$ for t real; hence, for $n > \mu$ and $k = j+1, j+2, \ldots$, $|x_k^{(n)}/(1+|x_k^{(n)}|) - x_k/(1+|x_k|)| < 2$. From (28), (29), and the definition of the distance function ρ in H, we obtain

$$\rho(f(p^{(n)}), f(p)) < \epsilon 5^{\frac{1}{2}} \qquad \text{for } n > \mu.$$

Consequently,

$$\lim_{n\to\infty} f(p^{(n)}) = f(p).$$

Suppose that $p^{(n)}$ is a sequence of elements of E_ω such that $\lim_{n\to\infty} f(p^{(n)}) = f(p)$ in H. From (27) and the properties of H, we obtain (28). But if $a/(1+|a|) = b$, then $|b| < 1$ and $a = b/(1-|b|)$. Hence if $a_1, a_2, \ldots$ is a sequence of real numbers such that

$$\lim_{n\to\infty} a_n/(1+|a_n|) = a/(1+|a|), \text{ then } \lim_{n\to\infty} a_n = a.$$

Therefore (28) gives

$$\lim_{n\to\infty} x_k^{(n)} = x_k \qquad k = 1, 2, \ldots.$$

In E_ω, this implies that

$$\lim_{n\to\infty} p^{(n)} = p.$$

Thus for $p^{(n)} \in E_\omega$ and $p \in E_\omega$ the relations

$$\lim_{n\to\infty} p^{(n)} = p (\text{in } E_\omega) \text{ and } \lim_{n\to\infty} f(p^{(n)}) = f(p) (\text{in } H)$$

are equivalent. This proves that f maps E_ω topologically on $f(E_\omega) = Q$.

Denote by H_ω (Fréchet) the set of all elements of E_ω with irrational coordinates. We shall prove that H_ω is homeomorphic with the set H_1 of all irrational numbers.

For x real, put

$$\phi(x) = \tfrac{1}{2} + \tfrac{1}{2}x/(1+|x|);$$

the function $\phi(x)$ obviously establishes a homeomorphism between the set of all real numbers and the open interval (0, 1) (its inverse function is $x(y) = (2y-1)/(1-|2y-1|)$ for $0 < y < 1$) and therefore also between the set H_1 of all irrational numbers and the set T_1 of all irrational numbers in the interval (0, 1). Now if, for $p = (x_1, x_2, \ldots) \in H_\omega$, we put

$$f(p) = (\phi(x_1), \phi(x_2), \ldots),$$

we obtain a homeomorphic mapping of the set H_ω on the set $T = f(H_\omega)$ of all elements of E_ω with irrational coordinates in the interval (0, 1). It will be sufficient, therefore to show that $T \, h \, T_1$.

Let $p = (x_1, x_2, \ldots)$ be a given element of T. The numbers $x_1, x_2, \ldots$ are therefore irrational and in the interval (0, 1); let

$$x_k = \frac{1}{n_1^{(k)}+} \frac{1}{n_2^{(k)}+} \cdots$$

be the development of the number x_k as a continued fraction. Employ the diagonal method to rearrange the double sequence $n_i^{(k)}$ into a single sequence $m_1, m_2, m_3, \ldots$. Put

$$f(p) = \frac{1}{m_1+} \frac{1}{m_2+} \frac{1}{m_3+} \cdots$$

It is easily shown that $T \, h \, T_1$. It is sufficient here to base the proof on the properties of sequences converging to a given limit in E_ω and on the following two known properties of infinite continued fractions: (i) For every irrational number x_0 and every positive integer k, there exists a positive number ϵ such that every irrational number x, which satisfies the inequality $|x - x_0| < \epsilon$, possesses a development as a continued fraction which is identical in the first k convergents with that of the number x_0 itself. (ii) For every irrational number x_0 and every positive number ϵ, there exists a positive integer k such that every irrational number x_1 whose development into a continued fraction has the first k convergents identical with the corresponding convergents in the development of x_0, satisfies the inequality $|x - x_0| < \epsilon$.

The relation $H_\omega \, h \, H_1$ may, therefore, be considered proved.

Translations of the space E_ω. Let $a_1, a_2, \ldots$ denote an infinite sequence of real numbers. Associate with each element $p = (x_1, x_2, \ldots) \in E_\omega$ the element

$$\phi(p) = (x_1 + a_1, x_2 + a_2, \ldots). \tag{30}$$

It is easily seen that $E_\omega \, h_\phi \, \phi(E_\omega)$. The transformation ϕ is called a *translation* of E_ω (by analogy with m-dimensional space). It follows readily from (30) that a translation of E_ω is an isometric mapping of E_ω on itself. Since by a suitable translation any element of E_ω can be transformed into any other, it may be said that E_ω is not only topologically but also metrically homogeneous.

Let now N be any set of elements of E_ω with cardinal less than that of the continuum. We shall show that there exists a translation of E_ω under which N maps into a certain subset of H_ω. To this end we first prove the following

LEMMA. *If Q is a set of real numbers of cardinal $< \mathbf{c}$, there exists a real number a such that for every number x of Q the number $x + a$ is irrational.*

Proof. Let $\mathbf{m}$ be the cardinal of Q; then $\mathbf{m} < \mathbf{c}$. Denote by S the set of all numbers of the form $r - x$, where r is a rational number and $x \in Q$. Hence the cardinal of S is $\leqslant \aleph_0 . \mathbf{m} < \mathbf{c}$. There exists, therefore, a real number a, which does not belong to S. The number $x + a$, for $x \in Q$, is clearly irrational (otherwise, $x + a = r$ would show $a \in S$, contrary to the definition of a). This proves the lemma.

If now $N \subset E_\omega$ and has cardinal $< \mathbf{c}$ then the set N_k of the kth coordinates of all elements of N has cardinal $< \mathbf{c}$ and so, by the above lemma, there exists a real number a_k such that, for every $x_k \in N_k$, the number $x_k + a_k$ is irrational. It follows that the translation (30) maps N on a certain subset of H_ω. From Theorem 75 and the relation $H_\omega \, h \, H_1$ we obtain

THEOREM 76. *Every separable metric space of cardinal less than that of the continuum is homeomorphic with a certain set of irrational numbers.*

We note that if the expression "separable" be omitted from the statement of Theorem 76 we obtain a theorem which is equivalent to the continuum hypothesis. For, if the continuum hypothesis is true, every metric space with cardinal less than that of the continuum is at most countable and so separable; it therefore satisfies the conditions of Theorem 76. If the hypothesis of the continuum is not true, then a metric space of cardinal $\aleph_1$ with $\rho(p, q) = 1$ for $p \neq q$, has cardinal less than that of the continuum but is not homeomorphic with any set of real numbers.

Let M denote a countable metric space. By Theorem 76, M is homeomorphic with a certain set P of irrational numbers. Let R be the set of all rational numbers. The set $P + R$ is countable, dense-in-itself, and has neither a first nor a last element when ordered according to increasing magnitude. By a theorem of Cantor,[59] $P + R$ is similar to R. But the similar mapping of the set $P + R$ on the set R is a homeomorphic mapping. The set P is therefore homeomorphic with a certain set of rational numbers. We thus have

COROLLARY 1. *Every countable metric space is homeomorphic with a certain set of rational numbers.*

This gives

COROLLARY 2. *The set of all rational numbers has the greatest dimensional type of all countable metric spaces.*

Every countable set of real numbers which is dense-in-itself and has no first nor last element is homeomorphic with the set of all rational numbers; we therefore obtain from Theorem 76

COROLLARY 3. *Every countable metric space which is dense-in-itself and has neither a first nor a last element is homeomorphic with the set of all rational numbers.*

Thus, for instance, the set of all rational points in Euclidean m-dimensional space (m any natural number), and the set of all rational points in Hilbert space (or in E_ω) are homeomorphic with the set of all rational numbers.

The sets of rational points in H and E_ω are therefore homeomorphic. The homeomorphism between them is not, however, established by the identity function since, for example, the sequence of points $p_n = (x_1^{(n)}, x_2^{(n)}, \ldots)$, where $x_n^{(n)} = 1$ for $n = 1, 2, \ldots$ and $x_k^{(n)} = 0$ for $k \neq n$, tends to the point $(0, 0, \ldots)$ in E_ω but has no limit in H.

In connection with Corollary 2 we note further that *among the dimensional types of non-countable metric spaces there is no least.* For a space of such a dimensional type would have to be homeomorphic with a subset of a non-countable isolated space (e.g. $\rho(p, q) = 1$ for $p \neq q$) and therefore would itself be isolated; on the other hand, it would have to be homeomorphic with a linear subset and, since it is isolated, it would have to be countable; this is impossible.

It can be shown that for every linear set of cardinal $\mathbf{c}$ there exists a set of cardinal $\mathbf{c}$ of smaller dimensional type.[60] This is not true, however, for arbitrary metric spaces of cardinal $\mathbf{c}$ (e.g. for isolated spaces).

Assuming the continuum hypothesis, it can be proved that there exists a greatest dimensional type among the dimensional types of metric spaces of cardinal $\mathbf{c}$. For if $\mathbf{c} = \aleph_1$, there exists[61] a metric space U of cardinal $\mathbf{c}$ such that every metric space of cardinal $\mathbf{c}$ is isometric (and therefore homeomorphic) with a certain subset of U.

We note in connection with Theorem 75 that Urysohn[62] has proved the existence of a separable metric space U such that every separable metric space is not only homeomorphic but even isometric with some subset of U.

The definition of Urysohn's space is rather complicated. Banach and Mazur have shown that the same property is possessed by the space (C) of all real functions continuous in the interval $0 \leqslant x \leqslant 1$ where the distance $\rho(f, g)$ between two functions f and g belonging to (C) is defined to be $\inf|f(x) - g(x)|$ for $0 \leqslant x \leqslant 1$. The proof given by Banach and Mazur[63] is based on the theory of functionals.

60. The 0-dimensional Baire space. The Cantor set. Let $M = M_1 \times M_2 \times \ldots$ denote the combinatorial product (§ **19**) of an infinite sequence of given metric spaces. The set M becomes a metric space if the distance ρ between two elements p and q of M is defined by

$$\rho(p, q) = \sum_{n=1}^{\infty} \frac{\rho_n(x_n, y_n)}{n!(1 + \rho_n(x_n, y_n))}$$

where ρ_n denotes the distance between two elements x_n, y_n in M_n. That M is a metric space is proved as in the case of E_ω (§ **59**).

Exactly as in the case of E_ω, it follows that for $p_n \in M$, $p_n = (x_1^{(n)}, x_2^{(n)}, \ldots)$, $n = 1, 2, \ldots$, where $x_k^{(n)} \in M_k$ (for any natural numbers n and k), and for $p \in M$, where $p = (x_1, x_2, \ldots)$,

$$\lim_{n\to\infty} p_n = p \text{ (in } M)$$

is equivalent to

$$\lim_{n\to\infty} x_k^{(n)} = x_k \text{ (in } M_k), \qquad k = 1, 2, \ldots.$$

E_ω is a special case of the product M when $M_n = R_1$ for all n, and where R_1 is the linear space. We may therefore write $E_\omega = R_1^{\aleph_0}$.

Similarly we write $H_\omega = H^{1\aleph_0}$; if the interval $(0 \leqslant x \leqslant 1)$ be denoted by I, then the fundamental cube[64] of E_ω, i.e. the set of all its elements $(x_1, x_2, \ldots)$ where $0 \leqslant x_k \leqslant 1$ for $k = 1, 2, \ldots$ may be denoted by $I^{\aleph_0}$.

Suppose now that each of the sets M_n $(n = 1, 2, \ldots)$ is the set of all integers (hence a metric space, since it is a subset of the linear space). In this case, the space M will consist of all infinite sequences of integers. Since an element of M_k is the limit of a sequence of elements of M_k if and only if from a certain stage in the sequence all terms are equal to x_k, it follows that the element $p = (x_1, x_2, \ldots) \in M$ is the limit of the sequence $p_n = (x_1^{(n)}, x_2^{(n)}, \ldots) \in M$ if and only if for every natural number m there exists a number μ_m such that

$$x_k^{(n)} = x_k \qquad \text{for } k = 1, 2, \ldots, m \text{ and } n > \mu_m.$$

The space M is the so-called 0-dimensional Baire space. It is homeomorphic with the set T_1 of all irrational numbers in the interval (0, 1,) and therefore with the set of all irrational numbers (§ **59**). To prove this, associate with each number

$$x = \frac{1}{n_1 +} \frac{1}{n_2 +} \cdots$$

of the set T_1 the element

$$f(x) = (k_{n_1}, k_{n_2}, \ldots)$$

of the set M, where $k_1, k_2, \ldots$ is the set of all different integers arranged in a sequence. The proof that $T_1 \, h_f \, M$ then follows from the properties of continued fractions and of limits in M. Hence *the 0-dimensional Baire space is homeomorphic with the set of all irrational numbers.*

Denote by P the set of all those elements of M which are infinite sequences composed of the two numbers 0 and 1. It is readily proved that P is compact and perfect. It will be shown that P is homeomorphic with the set C of all real numbers of the form

$$(31) \qquad 2(3^{-1}c_1 + 3^{-2}c_2 + \ldots),$$

where $c_n = 0$ or 1 for $n = 1, 2, \ldots$.

With each element $x = (c_1, c_2, \ldots)$ of the set P associate the number (31) of the set C and denote it by $f(x)$. The proof that $P\, h_f\, C$ then follows from Theorem 54, from the properties of limits in M, and from the fact that if the sequences $x = (c_1, c_2, \ldots)$ and $x' = (c'_1, c'_2, \ldots)$ first differ in their nth terms, then

$$|f(x) - f(x')| \leqslant \frac{1}{3^{n-1}}.$$

The set C (the so-called Cantor ternary set) is perfect and nowhere-dense. This set is obtained by dividing the closed interval (0, 1) into three equal parts and removing the interior of the middle interval (1/3, 2/3), by repeating this process with the remaining closed intervals (0, 1/3) and (2/3, 1) (i.e. removing the interior of the intervals (1/9, 2/9) and (7/9, 8/9)), and continuing in this manner indefinitely. If with every irrational number $x = 2^{-1}c_1 + 2^{-2}c_2 + \ldots$, expressed as a binary fraction, we associate the number (31), we obtain a homeomorphic mapping of T_1 on a subset of C. Since $T_1\, h\, M$ and $C\, h\, P$, where $P \subset M$, it follows that the *0-dimensional Baire space and the Cantor ternary set have the same dimensional type.*

It can be proved that the set T of all irrational numbers *has the greatest dimensional type of all linear sets whose dimensional type is less than that of the set of all real numbers*, that is, the inequality $dE < dR_1$ implies $dE \leqslant dT$.

For let E be a linear set such that $dE < dR_1$. Thus E cannot contain an interval for then $dE = dR_1$. There exists, therefore, a countable set $D \subset R_1 - E$ which is dense on R_1. By a theorem of Cantor, there exists a similar mapping ϕ of the set D on the set W of all rational numbers. But this mapping gives rise to a homeomorphic mapping ψ of R_1 on itself in such a manner that $\psi(x) = \phi(x)$ for $x \in D$. To obtain ψ it is sufficient to associate with each gap produced by a cut of D the gap produced by the corresponding cut of W. Since $E \subset R_1 - D$, we have $\psi(E) \subset \psi(R_1 - D) = T$; so E is homeomorphic with a certain subset of T. This gives $dE \leqslant dT$.

It may, however, be proved[65] that if X is a linear set such that $dX < dT$, there exists a set Y such that $dX < dY < dT$; hence there is no greatest among the dimensional types which are less than dT.

Furthermore, there exists a plane set whose dimensional type is greatest among all plane sets with dimensional types less than that of the plane (this is the set of all points of the plane with at least one irrational coordinate).

As to the sets D and W, we have seen that not only are they homeomorphic but there also exists a homeomorphic transformation of the straight line on itself under which the set D transforms into the set W. In particular, this relation holds for the set of all rational numbers and the set of all real algebraic numbers.

If two sets A and B contained in a space P are so related that under a homeomorphic mapping of P on itself A maps into B then $P - A$ maps into $P - B$. Consequently there exists a homeomorphic mapping of the straight line on itself under which the set of all irrational numbers maps on the set of all transcendental numbers.

In spite of the fact that the open interval (0, 1) and the set of all points on the line are homeomorphic, there exists no homeomorphic transformation, of the line on itself which would carry the first set into the second. Similarly there exists no such transformation for the set A consisting of the interval (0, 1) and the numbers 2 and 3 and the set B consisting of the interval (0, 1) and the numbers -1 and 2 (although $A \, h \, B$).

The set K of all points of a circle has obviously dimensional type $dK > dR_1$. It is easily seen that there is no dimensional type between dR_1 and dK (for if $E \subset K$ and $dE < dK$ then $K - E \neq 0$ and if $p \in K - E$ then $K - \{p\}$ is homeomorphic with R_1 and so, since $E \subset K - \{p\}$, we have $dE \leqslant dR_1$). There exist, however, plane sets Q such that $dK \neq dQ > dR_1$ and such that there is no dimensional type between dR_1 and dQ. This is true of the set Q consisting of the x-axis and the points on the y-axis with ordinates equal to $1/n$ for $n = 1, 2, \ldots$. Here neither $dQ \leqslant dK$ nor $dQ \geqslant dK$. Hence *among all dimensional types of plane sets which are* $> dR_1$ *there is no least* dS, that is, such that for every plane set X for which $dX > dR_1$ we have $dS \leqslant dX$; for, if there existed such sets, we would have $dR_1 < dS \leqslant dK$. From the property of the set K we could then conclude that $dS = dK$ and that $dK \leqslant dQ$; this is impossible.

Furthermore, there exist plane sets M (and even sets G_δ, as shown by Mazurkiewicz) such that neither $dM \leqslant dR_1$ nor $dM > dR_1$. Examples of such sets are rather complicated.

We note further that the Cantor ternary set is homeomorphic with the space S of all subsets of the set of natural numbers in which the limit of a sequence $\{E_n\}$ of sets is defined by the set-theoretic formula

$$\lim_{n \to \infty} E_n = \prod_{n=1}^{\infty} \sum_{k=n}^{\infty} E_k = \sum_{n=1}^{\infty} \prod_{k=n}^{\infty} E_k.$$

To obtain a homeomorphic mapping of S on C it is sufficient to map the null set of S on the number $0 \in C$ and every non-empty set $E \subset S$ on the number $f(E) = 2 \sum_{n \in E} 3^{-n}$.

Problems

1. Let $A \times B$ denote the metric product of the metric spaces A and B (§ **60**). Prove that if U and V are non-empty sets such that $U \subset A$ and $V \subset B$

then the set $U \times V$ is open in $A \times B$ if and only if U is open in A and V is open in B. Prove a similar result for closed sets and for connected sets.

2. Prove that the metric space $A \times B$ is dense-in-itself if and only if at least one of the spaces A and B is dense-in-itself.

3. Show that a metric product of two metric spaces is isolated if and only if each of the factor spaces is isolated. Prove a similar result for sets that are scattered, bounded, totally-bounded, compact-in-themselves, or separable.

61. Closed and compact sets as continuous images of the Cantor set.

THEOREM 77. *Two perfect, compact, and nowhere-dense linear sets are homeomorphic.*

Proof. Let E and E_1 be two compact and perfect sets which are nowhere dense in R_1. The set E is therefore contained in a certain finite interval (a, b) with endpoints in E; similarly E_1 is contained in a certain interval (c, d) with endpoints in E_1. Since E is perfect and nowhere-dense, the complement of E with respect to the interval (a, b) is the aggregate M of open intervals dense on (a, b). The complement of E_1 with respect to the interval (c, d) consists of the aggregate M_1 of open intervals dense on (c, d). The sets M and M_1 may be ordered as follows: of two intervals the one containing points with smaller abscissae shall precede the other. We thus have two ordered, countable, and dense sets without a first or last element; by Cantor's theorem, they are ordinally similar. Let ϕ denote the similar mapping of M on M_1. Every element p of E determines a cut in the set M; there will correspond to it a cut of M_1 which determines an element p_1 of E_1 under the mapping ϕ. Let p_1 be the image of p. This mapping is a homeomorphism between the sets E and E_1.

THEOREM 78. *Every non-empty closed and compact set contained in a metric space is a continuous image of the Cantor ternary set.*

Proof. Let T denote a given non-empty closed and compact set contained in a metric space with metric ρ. T is therefore the sum of a finite number of sets of arbitrarily small diameters (§ **52**). Hence we may write $T = M_1 + M_2 + \ldots + M_{s_1}$, where M_i is non-empty and $\delta(M_i) \leqslant 1$ for $i = 1, 2, \ldots, s_1$. Put $T_i = \bar{M}_i$ for $i = 1, 2, \ldots, s_1$; T_i is closed and compact and

(32) $$T = T_1 + T_2 + \ldots + T_{s_1},$$

(33) $$T_i \neq 0,$$

(34) $$\delta(T_i) \leqslant 1 \qquad \text{for } i = 1, 2, \ldots, s_1.$$

Similarly each of the sets T_i ($i = 1, 2, \ldots, s_1$) may be expressed as the sum of a finite number of closed and compact sets;

$$T_i = T_{i,1} + T_{i,2} + \ldots + T_{i,s_{2,i}},$$

$$T_{i,j} \neq 0 \text{ and } \delta(T_{i,j}) \leqslant 1/2 \qquad \text{for } j = 1, 2, \ldots, s_{2,i}.$$

Let s_2 be a natural number greater than each of the numbers $s_{2,1}, s_{2,2}, \ldots, s_{2,s_1}$; putting $T_{i,j} = T_{i,1}$ for $s_{2,i} < j \leqslant s_2$, we obtain, for $i = 1, 2, \ldots, s_1$,

$$T_i = T_{i,1} + T_{i,2} + \ldots + T_{i,s_2},$$

$$T_{i,j} \neq 0, \text{ and } \delta(T_{i,j}) \leqslant 1/2 \qquad \text{for } j = 1, 2, \ldots, s_2.$$

Continuing in this manner, we obtain an infinite sequence $s_1, s_2, s_3, \ldots$ of natural numbers (where we may assume that $s_k > 1$ for $k > 1$) such that for every finite set $n_1, n_2, \ldots, n_k$ of indices, where

$$n_i \leqslant s_i \qquad \text{for } i = 1, 2, \ldots, k, \tag{35}$$

the closed and compact set

$$T_{n_1,n_2,\ldots,n_k} \neq 0, \tag{36}$$

$$\delta(T_{n_1,n_2,\ldots,n_k}) \leqslant 1/k, \tag{37}$$

and

$$T_{n_1,n_2,\ldots,n_k} = T_{n_1,n_2,\ldots,n_k,1} + T_{n_1,n_2,\ldots,n_k,2} + \ldots + T_{n_1,n_2,\ldots,n_k,s_{k+1}}. \tag{38}$$

For each finite sequence $n_1, n_2, \ldots, n_k$ of indices satisfying (35) we now define a closed interval $E_{n_1,n_2,\ldots,n_k}$. Divide the closed interval (0, 1) into $2s_1$ equal intervals. Denote by $E_1, E_2, \ldots, E_{s_1}$ every second one of these intervals, end points included. Suppose we have the interval $E_{n_1,n_2,\ldots,n_k}$, where $n_1, n_2, \ldots, n_k$ is a sequence of indices satisfying (35); divide it into $2s_{k+1}$ equal intervals and denote every second one of these intervals by

$$E_{n_1,n_2,\ldots,n_k,1}, E_{n_1,n_2,\ldots,n_k,2}, \ldots, E_{n_1,n_2,\ldots,n_k,s_{k+1}}.$$

Put

$$S_k = \sum E_{n_1,n_2,\ldots,n_k}, \qquad k = 1, 2, \ldots, \tag{39}$$

where the summation extends over all finite sequences of indices $n_1, n_2, \ldots, n_k$, which satisfy (35). The sets (39) are non-empty, closed, and compact (since they are sums of a finite number of closed and compact sets). Moreover, $S_{k+1} \subset S_k$ for all k since, from the definition of the intervals $E_{n_1,n_2,\ldots,n_k}$, it follows that

$$E_{n_1,n_2,\ldots,n_k,n} \subset E_{n_1,n_2,\ldots,n_k} \qquad \text{for } n = 1, 2, \ldots, s_{k+1}. \tag{40}$$

Hence, by Theorem 29, the set

$$E = S_1 . S_2 . S_3 \ldots \tag{41}$$

is non-empty; it is obviously closed and compact. Furthermore, since $s_k > 1$

for $k > 1$, it follows readily that E is a perfect and nowhere-dense set in the closed interval $(0, 1)$.

Let x be a given number of the set E. From (41) $x \in S_1$ and so, from (39), $x \in (E_1 + E_2 + \ldots + E_{s_1})$, where the sets E_i $(i = 1, 2, \ldots, s_1)$ are disjoint. Hence there exists a unique index $n_1 \leqslant s_1$ such that $x \in E_{n_1}$. But $x \in S_2$ by (41); hence, from (40) and (39) and the fact that the terms of the sum S_1 are disjoint, we conclude the existence of an index $n_2 \leqslant s_2$ such that $x \in E_{n_1, n_2}$.

Continuing in this manner we obtain an infinite sequence $n_1, n_2, \ldots$ determined uniquely by the number $x \in E$ and such that

$$n_k \leqslant s_k \tag{42}$$

and

$$x \in E_{n_1, n_2, \ldots, n_k} \qquad \text{for } k = 1, 2, \ldots \tag{43}$$

Put

$$F(x) = T_{n_1} \cdot T_{n_1, n_2} \cdot T_{n_1, n_2, n_3} \cdots ; \tag{44}$$

the set $F(x)$ is a subset of T and is uniquely defined by $x \in E$.

It follows from (36) and (38) that (44) is a sequence of non-empty decreasing closed and compact sets and so, by Theorem 29, the set $F(x)$ is non-empty. Moreover, $\delta(F(x)) \leqslant 1/k$ for $k = 1, 2, \ldots$, by (37) and (44); hence $\delta(F(x)) = 0$. Consequently the set $F(x)$ consists of a single element which we denote by $f(x)$.

Thus to every number $x \in E$ corresponds an element $f(x) \in T$. We next show that $f(E) = T$. We already have $f(E) \subset T$; it remains to show that $T \subset f(E)$.

Let y denote a given element of T. There exists, by (32), at least one index $n_1 < s_1$ such that $y \in T_{n_1}$. By (38), for $k = 1$, there exists at least one index $n_2 \leqslant s_2$ such that $y \in T_{n_1, n_2}$. Continuing, we obtain an infinite sequence of indices $n_1, n_2, \ldots$, satisfying condition (42) and such that

$$y \in T_{n_1, n_2, \ldots, n_k} \qquad \text{for } k = 1, 2, \ldots. \tag{45}$$

Put

$$\phi(y) = E_{n_1} \cdot E_{n_1, n_2} \cdot E_{n_1, n_2, n_3} \cdots. \tag{46}$$

It follows from the definition of the sets $E_{n_1, n_2, \ldots, n_k}$ that (46) is the intersection of an infinite descending sequence of non-empty closed and compact sets with $\delta(E_{n_1, n_2, \ldots, n_k}) \leqslant 2^{-k}$ for $k = 1, 2, \ldots$; by Theorem 29, the set (46) consists of a single element $x = \phi(y)$. From (39), (41), and the definition of the function $f(x)$, it follows that $x \in E$ and $y = f(x)$; hence $y \in f(E)$. Consequently $T \subset f(E)$ and so $T = f(E)$.

Furthermore, the function f is continuous in E. For let x be an element of E and ϵ a given positive number. Choose a natural number p such that

$1/p < \epsilon$. Corresponding to the number x there exists a uniquely defined infinite sequence $n_1, n_2, \ldots$ of indices satisfying conditions (42) and (43). Put

$$\eta = \delta(E_{n_1,n_2,\ldots,n_p}). \tag{47}$$

Thus η is a definite positive number. Let x' be a number of the set E such that

$$|x - x'| < \eta. \tag{48}$$

From the definition of the intervals $E_{n_1,n_2,\ldots,n_k}$ and from (48), (47), (41), and (39), we obtain the relation

$$x' \in E_{n_1,n_2,\ldots,n_p}$$

and so, from the definition of the function f,

$$f(x') \in T_{n_1,n_2,\ldots,n_p}; \tag{49}$$

by (21), this gives

$$\delta(T_{n_1,n_2,\ldots,n_p}) \leqslant 1/p < \epsilon. \tag{50}$$

But, since $x \in E_{n_1,n_2,\ldots,n_p}$, $f(x) \in T_{n_1,n_2,\ldots,n_p}$; hence (49) and (50) give

$$\rho(f(x), f(x')) < \epsilon. \tag{51}$$

Thus for every $x \in E$ and every $\epsilon > 0$ there exists an $\eta > 0$ such that for $x' \in E$ the inequality (48) implies the inequality (51); this proves (§ **49**) that the function f is continuous in E.

We have therefore proved that the set T is a continuous image of the perfect and nowhere-dense set E. By Theorem 77, the set E is a continuous image of the Cantor ternary set; hence, by Theorem 21, the set T is a continuous image of the Cantor set. Theorem 78 is therefore proved.

It follows from Theorem 57 that the converse of Theorem 78 is also true; hence *a non-empty subset of a metric space is closed and compact if and only if it is a continuous image of the Cantor ternary set.* In particular, a non-empty metric space is compact if and only if it is a continuous image of the Cantor set.

Let Q denote the fundamental cube of E_ω. Then Q is closed and as we have shown previously (§ **59**), Q is compact in E_ω. The set Q is therefore a continuous image of the Cantor set C, i.e., $Q = f(C)$.

Let x be a number of the closed interval $(0, 1)$ not belonging to C; it is therefore in one of the open intervals whose removal gave rise to the set C (§ **60**) and whose endpoints a and b belong to C. If $f(a) = (a_1, a_2, \ldots)$ and $f(b) = (b_1, b_2, \ldots)$, assign to x the element $f(x) = (c_1, c_2, \ldots)$ of the set Q, where

$$c_k = a_k + \frac{x - a}{b - a}(b_k - a_k), \qquad k = 1, 2, \ldots. \tag{52}$$

It is easy to see that $f(x) \in Q$. For, since $a < x < b$, (52) gives, for

$k = 1, 2, \ldots$, either

$$a_k \leqslant c_k < b_k \text{ or } b_k \leqslant c_k < a_k;$$

since $(a_1, a_2, \ldots)$ and $(b_1, b_2, \ldots)$ are elements of Q, it follows that $|c_k| \leqslant 1$ for $k = 1, 2, \ldots$. Consequently $f(x) \in Q$. The function $f(x)$ is thus defined in the whole interval $I = [0 \leqslant x \leqslant 1]$ and $f(I) = Q$.

It follows from the continuity of the function f in the set C and the properties of limits in E_ω that, for every k, the kth coordinate of $f(x)$ is a continuous function of x in the set C. Since c_k is a linear function of x for $x \in I - C$, it is therefore continuous in the whole interval I. We deduce that $f(x)$ is continuous in I. Hence

THEOREM 79. *The fundamental cube of E_ω is a continuous image of the interval* $[0 \leqslant x \leqslant 1]$.

From Theorem 79 and the properties of E_ω we obtain

COROLLARY 1. *There exists an infinite sequence of functions $\phi_k(x)$ ($k = 1, 2, \ldots$) defined and continuous in the interval $I = [0 \leqslant x \leqslant 1]$, taking on values in that interval, and such that for every infinite sequence $c_1, c_2, \ldots$ of numbers in I there exists at least one number $x \in I$ such that*

$$\phi_k(x) = c_k \qquad \textit{for } k = 1, 2, \ldots.$$

Clearly the equations

$$x = \phi_1(t) \text{ and } y = \phi_2(t) \qquad \text{for } 0 \leqslant t \leqslant 1$$

define *a continuous curve filling a square.*[66] In general, the equations

$$x_1 = \phi_1(t), \quad x_2 = \phi_2(t), \quad \ldots, x_m = \phi_m(t) \qquad \text{for } 0 \leqslant t \leqslant 1,$$

define a continuous curve filling an m-dimensional parallelotope. Hence an m-dimensional parallelotope $[0 \leqslant x_1 \leqslant 1, 0 \leqslant x_2 \leqslant 1, \ldots, 0 \leqslant x_m \leqslant 1]$ is a continuous image of the interval $[0 \leqslant x \leqslant 1]$.

In order that a metric space be a continuous image of the closed interval $(0, 1)$ it is necessary and sufficient that it be a continuum or a single point and, if a continuum, that for every $\epsilon > 0$ it be the sum of a finite number of continua each of diameter $\leqslant \epsilon$. The necessity follows from the fact that a continuous image of a closed interval is a continuum or a point (§ **39**) and from Theorem 80 which is proved in § **63**. The proof of the sufficiency of the condition is rather complicated.[67] This condition may be replaced by another if the notion of local connectedness be introduced.

A set E is said to be *locally connected* at an element $p \in E$ if there exists a connected set of arbitrarily small diameter containing p and open in E. A set which is locally connected at every one of its points is called a *locally connected set;* a locally connected set need not be connected (for instance, the sum of two disjoint intervals). Thus *the necessary and sufficient condition for a metric space to be a continuous image of the interval* $[0 \leqslant x \leqslant 1]$ is that it be a point or a locally connected continuum.[68]

An example of a continuum which is not locally connected is given by the set E consisting of the points on the curve $y = \sin 1/x$, $0 < x \leqslant 1$ and the segment of the y-axis, $-1 \leqslant y \leqslant 1$. The set E is not locally connected at any point of this segment.

A homeomorphic image (contained in a metric space) of the interval $[0 \leqslant x \leqslant 1]$ is called a *simple arc*. Mazurkiewicz[69] has shown that if a metric space M is a continuous image of a closed interval then every pair of points of M can be joined in M by a simple arc; the converse need not be true (for example, the set consisting of E, defined above, and the interval $[-1 \leqslant x \leqslant 1]$). Knaster constructed an example of a rather complicated plane continuum which contains no simple arc (the idea of such an example was projected earlier by Janiszewski[69]).

It was proved in § **59** that E_ω is homeomorphic with a certain subset of its fundamental cube. It follows from Theorem 79 that E_ω is a continuous image of a certain set of real numbers. Hence, by Theorem 75, every separable metric space is a continuous image of some set of real numbers. On the other hand, a metric space which is a continuous image of a separable metric space is separable (for if $M_1 = f(M)$, where f is continuous, and if $M = \bar{P}$ then $M_1 = \overline{f(P)}$). We have therefore

COROLLARY 2. *A metric space is separable if and only if it is a continuous image of a set of real numbers.*

We note in connection with Theorem 76 that not every separable metric space is homeomorphic with a set of real numbers (for instance, the set of points in the plane).

62. Biuniform and continuous images of sets. In a metric space two sets which are (1, 1) continuous images of each other need not be homeomorphic. For instance,[70] the sets E_1 consisting of 0, the numbers $1/n$ and the numbers $2 + 1/n$ $(n = 1, 2, \ldots)$ and E_2 consisting of 0 and the numbers $2^{-m} + 2^{-m-n}$ $(m = 1, 2, \ldots; n = 1, 2, \ldots)$.

Two sets E_1 and E_2 (in a metric space) are said to have *the same γ-type, in symbols $\gamma E_1 = \gamma E_2$, if each is a* (1, 1) *continuous image of the other.* If a set E_1 is a (1, 1) continuous image of a set E_2 but the set E_2 is not a (1, 1) continuous image of the set E_1, then we write $\gamma E_1 < \gamma E_2$.

We have seen (§ **40**) that in a separable metric space a (1, 1) continuous image of a closed and compact set is bicontinuous and so a homeomorphic image. Consequently, if E_2 is closed and compact and $\gamma E_1 \leqslant \gamma E_2$, we must have $\gamma E_1 = \gamma E_2$. Hence there exists no set E such that $\gamma E < \gamma E_2$.

It follows that among the γ-types of all linear sets there is none that is smaller than each of the others. For, if E were such a set, we would have $\gamma E < \gamma E_1$, where $E_1 = \{0\} + \{1/n\}$ for $n = 1, 2, \ldots$ which, from the above, would be impossible, since E_1 is closed and compact. A similar argument establishes the same result for the γ-type of linear sets of cardinal $\mathbf{c}$.

However, among the γ-types of all countable metric spaces there is one which is greater than that of each of the others. This is clearly the γ-type of the set N of all natural numbers. Sets of type γN are obviously isolated countable sets. Hence sets whose γ-type is less than γN must be countable sets containing at least one limit point. Among the γ-types of these sets there is one which is greater than each of the others; namely the γ-type of the set E_1 defined above. But there is no greatest γ-type[71] among all γ-types which are less than γE_1.

It has been shown that there exist non-countable transfinite sequences of increasing (decreasing) γ-types of countable sets.[72] There exists also a non-countable family of countable sets whose γ-types are all incommensurable, i.e., no two can be related by one of the signs $=$, $<$, or $>$.

There are probably $\aleph_1$ different γ-types of all countable linear sets but this has not so far been proved without the continuum hypothesis. It can be proved,[73] however, that there are $2^{\mathbf{c}}$ different γ-types of all linear sets of cardinal $\mathbf{c}$.

The problem whether there exists for every family R of $\mathbf{c}$ linear sets, each of cardinal $\mathbf{c}$, a linear set whose γ-type is greater or equal to the γ-type of each set of R is not as yet solved. It is solved positively only for certain families R and in some cases with great difficulties.[74]

Continuity-types. Two sets E_1 and E_2 contained in a metric space are said to have the same c-type (continuity) if each is a continuous image of the other.[75] If E_1 is a continuous image of E^2 but not conversely then we write $cE_1 < cE_2$.

It can be shown that for every family of continuity-types which has cardinal $\leqslant \mathbf{c}$ there exists a continuity-type greater than each c-type of the family. The proof is rather difficult.[76] On the other hand, for every family (whatever its cardinal number) of metric spaces there exists a metric space M with c-type $\geqslant$ c-type of every space of the family. It is sufficient to take for M an isolated metric space of cardinal greater or equal to the cardinal of every space in the family.

There are $2^{\mathbf{c}}$ different continuity-types of linear sets and $\aleph_1$ continuity-types of scattered linear sets[77]; there is only one continuity-type of countable linear sets which are not scattered. Of two countable metric spaces at least one is always a continuous image of the other.[78] Hence their c-types are always commensurable. There exist, however, linear c-types which are not commensurable (for instance, the c-types of an open interval and the set consisting of two points). Furthermore, an example has been constructed of a family of $2^{\mathfrak{c}}$ subsets of R_3 no two of which have commensurable c-types.[79]

63. Uniform continuity.

THEOREM 80. *The function $f(p)$ is defined at all elements p of a closed and compact set E contained in a metric space with distance function ρ and takes on values in a metric space with distance function ρ_1. If f is continuous in E then to every number $\epsilon > 0$ corresponds a number $\eta > 0$ such that the relation*

$$\rho(p, q) < \eta \tag{53}$$

implies the inequality

$$\rho_1(f(p), f(q)) < \epsilon \tag{54}$$

for any pair of elements p and q belonging to E.

Proof. Let ϵ be a given positive number and p an element of E. Since f is continuous in E there exists (§ 49) a positive number $d(p)$ such that

$$\rho_1(f(p), f(q)) < \epsilon/2 \qquad \text{for } q \in E \,.\, S(p, d(p)). \tag{55}$$

For each element $p \in E$ put

$$Q(p) = S(p, \tfrac{1}{2}d(p)). \tag{56}$$

Let $M = \sum_{p \in E} Q(p)$; then M is an open covering of E and so, since E is closed and compact, there exists, by Theorems 68, 65, and 50, a finite number of sets of M, say $Q(p_1), Q(p_2), \ldots, Q(p_n)$, such that

$$E \subset Q(p_1) + Q(p_2) + \ldots + Q(p_n). \tag{57}$$

Let η be a positive number satisfying the inequalities

$$\eta \leqslant \tfrac{1}{2}d(p_i) \qquad \text{for } i = 1, 2, \ldots, n. \tag{58}$$

Let p and q be any two elements of E satisfying (53). On account of (57) there exists an index $k \leqslant n$ such that $q \in Q(p_k)$ and so, by (56), $q \in S(p_k, \frac{1}{2}d(p_k))$; hence

$$\rho(p_k, q) < \tfrac{1}{2}d(p_k) \tag{59}$$

and since, from (53) and (58), $\rho(p, q) < \eta \leqslant \frac{1}{2}d(p_k)$, we have

$$\rho(p_k, p) \leqslant \rho(p_k, q) + \rho(q, p) < d(p_k).$$

Thus $p \in E \,.\, S(p_k, d(p_k))$ and so, by (55),

$$\rho_1(f(p_k), f(p)) < \epsilon/2. \tag{60}$$

By (59), $q \in E \,.\, S(p_k, d(p_k))$ and so, from (55),

$$\rho_1(f(p_k), f(q)) < \epsilon/2. \tag{61}$$

The inequalities (60) and (61) give the inequality (54). This theorem may also be stated in the form: *a function continuous in a closed and compact set is uniformly continuous in this set.*

It is easy to see that neither the condition that the set E be closed nor the condition that it be compact can be omitted from Theorem 80.

For the function $f(x) = 1/x$ is continuous in the set $E = [0 < x < 1]$ which is compact in R_1 but $f(x)$ is not uniformly continuous in E because, for every positive $\delta < 1$, we have $(\delta^{\frac{1}{2}} + \delta) - \delta^{\frac{1}{2}} = \delta$ but

$$1/\delta^{\frac{1}{2}} - 1/(\delta^{\frac{1}{2}} + \delta) = \delta/\delta^{\frac{1}{2}}(\delta^{\frac{1}{2}} + \delta) > \delta/\delta^{\frac{1}{2}}(\delta^{\frac{1}{2}} + \delta^{\frac{1}{2}}) = 1/2.$$

Similarly, the function $f(x) = x^2$ is continuous but not uniformly continuous in the set of all real numbers which is closed. For, whatever the positive number δ, we have

$$|(1/\delta + \delta/2) - 1/\delta| < \delta,$$

but

$$(1/\delta + \delta/2)^2 - (1/\delta)^2 = 1 + \delta^2/4 > 1.$$

Examples

1. Show that if every real continuous function defined in a set E contained in a metric space is uniformly continuous in E then E is closed but not necessarily compact.

Proof. If E is not closed and $a \in E' - E$ then the function $f(p) = 1/\rho(p, a)$ is continuous in E but not uniformly continuous. On the other hand, every continuous function defined in the set E of all natural numbers is uniformly continuous in E but E is not compact.

2. Prove that a function which is uniformly continuous in a compact set is bounded in that set.

The proof follows from the theorem on total boundedness of a compact set (§ **51**) and from the definition of uniform continuity.

3. A function f is defined in a metric space M with distance function ρ and takes on values in a metric space M_1 with distance function ρ_1. Show that f is uniformly continuous in M if and only if for every pair of infinite sequences $x_1, x_2, \ldots$ and $y_1, y_2, \ldots$ of elements of E for which

$$\lim_{n\to\infty} \rho(x_n, y_n) = 0$$

we have

$$\lim_{n\to\infty} \rho_1(f(x_n), f(y_n)) = 0.$$

Proof. Suppose f is uniformly continuous in E; then for every $\epsilon > 0$ there exists a $\delta > 0$ such that $\rho_1(f(x), f(y)) < \epsilon$ for $x \in E$, $y \in E$, and $\rho(x, y) < \delta$.

If $x_n \in E$ and $y_n \in E$ for $n = 1, 2, \ldots$, and $\lim_{n\to\infty} \rho(x_n, y_n) = 0$, then there exists an index μ such that $\rho(x_n, y_n) < \delta$ for $n > \mu$ and so $\rho_1(f(x_n), f(y_n)) < \epsilon$ for $n > \mu$. Since ϵ is arbitrary this gives

$$\lim_{n\to\infty} \rho_1(f(x_n), f(y_n)) = 0.$$

Thus the condition is necessary.

Suppose now that f is not uniformly continuous in E. Then there exists a number $\epsilon > 0$ such that for every natural n there exist points $x_n \in E$, $y_n \in E$ for which $\rho(x_n, y_n) < 1/n$ but $\rho_1(f(x_n), f(y_n)) \geqslant \epsilon$. This gives

$$\lim_{n\to\infty} \rho(x_n, y_n) = 0 \quad \text{but} \quad \lim_{n\to\infty} \rho_1(f(x_n), f(y_n)) \neq 0.$$

Hence the condition is sufficient.

64. Uniform convergence of a sequence of functions. Let E denote a given set contained in a metric space M with distance function ρ; let $f_n(p)$ $(n = 1, 2, \ldots)$ be an infinite sequence of functions defined in E and taking on values in a metric space M_1 with distance function ρ_1. The sequence $\{f_n(p)\}$ is said to *converge uniformly* in E to the function $f(p)$ defined in the set E and taking on values in M_1 if, for every $\epsilon > 0$ and $p \in E$, there exists an index μ (depending on ϵ but not on p) such that $\rho_1(f_n(p), f(p)) < \epsilon$ for $n > \mu$.

THEOREM 81. *The limit of an infinite uniformly convergent sequence of continuous functions in a set E contained in a metric space M and taking on values in a metric space M_1 is continuous in E.*

Proof. Let $\{f_n(p)\}$ denote an infinite sequence of functions continuous in E and converging uniformly in that set to the function $f(p)$. Let p_0 be an element of E and ϵ an arbitrary positive number. Since the given sequence converges uniformly in E there exists an index n such that

$$\rho_1(f_n(p), f(p)) < \epsilon/3 \qquad \text{for } p \in E \tag{62}$$

and so, in particular,

$$\rho_1(f_n(p_0), f(p_0)) < \epsilon/3. \tag{63}$$

Since $f_n(p)$ is continuous in E there exists, for the given ϵ, a number $\delta > 0$ such that

$$\rho_1(f_n(p), f_n(p_0)) < \epsilon/3 \text{ for } p \in E \text{ and } \rho(p, p_0) < \delta. \tag{64}$$

Relations (62), (63), (64), and the triangle law give

$$\rho_1(f(p), f(p_0)) < \epsilon \text{ for } p \in E \text{ and } \rho(p, p_0) < \delta. \tag{65}$$

Thus for every $\epsilon > 0$ there exists a $\delta > 0$ such that (65) is true; this proves the continuity of the function f in E.

THEOREM 82.[80] *An infinite sequence $\{f_n(p)\}$ of functions continuous in a closed and compact set contained in a metric space M and taking on values in a metric space M_1 converges uniformly in that set to a function $f(p)$ if and only if the condition*

$$(66) \qquad \lim_{n\to\infty} p_n = p_0, \qquad p_n \in E,\ n = 0, 1, 2, \ldots,$$

implies the relation

$$(67) \qquad \lim_{n\to\infty} f_n(p_n) = f(p_0).$$

Proof. Suppose that the sequence $\{f_n(p)\}$ of functions continuous in E converges uniformly in E to the function $f(p)$. Let p_n, $n = 0, 1, 2, \ldots$, denote an infinite sequence satisfying (66). Let ϵ be an arbitrary positive number. Since, by Theorem 81, f is continuous in E we have $\lim_{n\to\infty} f(p_n) = f(p_0)$ and so there exists an index μ such that

$$(68) \qquad \rho_1(f(p_n), f(p_0)) < \epsilon/2 \qquad \text{for } n > \mu,$$

where ρ_1 is the distance function in M_1. But the sequence $f_n(p)$ converges uniformly to $f(p)$ in E; hence there exists an index ν such that

$$\rho_1(f_n(p),\ f(p)) < \epsilon/2 \qquad \text{for } p \in E \text{ and } n > \nu$$

and so, in particular,

$$(69) \qquad \rho_1(f_n(p_n), f(p_n)) < \epsilon/2 \qquad \text{for } n > \nu.$$

Now (68), (69), and the triangle law give

$$(70) \qquad \rho_1(f_n(p_n), f(p_0)) < \epsilon \qquad \text{for } n > \mu + \nu.$$

Thus corresponding to every positive ϵ there exists an index $\mu + \nu$ such that (70) holds; this gives (67) and with it the necessity of the condition follows.

Next suppose that E is a closed and compact set and $f_n(p)$, $n = 1, 2, \ldots$, an infinite sequence of functions continuous in E for which condition (66) always implies (67). We first show that the function f is continuous in E. Suppose that f is not continuous in E at p_0. Then there exists a number $\epsilon > 0$ and an infinite sequence q_k, $k = 1, 2, \ldots$, of elements of E such that

$$(71) \qquad \lim_{k\to\infty} q_k = p_0 \text{ and } \rho_1(f(q_k), f(p_0)) \geqslant \epsilon \qquad \text{for } k = 1, 2, \ldots.$$

The sequence $p_n = q_k$ for $n = 1, 2, \ldots$ and the fact that (66) implies (67) give

$$\lim_{n\to\infty} f_n(q_k) = f(q_k), \qquad k = 1, 2, \ldots;$$

hence for every natural k there exists an index $l_k > k$ such that

$$\rho_1(f_{l_k}(q_k), f(q_k)) < \epsilon/2$$

and so, from (71),

$$(72) \qquad \rho_1(f_{l_k}(q_k), f(p_0)) > \epsilon/2, \qquad k = 1, 2, \ldots.$$

Since $l_k > k$, $\lim_{k\to\infty} l_k = +\infty$; we may therefore remove from the sequence $l_k\ (k = 1, 2, \ldots)$ an infinite increasing sequence $l_{k_i}\ (i = 1, 2, \ldots)$. Put $p_n = q_{k_i}$ for $n = l_{k_i}$ and $p_n = p_0$ for n not a term of the sequence $l_{k_i}\ (i = 1, 2, \ldots)$. It follows from (71) that $\lim_{n\to\infty} p_n = p_0$ and so (66) holds; from (72), we get

$$\rho_1(f_{l_{k_i}}(p_{l_{k_i}}), f(p_0)) > \epsilon/2$$

which contradicts (67). The function f is therefore continuous in E.

Next, assume that the sequence $\{f_n(p)\}$ does not converge uniformly to the function $f(p)$ in the closed and compact set E. Then there exists a positive number ϵ such that to every natural l corresponds an index $k_l > l$ and a point $q_l \in E$ for which

(73) $$\rho_1(f_{k_l}(q_l), f(q_l)) \geqslant \epsilon, \qquad l = 1, 2, \ldots.$$

Since E is closed and compact there exists an infinite increasing sequence of indices $l_i (i = 1, 2, \ldots)$ such that

$$\lim_{i\to\infty} q_{l_i} = p_0 \in E,$$

whether the terms of the sequence $\{q_l\}$ are different or not.

Furthermore, since f is continuous in E, there exists an index μ such that

$$\rho_1(f(q_{l_i}), f(p_0)) < \epsilon/2 \qquad \text{for } i > \mu,$$

and so, from (73),

(74) $$\rho_1(f_{k_{l_i}}(q_{l_i}), f(p_0)) > \epsilon/2 \qquad \text{for } i > \mu.$$

Since $k_{l_i} > l_i$ we have $\lim_{i\to\infty} k_{l_i} = +\infty$; thus there exists an infinite increasing subsequence $k_{l_{i_s}} (s = 1, 2, \ldots)$ of the sequence $\{k_{l_i}\}$. Put $p_n = q_{l_{i_s}}$ for $n = k_{l_{i_s}}$ and $p_n = p_0$ for n not a term of the sequence $k_{l_{i_s}} (s = 1, 2, \ldots)$. Then $p_{k_{l_{i_s}}} = q_{l_{i_s}}$ for $s = 1, 2, \ldots$ and $\lim_{n\to\infty} p_n = p_0$. Thus (66) is satisfied; from (74),

$$\rho_1(f_{k_{l_{i_s}}}(p_{k_{l_{i_s}}}), f(p_0)) > \epsilon/2 \qquad \text{for } i > \mu,$$

contrary to (67). This proves the theorem.

*It is necessary that the set E in Theorem 82 be closed and compact. For suppose that E is either not closed or not compact. In either case there exists an infinite sequence $Q = \{q_1, q_2, \ldots\}$ of different points of E such that $Q' . E = 0$. Put, for $p \in E, f_n(p) = 0$ if $\rho(p, q_n) \geqslant 1/n, f_n(p) = 1 - n\rho(p, q_n)$ if $\rho(p, q_n) < 1/n$; also set $f(p) = 0$ for all $p \in E$.

It follows readily that the functions $f_n(p)$, $n = 1, 2, \ldots$, are continuous in the set E. If p_0 is any point of E then, since $Q' . E = 0$, there exists a positive number δ such that

(75) $$\rho(p, p_0) < \delta, p \in E, \text{ implies that } p \notin Q.$$

Thus if $\lim_{n\to\infty} p_n = p_0$, $p_n \in E$, $n = 1, 2, \ldots$, then, for sufficiently large n, $S(p_n, 1/n) \subset S(p_0, \delta)$. If $\rho(p_n, q_n) < 1/n$, then $q_n \in S(p_0, \delta)$, that is, $\rho(q_n, p_0) < \delta$, contrary to (75). Consequently, for n sufficiently large, $\rho(p_n, q_n) \geqslant 1/n$ and so $f_n(p_n) = 0$. This gives $\lim_{n\to\infty} f_n(p_n) = 0 = f(p_0)$. Thus relation (66) implies relation (67) but the sequence $\{f_n(p)\}$ does not converge uniformly to $f(p)$ in E; for $f_n(q_n) = 1$ while $f(q_n) = 0$, $n = 1, 2, \ldots$. Hence Theorem 82 does not hold in E.

On the other hand, as seen from the proof of Theorem 82, the condition of the theorem is necessary for every set E even if E is not closed nor compact.*

65. The (C) **space of all functions continuous in the interval** $[0, 1]$. Let (C) denote the set of all real functions $f(t)$ defined and continuous in the interval $I = [0 \leqslant t \leqslant 1]$. If $f(t)$ and $g(t)$ are two given functions of the set (C), then the set of all numbers $|f(t) - g(t)|$, where $t \in I$, is obviously bounded and so has an upper bound which we denote by

$$r(f, g) = \sup_{0 \leqslant t \leqslant 1} |f(t) - g(t)|. \tag{76}$$

It is easy to prove that the function (76) satisfies the three distance axioms; it may therefore be employed to define distance in the set (C). (C) thereby becomes a metric space, the so-called *space of all continuous functions in the interval* $0 \leqslant t \leqslant 1$. *The space* (C) *is separable.* This may be easily deduced from Weierstrass's theorem that a function continuous in a finite interval is the limit of a uniformly convergent sequence of polynomials in this interval. It may also be proved directly in the following manner. For a given natural n denote by P_n the set of all functions defined in I as follows: for $t = k/n$, $k = 0, 1, \ldots n$, $f(t)$ is rational, and for $(k - 1)/n \leqslant t \leqslant k/n$, $k = 1, 2, \ldots n$, $f(t)$ is linear. The sets P_n are obviously countable; consequently the set $P = P_1 + P_2 + \ldots$ is countable. We shall show that P is dense on (C).

Let f denote a given function belonging to (C) and ϵ a given positive number. By Theorem 80, there exists a number $\eta > 0$ such that

$$|f(x) - f(y)| < \epsilon/5 \text{ for } |x - y| < \eta,\ x \in I, y \in I. \tag{77}$$

Choose n such that $1/n < \eta$. Let k denote one of the numbers $0, 1, 2, \ldots, n$. There exists a rational number w_k such that

$$|f(k/n) - w_k| < \epsilon/5. \tag{78}$$

Let $g(t)$ be a function which has $g(k/n) = w_k$ for $k = 0, 1, 2, \ldots, n$, and which is linear in each of the intervals $((k - 1)/n, k/n)$, $k = 1, 2, \ldots, n$; clearly $g(t) \in P$.

Next, let t be any number of the interval I; then there exists, for every n, a number k of the sequence $0, 1, 2, \ldots, n$, such that $(k - 1)/n \leqslant t \leqslant k/n$. Since $g(t)$ is a linear function in the interval $((k - 1)/n, k/n)$, and since

$g((k-1)/n) = w_{k-1}$, $g(k/n) = w_k$, we have

$$(79) \qquad |g(t) - g(k/n)| \leqslant |w_{k-1} - w_k|.$$

Since $1/n < \eta$, (77) gives

$$(80) \qquad |f(k/n) - f((k-1)/n)| < \epsilon/5$$

and from (78)

$$(81) \qquad |f((k-1)/n) - w_{k-1}| < \epsilon/5.$$

The inequalities (80), (81), and (78) give

$$(82) \qquad |w_{k-1} - w_k| < 3\epsilon/5.$$

But

$$(83) \quad f(t) - g(t) = f(t) - f(k/n) + f(k/n) - g(k/n) + g(k/n) - g(t);$$

from (77) and the fact that $1/n < \eta$, we have

$$(84) \qquad |f(t) - f(k/n)| < \epsilon/5;$$

consequently (84), (78) (since $g(k/n) = w_k$), (79), (82), and (83) give the inequality

$$|f(t) - g(t)| < \epsilon.$$

Since this inequality holds for every $t \in I$ we have $r(f, g) < \epsilon$. Thus for every element f of the space (C) there exists an element g of P at a distance less than ϵ from f. Since ϵ is arbitrary, it follows that $(C) \subset \bar{P}$.

THEOREM 83 (Banach-Mazur). *Every separable metric space is isometric with a subset of the space (C).*

Proof. Let M be a given separable space with distance function ρ and $Q = \{p_1, p_2, \ldots\}$ a subset of M such that $M \subset \bar{Q}$. The sequence $\{p_n\}$ may be assumed to be infinite, repeating one term an infinite number of times if necessary. Put

$$(85) \qquad \gamma_n(p) = \rho(p, p_n) - \rho(p, p_1) \qquad \text{for } p \in Q,\ n = 1, 2, \ldots.$$

Since, by the triangle law,

$$-\rho(p_1, p_n) \leqslant \rho(p, p_n) - \rho(p, p_1) \leqslant \rho(p_1, p_n),$$

we obtain from (85)

$$(86) \qquad |\gamma_n(p)| \leqslant \rho(p_1, p_n) \qquad \text{for } p \in Q,\ n = 1, 2, \ldots.$$

Let $\{\phi_k\}$ $(k = 1, 2, \ldots)$ be an infinite sequence of functions satisfying Corollary 1 to Theorem 79 (§ 61). It follows from the properties of the sequence $\{\phi_k\}$ and from inequality (86) that there exists for every natural number k at least one real number t_k such that $0 \leqslant t_k \leqslant 1$ and

$$(87) \qquad \gamma_n(p_k) = \rho(p_1, p_n)(2\phi_n(t_k) - 1), \qquad n = 1, 2, \ldots.$$

Put $T = \{t_1, t_2, \ldots\}$ and

(88) $\quad f_n(t) = \rho(p_1, p_n)(2\phi_n(t) - 1)$ for $t \in \bar{T}$ and $n = 1, 2, \ldots.$

The function $f_n(t)$ is continuous in the closed set $\bar{T}$. Its definition may be extended to the whole of the interval $0 \leqslant t \leqslant 1$ by assuming $f_n(t)$ to be linear in every open interval contained in the complement of $\bar{T}$ with respect to the interval $[0, 1]$; if t' and t'' are the lower and upper bounds respectively of the numbers in the set $\bar{T}$, put $f_n(t) = f_n(t')$ for $0 \leqslant t < t'$ and $f_n(t) = f_n(t'')$ for $t'' < t \leqslant 1$. The function $f_n(t)$ is obviously continuous in $[0, 1]$; it is therefore an element of the space (C).

It follows that

(89) $$\rho(p_i, p_k) = r(f_i, f_k), \qquad i = 1, 2, \ldots; k = 1, 2, \ldots,$$

where r denotes distance in the space (C).

For (87) and (88) give

(90) $$\gamma_i(p_k) = f_i(t_k).$$

Hence, from (85),

$$\rho(p_i, p_k) = \gamma_i(p_k) - \gamma_k(p_k) = f_i(t_k) - f_k(t_k);$$

since $0 \leqslant t_k \leqslant 1$, we have

$$|f_i(t_k) - f_k(t_k)| \leqslant \sup_{0 \leqslant t \leqslant 1} |f_i(t) - f_k(t)|;$$

consequently, from (76),

(91) $$\rho(p_i, p_k) \leqslant r(f_i, f_k).$$

But, for any three natural numbers k, i, j, we find from (90) and (85) that

$$f_i(t_j) - f_k(t_j) = \gamma_i(p_j) - \gamma_k(p_j) = \rho(p_j, p_i) - \rho(p_j, p_k)$$

and so, since

$$|\rho(p_j, p_i) - \rho(p_j, p_k)| \leqslant \rho(p_i, p_k),$$

we have

$$|f_i(t_j) - f_k(t_j)| \leqslant \rho(p_i, p_k) \text{ for } i = 1, 2, \ldots; k = 1, 2, \ldots; j = 1, 2, \ldots.$$

Thus

(92) $$|f_i(t) - f_k(t)| \leqslant \rho(p_i, p_k) \qquad \text{for } t \in T.$$

Since the functions f_i and f_k are continuous in the interval $I = [0 \leqslant t \leqslant 1]$ and since $\bar{T} \subset I$, it may be concluded that the inequality (92) holds in $\bar{T}$; furthermore, it follows from the definition of the functions f_n in the set $I - \bar{T}$ that (92) holds in the whole interval I; hence, from (76), we obtain

(93) $$r(f_i, f_k) \leqslant \rho(p_i, p_k).$$

Relations (91) and (93) give (89).

Next let p denote an element of M. Since $M \subset \bar{Q}$ there exists, by Theorem 53, an infinite sequence $n_1, n_2, \ldots$ of natural numbers such that

$$p = \lim_{k\to\infty} p_{n_k}. \tag{94}$$

Thus $\lim_{k\to\infty} \rho(p, p_{n_k}) = 0$ and so, for every positive number ϵ, there exists an index $\mu = \mu(\epsilon)$ such that

$$\rho(p_{n_i}, p_{n_k}) < \epsilon \qquad \text{for } i > \mu \text{ and } k > \mu;$$

from (89),

$$r(f_{n_i}, f_{n_k}) < \epsilon \qquad \text{for } i > \mu \text{ and } k > \mu.$$

This and (1) give

$$|f_{n_i}(t) - f_{n_k}(t)| < \epsilon \qquad \text{for } i, k, > \mu,\ 0 \leqslant t < 1;$$

consequently the sequence $f_{n_k}(t)$, $k = 1, 2, \ldots$, converges uniformly in the interval I. Put

$$f^{(p)}(t) = \lim_{k\to\infty} f_{n_k}(t), \qquad 0 \leqslant t \leqslant 1; \tag{95}$$

$f^{(p)}(t)$ is clearly continuous in I and so $f^{(p)}(t) \in (C)$. Now let $m_1, m_2, \ldots$ be an infinite sequence of indices such that

$$p = \lim_{k\to\infty} p_{m_k};$$

from (94),

$$\lim_{k\to\infty} \rho(p_{m_k}, p_{n_k}) = 0$$

and so, from (89),

$$\lim_{k\to\infty} r(f_{m_k}, f_{n_k}) = 0.$$

Therefore (95) gives

$$f^{(p)}(t) = \lim_{k\to\infty} f_{m_k}(t), \qquad 0 \leqslant t \leqslant 1.$$

Consequently, the function $f^{(p)}(t)$ depends only on the point p of the space M and not on the sequence $\{p_{n_k}\}$ satisfying (94).

With each element p of the space M associate a certain element $f^{(p)}$ of the space (C). Next, let p and q be given elements of M. Then there exists an infinite sequence $\{n_k\}$ of natural numbers for which (94) holds and a sequence $\{h_k\}$ such that

$$q = \lim_{k\to\infty} p_{h_k} \tag{96}$$

and

$$f^{(q)}(t) = \lim_{k\to\infty} f_{h_k}(t), \qquad 0 \leqslant t \leqslant 1, \tag{97}$$

and where the sequence $\{f_{h_k}(t)\}$ converges uniformly in the interval I. Now (95) and (97) give

$$f^{(p)}(t) - f^{(q)}(t) = \lim_{k\to\infty} (f_{n_k}(t) - f_{h_k}(t)), \qquad 0 \leqslant t \leqslant 1,$$

with the sequence on the right uniformly convergent in I. Hence, given $\epsilon > 0$, there exists, by (94) and (96), an index k such that

$$|\rho(p, q) - \rho(p_{n_k}, p_{h_k})| < \epsilon \tag{98}$$

and

$$|f^{(p)}(t) - f^{(q)}(t) - (f_{n_k}(t) - f_{h_k}(t))| < \epsilon, \qquad 0 \leqslant t \leqslant 1. \tag{99}$$

Hence (99) and (76) give

$$|r(f^{(p)}, f^{(q)}) - r(f_{n_k}, f_{h_k})| \leqslant \epsilon \tag{100}$$

and since, from (89),

$$r(f_{n_k}, f_{h_k}) = \rho(p_{n_k}, p_{h_k}),$$

(98) and (100) give

$$|r(f^{(p)}, f^{(q)}) - \rho(p, q)| < 2\epsilon;$$

since ϵ is arbitrary we have

$$r(f^{(p)}, f^{(q)}) = \rho(p, q).$$

The space M is therefore isometric with the set of functions $f^{(p)}$, $p \in M$, of the space (C). This proves Theorem 83.

A separable metric space U is called *universal* if every separable metric space is isometric with some subset of U. Theorem 83 may therefore be stated as follows: *Space* (C) *is a universal separable metric space.* The proof of this theorem given by Banach[81] is based on the theory of linear functionals; it is therefore less elementary than the one given above.[82]

Note that Hilbert space is not a universal separable metric space for, as shown in § **51**, there exists a metric space consisting of four elements which cannot be embedded in Hilbert space. Nor is the space E_ω a universal space for separable metric spaces because, the distance between any two of its elements being $< e - 1 < 2$, it is impossible for the set consisting of two elements whose distance is 2 to be embedded in E_ω.

*The first proof of the existence of universal separable metric spaces was given by Urysohn.[83] The universal space U constructed by Urysohn has also the property (called by Urysohn metric homogeneity) that, if A and B are any two finite isometric subsets of U, then there exists an isometric mapping of U on itself which maps A on B. The space (C) does not possess this property.

For if f_1 denotes the function equal to zero for $0 \leqslant t \leqslant 1$, f_2 the function having the constant value unity, and f_3 the function $f_3(t) = t$, then $r(f_1, f_2) = r(f_1, f_3) = 1$, i.e., the sets $\{f_1, f_2\}$ and $\{f_1, f_3\}$ are isometric but there exists no isometric mapping of the space (C) on itself which maps the elements f_1 and f_2 on f_1 and f_3 or on f_3 and f_1. This follows from the fact that f_1 and f_2 have only one mid-point, i.e., an element f such that $r(f, f_1) = r(f, f_2)$

$= \frac{1}{2}r(f_1, f_2)$; this is the function f having the constant value 1/2. But, for f_1 and f_3 there exists more than one such element; in fact, there are non-countably many of them, namely, each of the functions $f(t) = (1/2 - a)t + a$ where a is a constant such that $0 \leqslant a \leqslant 1/2$. Under an isometric mapping, however, different mid-points of two elements map into different mid-points of the images of the two elements.*

The set T of all bounded continuous functions of a real variable becomes a metric space if the distance function r is defined by the relation

$$r(f, g) = \sup_{-\infty<t<+\infty} |f(t) - g(t)|.$$

The space T contains a subset which is isometric with the space (C). This subset is obtained by associating with each function $f \in (C)$ a function $\phi \in T$ defined as follows: $\phi(t) = f(t)$ for $0 \leqslant t \leqslant 1$, $\phi(t) = f(0)$ for $-\infty < t < 0$, and $\phi(t) = f(1)$ for $1 < t < +\infty$.

However, the space T is not even separable at any one of its points. For let $f \in T$ and $\eta > 0$. For every set N of natural numbers, denote by f_N the function $f_N(t)$ defined by the relation $f_N(k) = f(k) + \eta/2$ for $k \in N$, $f_N(k) = f(k)$ for k an integer not in N, and let f_N be a linear function in each of the intervals $(k, k + 1)$ where k is an integer. Obviously $r(f_N, f) \leqslant \eta/2$ and so $f_N \in S(f, \eta)$. But for two different sets N_1 and N_2 of natural numbers we have $r(f_{N_1}, f_{N_2}) = \eta/2$; hence the set of all elements f_N of the space T, where N is any set of natural numbers, is an isolated set of cardinal $\mathbf{c}$ and is contained in the sphere $S(f, \eta)$. Since η is arbitrary, this proves that T is not separable at the point f.

It can be proved that T has the same dimensional type as the Fréchet space D_ω which consists of all infinite bounded sequences $\{x_n\}$ of real numbers and in which the distance between two points $x = \{x_n\}$ and $y = \{y_n\}$ is given by $\rho(x, y) = \sup |x_n - y_n|$ for $n = 1, 2, \ldots$. Moreover each space can be embedded in the other.[84]

66. The space of all bounded closed sets of a metric space. Let M denote a given metric space with distance function ρ and F the family of all non-empty bounded and closed sets of M. for $p \in M$ and $E \in F$ put

$$\rho_1(p, E) = \inf_{q \in E} \rho(p, q). \tag{101}$$

For $p \in E$ we have $\rho_1(p, E) = 0$ but if $p \notin E$, since E is closed, there exists a natural number n such that $E . S(p, 1/n) = 0$; hence $\rho(p, q) \geqslant 1/n$ for $q \in E$, and so, from (101) $\rho_1(p, E) \geqslant 1/n > 0$.

If $E \in F$ and $T \in F$ then $E + T$ is bounded (§ **52**); so there exists a finite number a such that $\rho(p, q) < a$ for $p \in T$ and $q \in E$ and hence, from (101), $\rho_1(p, E) < a$ for $p \in T$ and $\sup_{p \in T} \rho_1(p, E) \leqslant a$. Similarly $\sup_{q \in E} \rho_1(q, T) \leqslant a$.

We now define, for $E \in F$ and $T \in F$,

$$r(E, T) = \max\left(\sup_{p \in E} \rho_1(p, T), \sup_{p \in T} \rho_1(p, E)\right); \tag{102}$$

hence $r(E, T)$ is a finite real number $\geqslant 0$.

The function r, defined in the set $F \times F$, satisfies the distance axioms. The axiom of symmetry is evident from the definition. Furthermore, if $E \in F$, $T \in F$, and $E \neq T$, there exists at least one element in one of the sets but not in the other, $p_0 \in E - T$ say. Since $p_0 \notin T$, we have $\rho_1(p_0, T) > 0$; hence $\sup_{p \in E} \rho_1(p, T) \geqslant \rho_1(p_0, T) > 0$ and so, from (102), $r(E, T) > 0$. If $E = T$ then, for $p \in E$, $\rho_1(p, T) = \rho_1(p, E) = 0$; hence $r(E, T) = 0$. This gives the second distance axiom (identity). It remains to prove the triangle axiom. We note first that if $p_1 \in E_1$, $p_2 \in E_2$, $p_3 \in E_3$, where E_1, E_2, and E_3 are all in F, then

$$\rho(p_1, p_3) \leqslant \rho(p_1, p_2) + \rho(p_2, p_3)$$

which gives

$$\inf_{p_3 \in E_3} \rho(p_1, p_3) \leqslant \rho(p_1, p_2) + \inf_{p_3 \in E_3} \rho(p_2, p_3) \quad \text{for } p_1 \in E_1, p_2 \in E_2. \tag{103}$$

Hence (103) and (101) give

$$\rho_1(p_1, E_3) \leqslant \rho(p_1, p_2) + \rho_1(p_2, E_3) \quad \text{for } p_1 \in E_1 \text{ and } p_2 \in E_2. \tag{104}$$

From (102),

$$\rho_1(p_2, E_3) \leqslant r(E_2, E_3) \quad \text{for } p_2 \in E_2;$$

hence (104) gives

$$\rho_1(p_1, E_3) \leqslant \rho(p_1, p_2) + r(E_2, E_3), \quad p_1 \in E_1, p_2 \in E_2,$$

and therefore

$$\rho_1(p_1, E_3) \leqslant \inf_{p_2 \in E_2} \rho(p_1, p_2) + r(E_2, E_3) \quad \text{for } p_1 \in E_1. \tag{105}$$

But, from (101) and (102), we have, for $p_1 \in E_1$, $\inf_{p_2 \in E_2} \rho(p_1, p_2) = \rho_1(p_1, E_2) \leqslant r(E_1, E_2)$; hence (105) gives

$$\rho_1(p_1, E_3) \leqslant r(E_1, E_2) + r(E_2, E_3) \quad \text{for } p_1 \in E_1. \tag{106}$$

Interchanging E_1 and p_1 with E_3 and p_3, we obtain

$$\rho_1(p_3, E_1) \leqslant r(E_3, E_2) + r(E_2, E_1). \tag{107}$$

Since the function r is symmetric, relations (106), (107), and (102) give

$$r(E_1, E_3) \leqslant r(E_1, E_2) + r(E_2, E_3);$$

hence r satisfies the triangle axiom. The function r may therefore be employed to define distance in the space F.

In particular, let M denote a separable metric space with distance function ρ and F the family of all non-empty closed and compact (hence bounded) subsets of M. Let r, defined by (101) and (102), be the distance function in F.

Since M is separable there exists a sequence $\{q_n\} = Q$ such that $Q \subset M \subset \bar{Q}$. Let ϵ be a given positive number. If $E \in F$, there exists, by Theorem 67, a finite sequence $p_1, p_2, \ldots p_m$ of elements of E such that every element p of E is at a distance less than ϵ from at least one of the elements of the sequence. On the other hand, there exists for $i = 1, 2, \ldots, m$, an element $q_{n_i} \in Q$ such that $\rho(q_{n_i}, p_i) < \epsilon$. Put $S = \{q_{n_1}, q_{n_2}, \ldots, q_{n_m}\}$. Obviously $S \in F$. It follows that $r(S, E) \leqslant 2\epsilon$. For, since $\rho(q_{n_i}, p_i) < \epsilon$ and $p_i \in E$, for $i = 1, 2, \ldots, m$, we have

$$\rho_1(q_{n_i}, E) = \inf_{q \in E} \rho(q_{n_i}, q) < \epsilon, \qquad i = 1, 2, \ldots, m.$$

Hence

$$\sup_{p \in S} \rho_1(p, E) = \sup_{p \in S} \inf_{q \in E} \rho(p, q) \leqslant \epsilon.$$

Next assume that $p \in E$. Then there exists an index $i \leqslant m$ such that $\rho(p, p_i) < \epsilon$; since $\rho(q_{n_i}, p_i) < \epsilon$, we have $\rho(p, q_{n_i}) < 2\epsilon$ and so

$$\rho_1(p, S) = \inf_{q \in S} \rho(p, q) < 2\epsilon.$$

This gives

$$\sup_{p \in E} \rho_1(p, S) \leqslant 2\epsilon.$$

It therefore follows from (102) that $r(S, E) \leqslant 2\,\epsilon$.

This proves that the aggregate of all finite subsets of the set Q is dense on F. Since Q is countable this aggregate is countable. Consequently F is separable, that is, *the space of all non-empty closed and compact subsets of a separable metric space is separable.*

In particular the space F_2 of all non-empty closed and compact subsets of the plane is separable and therefore isometric with a subset of the space (C). It can be shown that *the spaces F_2 and (C) have the same dimensional type.* It is sufficient to show that (C) is homeomorphic with a subset of F_2, namely, the subset consisting of the graphs of all the functions in (C), in other words, the subset consisting of all the closed sets $E_{(x,y)}[y = f(x)]$ in the plane, where $f \in (C)$. The proof is left to the reader.

Similarly, the space F_1 of all non-empty closed and compact linear sets has the same dimensional type as the space (C).[85] For, in virtue of Theorems 83 and 75 and the fact that E_ω is homeomorphic with the subset E_1 composed of all the elements with coordinates > 0 and < 1 (§ **59**), it is obviously sufficient to show that E_1 is homeomorphic with a subset of F_1. Hence, for $p = (x_1, x_2, \ldots) \in E_1$, denote by $f(p)$ the closed and compact set consisting of the number 0 and the numbers $(1 + x_n)2^{-n}$ where $n = 1, 2, \ldots$; let T_1 be the set of all sets $f(p)$ for $p \in E_1$. It is easily verified that the function f establishes a homeomorphic mapping of E_1 on T_1.

*It can be shown that Hilbert space cannot be embedded in the space F_2 (and therefore not in F_1). Consequently *neither the space F_1 nor the space F_2 is a universal space with respect to separable metric spaces.* For Hilbert space contains a countable subset whose elements are at a distance 1 from each other whereas it can be shown that F_2 does not contain such a subset, although both F_1 and F_2 contain countable subsets whose elements are at a distance $\geqslant 1$ from each other; these are the subsets composed of all unit sets such that all coordinates of the elements in each set are integers.*

67. Sets F_σ and G_δ. A set which is the sum of a countable aggregate of closed sets is called (Hausdorff) an F_σ; its complement, that is, the intersection of a countable aggregate of open sets is called a G_δ. The sum of a countable aggregate of sets F_σ is clearly an F_σ and the intersection of a countable aggregate of sets G_δ is a G_δ. Furthermore, it follows readily from Theorems 35 and 36 that the intersection of a finite number of sets F_σ is an F_σ and the sum of a finite number of sets G_δ is a G_δ. But, as will be seen later, the intersection of a countable aggregate of sets F_σ need not be an F_σ and consequently the sum of a countable aggregate of sets G_δ need not be a G_δ.

In a countable metric space every subset is both an F_σ and a G_δ; this is also true in an isolated space where every subset is both closed and open. It is not known, however, whether there exists a non-countable separable metric space in which every set is an F_σ. It would follow from the continuum hypothesis that such a space could not exist.

THEOREM 84. *Every closed set contained in a metric space is a set G_δ.*

Proof. Let F denote a given closed set contained in a metric space M with distance function ρ. Put

$$\Gamma_n = \sum_{p \in F} S(p, 1/n);$$

the set Γ_n is open. Then

$$F = \Gamma_1 . \Gamma_2 . \Gamma_3 . \dots \tag{108}$$

For F is obviously contained in Γ_n for $n = 1, 2, \dots$. Let $q \in \Gamma_n$ for $n = 1, 2, \dots$. Then there exists an element $p_n \in F$ such that $q \in S(p_n, 1/n)$ and so $\rho(p_n, q) < 1/n$. Since this inequality holds for $n = 1, 2, \dots, q \in F$ or $q \in F'$; hence, in either case q is an element of F since F is closed. Thus (108) follows and Theorem 84 is proved. Passing to complements we obtain from Theorem 84

THEOREM 85. *Every open set contained in a metric space is an F_σ.*

We note that Theorems 84 and 85 may not hold in some topological spaces. For instance, if K is a non-countable topological space in which $\bar{E} = E$ for E finite, and $\bar{E} = K$ for E infinite, then all finite sets contained in K are closed but they are not sets G_δ. For it is easy to see that a non-empty $G_\delta \subset K$ differs from K by an at-most-countable set and so must itself be non-countable.

Problems

1. R is a family of closed sets of a given metric space M such that of any two sets of R one is contained in the other. Show that the sum S of all sets of R is an F_σ.

Solution. Let M have distance function ρ. Consider two cases:

(i) There exists an infinite sequence $E_1, E_2, \ldots$ of sets of R such that $S = E_1 + E_2 + \ldots$. In this case S is certainly an F_σ.

(ii) For every infinite sequence $E_1, E_2, \ldots$ of the family R we have $S - (E_1 + E_2 + \ldots) \neq 0$ and so there exists a point $p \in S - (E_1 + E_2 + \ldots)$. Since $p \in S$ there exists a set $E \in R$ such that $p \in E$. But $p \notin E_n$; hence $E - E_n \neq 0$ for $n = 1, 2, \ldots$. Since of any two sets of R one is contained in the other, we have $E_n \subset E$ for $n = 1, 2, \ldots$. Thus for every infinite sequence $E_1, E_2, \ldots$ of R there exists a set E of R such that $E_n \subset E$ for $n = 1, 2, \ldots$. We next show that in case (ii) S is closed. Let $p \in S'$; then there exists for every natural n a point $p_n \in S$ such that $\rho(p_n, p) < 1/n$. Since $p_n \in S$ there exists a set E_n of R such that $p_n \in E_n$. As shown above, there exists a set E of R such that $E_n \subset E$ for $n = 1, 2, \ldots$. Hence $p_n \in E$ for $n = 1, 2, \ldots$ and, since $p = \lim_{n \to \infty} p_n$ and E is closed, we have $p \in E$ and therefore $p \in S$. Consequently $S' \subset S$, that is, S is closed and so is an F_σ.

2. Prove that the sum of a transfinite ascending sequence of closed sets of type Ω is closed.

The proof follows from the result of problem 1, case (ii).

68. The straight line as the sum of $\aleph_1$ ascending sets G_δ. The Hausdorff Theorem[86] that *the set of all real numbers is the sum of an ascending transfinite sequence of type Ω of sets G_δ* which was proved without the aid of the continuum hypothesis is, of all known theorems of this type, the closest to this hypothesis.

It follows from the continuum hypothesis ($\mathbf{c} = \aleph_1$) that the line is the sum of $\aleph_1$ increasing sets F_σ. This theorem cannot be proved, at the present state of our knowledge, without the continuum hypothesis.

Results closely approaching Hausdorff's theorem were published by Lusin[87] in 1947. We give here a modification of Hausdorff's proof.

A and B are two given sets. We say that A *is almost contained in* B and write[88]

$$A \underset{*}{\subset} B \text{ if } \overline{\overline{A - B}} < \aleph_0.$$

If $A \underset{*}{\subset} B$ but not $B \underset{*}{\subset} A$ (that is, $\overline{\overline{A - B}} < \aleph_0$ and $\overline{\overline{B - A}} \geqslant \aleph_0$, in other words, A is almost contained in B but differs from it by an infinite number of elements), then we write

$$A \prec B.$$

The relation $*\subset$ is transitive (for, if $A * \subset B$ and $B * \subset C$, then $\overline{\overline{A - B}} < \aleph_0$ and $\overline{\overline{B - C}} < \aleph_0$; hence, from $A - C \subset (A - B) + (B - C)$, we find $\overline{\overline{A - C}} \leqslant \overline{\overline{A - B}} + \overline{\overline{B - C}} < \aleph_0$) and reflexive. The relation $\prec$ is asymetric and transitive; it therefore partially orders the sets. Furthermore, if for a natural number m, $A_k * \subset B$ for $k = 1, 2, \ldots, m$, then $A_1 + A_2 + \ldots + A_m * \subset B$; and if $A * \subset B_k$ for $k = 1, 2, \ldots, m$, then $A * \subset B_1 . B_2 \ldots B_m$, since $A - B_1 . B_2 \ldots B_m = (A - B_1) + (A - B_2) + \ldots + (A - B_m)$. From $A_1 \prec B$ and $A_2 \prec B$ it need not follow that $A_1 + A_2 \prec B$; for instance, take $A_1 = \{1, 3, 5, \ldots\}$, $A_2 = \{2, 4, 6, \ldots\}$, $B = \{1, 2, 3, \ldots\}$.

We shall prove, however,

LEMMA 1. *If, for a natural number* m, $A_1 \prec A_2 \prec \ldots \prec A_m \prec B$, *then* $A_1 + A_2 + \ldots + A_m \prec B$.

Proof. Since relation $\prec$ is transitive and $A \prec B$ implies $A * \subset B$, it follows from $A_1 \prec A_2 \prec \ldots \prec A_m \prec B$ that $A_k * \subset B$ for $k = 1, 2, \ldots, m$ and so $A_1 + A_2 + \ldots + A_m * \subset B$. It remains to show that the cardinal of $B - (A_1 + A_2 + \ldots + A_m)$ is $\geqslant \aleph_0$. We have the identity

$$B - (A_1 + A_2 + \ldots + A_m) = (B - A_m) - [(A_1 + A_2 + \ldots + A_m) - A_m].$$

Since $A_m \prec B$, we have $\overline{\overline{B - A_m}} \geqslant \aleph_0$; since $A_1 \prec A_2 \prec \ldots \prec A_m$, we have $A_1 + A_2 + \ldots + A_m * \subset A_m$ and so the cardinal of $(A_1 + A_2 + \ldots + A_m) - A_m$ is $< \aleph_0$. Consequently the cardinal of $B - (A_1 + A_2 + \ldots + A_m)$ is $\geqslant \aleph_0$.

If R denotes a given family of sets and B a given set we shall write[89]

$$R \gamma B$$

to express the fact that for every finite set S there exists only a finite number of sets A of the family R such that $A - B \subset S$.

From the relation $A - B \subset (A - C) + (C - B)$ it follows that if $R \gamma B$ and $C * \subset B$ then $R \gamma C$.

LEMMA 2. *If* $A_1, A_2, \ldots$ *is an infinite sequence of infinite sets such that* $A_1 \prec A_2 \prec \ldots \prec B$, *then there exists a set* A *such that* $A_1 \prec A_2 \prec \ldots \prec A \prec B$ *and* $\{A_1, A_2, \ldots\} \gamma A$.

Proof. Let k be a given natural number. Since $A_1 \prec A_2 \prec \ldots \prec A_k$, $A_1 + A_2 + \ldots + A_{k-1} \prec A_k$, by Lemma 1. So the cardinal of $A_k - (A_1 + A_2 + \ldots + A_{k-1})$ is $\geqslant \aleph_0$, since $A_k \prec B$, i.e., $\overline{\overline{A_k - B}} < \aleph_0$. Thus from the relation

$$A_k \,.\, B - (A_1 + A_2 + \ldots + A_{k-1})$$
$$= (A_k - (A_1 + A_2 + \ldots + A_{k-1})) - (A_k - B)$$

we find that the cardinal of $A_k \,.\, B - (A_1 + A_2 + \ldots + A_{k-1})$ is $\geqslant \aleph_0$. Hence there exists an element $a_k \in A_k \,.\, B - (A_1 + A_2 + \ldots + A_{k-1})$; for $k \neq l$, we have $a_k \neq a_l$. Put $A = B - \{a_1, a_2, \ldots\}$; obviously $A \subset B$ and $\overline{\overline{B - A}} = \aleph_0$; hence $A \prec B$. On the other hand, for any natural number k we have

$$A_k - A = (A_k - B) + A_k(B - A) = (A_k - B) + A_k\{a_1, a_2, \ldots\}$$
$$= (A_k - B) + A_k\{a_1, a_2, \ldots, a_k\}$$

since $a_n \notin A_k$ for $n > k$. Since $\overline{\overline{A_k - B}} < \aleph_0$, it follows that $\overline{\overline{A_k - A}} < \aleph_0$. But since $A_k \prec A_{k+1}$ we have $\overline{\overline{A_{k+1} - A_k}} \geqslant \aleph_0$ while $\overline{\overline{A_{k+1} - A}} < \aleph_0$. Consequently, we conclude from

$$A - A_k \supset (A_{k+1} - A_k) - (A_{k+1} - A)$$

that $\overline{\overline{A - A_k}} \geqslant \aleph_0$. Hence $A_k \prec A$ for $k = 1, 2, \ldots$. Finally, we prove that $\{A_1, A_2, \ldots\} \,\gamma\, A$.

Let S be a given finite set. If $A_k - A \subset S$ for infinitely many values of k we would have an infinite sequence $k_1 < k_2 < \ldots$ of indices such that $A_{k_i} - A \subset S$ for $i = 1, 2, \ldots$. But $a_{k_i} \in A_{k_i} - A$ for $i = 1, 2, \ldots$; hence $a_{k_i} \in S$ for $i = 1, 2, \ldots$. This is impossible since S is finite and the elements $a_{k_1}, a_{k_2}, \ldots$ are all different. Consequently $\{A_1, A_2, \ldots\} \,\gamma\, A$; this proves the lemma.

LEMMA 3. *If R_1 and R_2 are two at-most-countable families of infinite sets ordered by the relation $\prec$ and such that $A \prec B$ for $A \in R_1$, $B \in R_2$, then there exists a set E such that $A \prec E \prec B$ for all $A \in R_1$ and all $B \in R_2$.*

Proof. If there exists in the family R_1 (ordered by the relation $\prec$) a last set A_1 and in the family R_2 a first set B_1, then $A_1 \prec B_1$ and so $\overline{\overline{B_1 - A_1}} \geqslant \aleph_0$; hence $B_1 - A_1$ is the sum of two infinite disjoint sets X and Y. Put $E = A_1 + X$; this gives

$$B_1 - E = (B_1 - A_1) - X = Y$$

and $E - A_1 = X$; hence $\overline{\overline{B_1 - E}} \geqslant \aleph_0$ and $\overline{\overline{E - A}} \geqslant \aleph_0$. On the other hand, $A_1 - E = 0$ and

$$E - B_1 = (A_1 + X) - B_1 = A_1 - B_1$$

and so, from $A_1 \prec B_1$, we have $\overline{\overline{B_1 - E}} < \aleph_0$. Therefore $A_1 \prec E \prec B_1$ and, since A_1 is the last set of R_1 and B_1 the first of R_2, $A \prec E \prec B$ for $A \in R_1$ and $B \in R_2$.

Suppose next that R_1 has no last set but R_2 has a first set B_1; then, since R_1 is countable, there exists an infinite sequence $A_1 \prec A_2 \prec \ldots$ of sets of R_1 co-final[90] with R_1, where $A_1 \prec A_2 \prec \ldots \prec B_1$; hence, by Lemma 2, there exists a set E such that $A_1 \prec A_2 \prec \ldots \prec E \prec B_1$. If $A \in R_1$, since the sequence $A_1 \prec A_2 \prec \ldots$ is co-final with R_1, there exists a natural number k such that $A \prec A_k$; since $A_k \prec E$ we have $A \prec E$. If $B \in R_2$, $B \neq B_1$, we have $B_1 \prec B$ and so $E \prec B$. Consequently $A \prec E \prec B$ for $A \in R_1$ and $B \in R_2$.

Consider now the case where R_1 has a last set A_1 but R_2 has no first set. Denote by R' the family of complements of all sets of the family $R \subset R_1 + R_2$ with respect to the sum S of all sets of the family $R_1 + R_2$. Obviously $A \prec B$ for $A \in R_1$, $B \in R_2$, implies $B' \prec A'$ for $B' \in R'_2$, $A' \in R'_1$; also the family R'_2 has no last set but R'_1 has a first set. But in this case, as proved above, there exists a set E' such that $B' \prec E' \prec A'$ for $B' \in R'_2$ and $A' \in R'_1$. Denoting by A, B, E, the complements with respect to S of the sets A', B', and E' respectively, we obtain $A \prec E \prec B$ for $A \in R_1$ and $B \in R_2$.

Finally consider the case where the family R_1 has no last set and the family R_2 no first. Let $R_1 = \{E_1, E_2, \ldots\}$, $R_2 = \{H_1, H_2, \ldots\}$ where the sets E_k and H_k $(k = 1, 2, \ldots)$ may not be ordered by the relation $\prec$ with increasing indices. Put

$$E = E_1 . H_1 + E_2 . H_1 . H_2 + E_3 . H_1 . H_2 . H_3 + \ldots + E_i . H_1 . H_2 \ldots H_i + \ldots .$$

Since $E_i \prec H_k$ for all i and k, we have $E_i * \subset E_i . H_1 . H_2 \ldots . H_i$ for $i = 1, 2, \ldots$, and so $E_i * \subset E$ for $i = 1, 2, \ldots$; again $H_1 . H_2 \ldots H_i \subset H_k$ for $i \geqslant k$; hence

$$E \subset E_1 + E_2 + \ldots + E_{k-1} + H_k * \subset H_k.$$

Consequently $E * \subset H_k$ for $k = 1, 2, \ldots$. We thus have $A * \subset E * \subset B$ for $A \in R_1$ and $B \in R_2$. If now $A \in R_1$ and $B \in R_2$ it follows from the hypothesis concerning the families R_1 and R_2 that there exist sets $A_1 \in R_1$ and $B_1 \in R_2$ such that $A \prec A_1$ and $B \prec B_1$; hence $A \prec A_1 * \subset E * \subset B_1 \prec B$. This gives $A \prec E \prec B$ and completes the proof of Lemma 3.

A section $R(A)$ of R is the family of all those sets of R which $\prec A$.

LEMMA 4. *If R is a countable ordered family R of sets of natural numbers such that $A \prec B$ for all $A \in R$ and, for every section $R(A)$ of R, we have $R(A) \gamma B$, then there exists a set E such that $A \prec E \prec B$ for all $A \in R$; further, $R \gamma E$.*

Proof. Suppose that $R \gamma B$. Put $R_1 = R$, $R_2 = \{B\}$; by Lemma 3 there exists a set E such that $A \prec E \prec B$ for all $A \in R$. Since $R \gamma B$ and $E \prec B$ we have $R \gamma E$. Hence Lemma 4 holds. We may therefore assume that $R \gamma B$ is not true.

If there is a last set A_1 in R_1, then $R(A_1) \gamma B$ and so $R \gamma B$ contrary to assumption. Hence R does not possess a last set and so, since it is countable, it is of type co-final with ω.

Since $R \gamma B$ is not true there exists a finite set S_1 such that for infinitely many sets A of the family R we have $A - B \subset S_1$. But, by the hypothesis of the Lemma, $R(A_1) \gamma B$ for every $A_1 \in R$; hence for every $A_1 \in R$ there exists only a finite number of sets $A \prec A_1$ such that $A - B \subset S_1$. Thus, the sets A in R such that $A - B \subset S_1$ form an ordered set (by the relation $\prec$) of type ω which is co-final with R. Let these sets be denoted by $A_1^{S_1}, A_2^{S_1}, \ldots$. Then

$$A_1^{S_1} \prec A_2^{S_1} \prec \ldots \prec B.$$

Hence there exists, by Lemma 2, a set A^{S_1} such that

$$\{A_1^{S_1}. A_2^{S_1}, \ldots\} \gamma A^{S_1}$$

and

$$A_1^{S_1} \prec A_2^{S_1} \prec \ldots \prec A^{S_1} \prec B.$$

Since the sequence $A_1^{S_1}, A_2^{S_1}, \ldots$ is co-final with R we conclude that

$$A \prec A^{S_1} \prec B \text{ for all } A \in R.$$

Since R is a family of sets of natural numbers it may be assumed that the finite set S_1 consists of natural numbers; hence there exists a natural number m such that $S_1 \subset \{1, 2, \ldots m\}$. Now put $S_2 = \{1, 2, \ldots, m, m + 1\}$; we conclude as before that there exist infinitely many sets $A \in R$, that is, sets $A \prec A^{S_1}$ for which $A - B \subset S_2$, and that these sets A form an ordered set $A_1^{S_2} \prec A_2^{S_2} \prec \ldots$ of type ω co-final with R. Since $A_n^{S_2} \prec A^{S_1}$ for $n = 1, 2, \ldots$ we conclude similarly that there exists a set A^{S_2} such that

$$A_1^{S_2} \prec A_2^{S_2} \prec \ldots \prec A^{S_2} \prec A^{S_1} \text{ and } \{A_1^{S_2}, A_2^{S_2}, \ldots\} \gamma A^{S_2}.$$

Repeating this process and setting $S_k = \{1, 2, \ldots m + k - 1\}$, we deduce the existence of an infinite sequence $A^{S_1}, A^{S_2}, \ldots$, of sets such that

$$A \prec \ldots \prec A^{S_k} \prec A^{S_{k-1}} \prec \ldots \prec A^{S_1} \prec B \text{ for all } A \in R, k = 1, 2, \ldots$$

and where, for every natural k, there exists an infinite sequence of sets $A_1^{S_k}, A_2^{S_k}, \ldots$ co-final with R and such that

$$\{A_1^{S_k}, A_2^{S_k}, \ldots\} \gamma A^{S_k}.$$

By Lemma 3 there exists a set E such that

$$A \prec E \prec \ldots \prec A^{S_k} \prec A^{S_{k-1}} \prec \ldots \prec A^{S_1} \prec B \text{ for } A \in R.$$

It follows that $R \gamma E$. Otherwise there would exist a finite set S such that $A - E \subset S$ for infinitely many sets $A \in R$. For all such sets A the relation $A - B \subset (A - E) + (E - B)$ implies that $A - B \subset S + E - B$; and since $\overline{\overline{E - B}} < \aleph_0$ there exists a natural number k such that $A - B \subset S_k$. Thus the sets A of R for which $A - E \subset S$ are members of the sequence $A_1^{S_k}, A_2^{S_k}, \ldots$; so, for infinitely many different indices i, we have $A_i^{S_k} - E \subset S$. Since

$$A_i^{S_k} - A^S \subset (A_i^{S_k} - E) + (E - A^{S_k}) \subset S + (E - A^{S_k}) = T$$

and

$$E \prec A^{S_k}$$

we find that T is finite. Hence $A_i^{S_k} - A^S$ is contained in a finite set for infinitely many values of i, contrary to the fact that $\{A_1^{S_k}, A_2^{S_k}, \ldots\} \gamma A^{S_k}$. Therefore $R \gamma E$ and Lemma 4 is proved.

We next define by transfinite induction two transfinite sequences $\{A_\xi\}$, and $\{B_\xi\}$, $\xi < \Omega$, of natural numbers. Let A_1 and B_1 be any two infinite sets of natural numbers such that $A_1 \prec B_1$; for instance, $A_1 = \{1, 3, 5, \ldots\}$ and $B_1 = \{1, 2, 3, \ldots\}$. Let α be a given ordinal number, $1 < \alpha < \Omega$, and suppose that all sets A_ξ and B_ξ, $\xi < \alpha$, are already defined in such a way that

(109) $$A_\xi \prec A_\eta \prec B \prec B_\xi, \qquad \xi \prec \eta \prec \alpha$$

and

(110) $$\{A_\xi\}_{\xi<\beta} \gamma B_\beta \qquad \text{for all } \beta < \alpha.$$

This is certainly true for $\alpha = 2$ since $\{A_1\} \gamma B_1$.

By Lemma 3 there exists a set E such that $A_\xi \prec E \prec B_\xi$ for $\xi < \alpha$. Since $E \prec B_\beta$ for $\beta < \alpha$ we have, from (110), $\{A_\xi\}_{\xi<\beta} \gamma E$ for every $\beta < \alpha$. By Lemma 4 there exists a set B_α such that $A_\xi \prec B_\alpha \prec E$ for $\xi < \alpha$ and $\{A_\xi\}_{\xi<\alpha} \gamma B_\alpha$. By Lemma 3 there exists a set A_α such that $A_\xi \prec A_\alpha \prec B_\alpha$ for $\xi < \alpha$.

Hence from (109)

(111) $$A_\xi \prec A_\eta \prec B_\eta \prec B_\xi \text{ for } \xi < \eta \leqslant \alpha \text{ and } \{A_\xi\}_{\xi<\alpha} \gamma B_\alpha.$$

The sets A_ξ and B_ξ are thus defined by transfinite induction for all $\xi < \Omega$.

We next show that there exists no set E such that

(112) $$A_\xi * \subset E * \subset B_\xi \qquad \text{for all } \xi < \Omega.$$

Suppose that such a set E exists. Now from (112) $\overline{\overline{A_\xi - E}} < \aleph_0$ for all $\xi < \Omega$; so, for every $\xi < \Omega$, there exists a natural number $k = k_\xi$ such that $A_\xi - E \subset \{1, 2, \ldots, k_\xi\}$. Consequently there exists a sequence $\xi_1 < \xi_2 < \ldots$

with this property. Since $\xi_i < \Omega$ for $i = 1, 2, \ldots$ there exists an ordinal number $\alpha < \Omega$ such that $\xi_i < \alpha$ for $i = 1, 2, \ldots$. Thus $A_{\xi_i} - E \subset \{1, 2, \ldots k\}$ for $\xi_1 < \xi_2 < \ldots < \alpha$ contrary to the relation $\{A_\xi\}_{\xi<\alpha} \gamma E$ which follows from $E_* \subset B_\alpha$ and $\{A_\xi\}_{\xi<\alpha} \gamma B_\alpha$. Hence there exists no set E which satisfies (112), that is, the sequences $\{A_\xi\}$, $\xi < \Omega$, and $\{B_\xi\}$, $\xi < \Omega$, give rise to a gap[91] of type $(\Omega\,\Omega^*)$. There exist, therefore, sequences of natural numbers which give rise to a gap of type $(\Omega\,\Omega^*)$.

For every infinite set E of natural numbers denote by $f(E)$ the real number

$$x = 2\sum_{n=1}^{\infty} x_n/3^n \tag{113}$$

where $x_n = 1$ for $n \in E$ and $x_n = 0$ for $n \notin E$. It follows readily that if

$$a = 2\sum_{n=1}^{\infty} a_n/3^n, \quad b = 2\sum_{n=1}^{\infty} b_n/3^n, \quad \text{and } a \leqslant b$$

then, for a given natural number n, the set F_n of all real numbers (113) for which $a_n \leqslant x_n \leqslant b_n$ is a closed set. If A and B are infinite sets of natural numbers such that $A_* \subset B$ and if we put $f(A) = a$, $f(B) = b$, then it can be shown that the set $f(E)$ of all real numbers where $A_* \subset E_* \subset B$, is the set

$$\sum_{n=1}^{\infty} F_n \,.\, F_{n+1} \,.\, F_{n+2} \ldots,$$

that is, an F_σ.

For $\alpha < \Omega$, denote by Q_α the set of all numbers $f(E)$ such that $A_{\alpha *} \subset E_* \subset B_\alpha$, where $\{A_\xi\}$, $\xi < \Omega$, and $\{B_\xi\}$, $\xi < \Omega$, are transfinite sequences previously defined with a gap of type $(\Omega\,\Omega^*)$; the sets Q_α are sets F_σ and

$$\prod_{\alpha<\Omega} Q_\alpha = 0$$

since there is no set E with the property (112). It follows from (111) that, for $\xi < \eta < \Omega$, $Q_\eta \subset Q_\xi$; also $f(A_\xi) \in Q_\xi - Q_\eta$. Hence $Q_\xi \neq Q_\eta$. Consequently $\{Q_\xi\}$, $\xi < \Omega$, is a transfinite sequence of type Ω of descending linear sets F_σ whose intersection is the null set. This leads at once to the result that the complements P_ξ of the sets Q_ξ with respect to the set of all real numbers form a transfinite ascending sequence of sets G_δ. The theorem of Hausdorff is therefore proved.

Putting

$$X_\alpha = P_\alpha - \sum_{\xi<\alpha} P_\xi \text{ for } \alpha < \Omega,$$

we obtain a decomposition of the straight line[92]

$$X = \sum_{\alpha<\Omega} X_\alpha$$

into a sum of $\aleph_1$ disjoint non-empty sets $F_{\sigma\delta}$.

69. Hausdorff's sets P^{α} and Q^{α}. Following Hausdorff,[93] denote an open set by P^1, a closed set by Q^1, and define by transfinite induction, for $1 < \alpha < \Omega$, sets P^{α} and Q^{α} as follows:

$$P^{\alpha} = \sum_{n=1}^{\infty} Q^{\xi_n}, \quad Q^{\alpha} = \prod_{n=1}^{\infty} P^{\xi_n}, \qquad \xi_n < \alpha, n = 1, 2, \ldots.$$

Thus a set P^2, being the sum of a countable aggregate of sets Q^1, is a set F_{σ} (and conversely). Similarly a set Q^2, being the intersection of a countable aggregate of sets P^1, is a G_{δ} (and conversely). A set P^3 is the sum of a countable aggregate of sets Q^2, i.e., of sets G_{δ}; hence P^3 is a so-called set $G_{\delta\sigma}$ (and conversely). Similarly Q^3 is the intersection of a countable aggregate of sets F_{σ} and hence a so-called set $F_{\sigma\delta}$.

We next deduce several properties of the sets P^{α} and Q^{α}.

Property 1. *In a metric space a set P^{α} (Q^{α}) is also a set P^{β} (Q^{β}) for $\alpha < \beta < \Omega$.*

Proof. Since a set P^1 is open it is an F_{σ} and so a P^2; similarly, since Q^1 is closed, it is a G_{δ} and therefore a Q^2. It remains to prove property 1 for $\alpha \geqslant 2$.

Let $2 \leqslant \alpha < \beta$ and let E be a set P^{α}; then $E = E_1 + E_2 + \ldots$, where E_n is a Q^{ξ_n} and $\xi_n < \alpha$ for $n = 1, 2, \ldots$; certainly $\xi_n < \beta$ for $n = 1, 2, \ldots$ and consequently E is a P^{β}.

Property 2. *In a metric space the sum (intersection) of a finite or countable aggregate of sets $P^{\alpha}(Q^{\alpha})$ is a set $P^{\alpha}(Q^{\alpha})$ for $\alpha < \Omega$.*

Proof. For $\alpha = 1$, property 2 follows from Theorems 2 and 4. For $\alpha > 1$, property 2 comes from the definition of the sets P^{α} and Q^{α} and the fact that an infinite double sequence can be arranged as a single sequence.

Note that property 2 holds in all (V)spaces, but property 1 may not hold in some topological spaces (cf. the remark at the end of § 65).

Property 3. *The complement of a set $P^{\alpha}(Q^{\alpha})$ with respect to the metric space in which it is contained is a set $Q^{\alpha}(P^{\alpha})$.*

Proof. For $\alpha = 1$ property 3 follows from the definition of the sets P^1 and Q^1. Let α be a given ordinal number where $1 < \alpha < \Omega$ and suppose that property 3 holds for all ordinal numbers $\xi < \alpha$. Let E denote a set P^{α}; then $E = E_1 + E_2 + \ldots$, where E_n is a set Q^{ξ_n} and $\xi_n < \alpha$ for n $= 1, 2, \ldots$. Since property 3 is assumed to hold for $\xi < \alpha$, CE_n is a set P^{ξ_n} and so, from $CE = CE_1 \,.\, CE_2 \,.\, CE_3 \ldots$, we find that CE is a set Q^{α}.

On the other hand, let E denote a set Q^{α}; then $E = E_1 \,.\, E_2 \,.\, E_3 \ldots$, where E_n is a set P^{ξ_n} and $\xi_n < \alpha$. Since property 3 holds for $\xi < \alpha$, CE_n is a set Q^{ξ_n} and, since $CE = CE_1 + CE_2 + \ldots$, it follows that CE is a set P^{α}. Property 3 is thus established by transfinite induction.

Property 4. *In a metric space the intersection (sum) of a finite number of sets $P^{\alpha}(Q^{\alpha})$ is a set $P^{\alpha}(Q^{\alpha})$ for $\alpha < \Omega$.*

Proof. It is obviously sufficient to establish property 4 for two sets and then extend the result by ordinary induction to any finite number of sets.

For $\alpha = 1$ property 4 follows from Theorems 35 and 36. Let α denote an ordinal number, where $1 < \alpha < \Omega$, and let E and T be two sets P^α. Hence $E = E_1 + E_2 + \ldots, T = T_1 + T_2 + \ldots$, where E_n is a set Q^{ξ_n} for $\xi_n < \alpha$, T_n is a set Q^{η_n} for $\eta_n < \alpha$ and $n = 1, 2, \ldots$. Thus

$$E \,.\, T = \sum_{m=1}^{\infty} \sum_{n=1}^{\infty} E_m \,.\, T_n.$$

Let $\zeta_{mn} = \max(\xi_m, \eta_n)$; since $\xi_m < \alpha$ for $m = 1, 2, \ldots$ and $\eta_n < \alpha$ for $n = 1, 2, \ldots$, we have $\zeta_{mn} < \alpha$ for all m and n. Since $\xi_m \leqslant \zeta_{mn}$ and $\eta_n \leqslant \zeta_{mn}$, it follows from property 1 that E_m and T_n are each sets $Q^{\zeta_{mn}}$. The set $E_m \,.\, T_n$, as the intersection of two sets $Q^{\zeta_{mn}}$, is by property 2 a set $Q^{\zeta_{mn}}$. Since $\zeta_{mn} < \alpha$ for all natural m and n the set $E \,.\, T$ is a set P^α.

Next let E and T be two sets Q^α. By property 3, the sets CE and CT are sets P^α and so, from the first part of the proof, we deduce that the set

$$CE \,.\, CT = C(E + T)$$

is a set P^α; hence $E + T$ is a set Q^α by property 3. This establishes property 4.

It is easily seen that property 3 holds in all (V)spaces while property 4 holds in all topological spaces but may not hold in some (V)spaces.

Property 5. *In a metric space every set $P^\alpha(Q^\alpha)$ is a set $Q^{\alpha+1}(P^{\alpha+1})$ for $\alpha < \Omega$.*

Proof. For $\alpha = 1$ property 5 follows from Theorems 84 and 85. Let α be an ordinal number such that $1 < \alpha < \Omega$; if E is a set P^α, it is sufficient to write $E = E \,.\, E \,.\, E \ldots$ and then, from the definition of sets $Q^{\alpha+1}$, property 5 follows. If E is a set Q^α then, writing $E = E + E + \ldots$, it follows from the definition of sets $P^{\alpha+1}$ that E is such a set.

From properties 5 and 2 we obtain

Property 6. *In a metric space the sum (intersection) of a finite or countable aggregate of sets $P^\alpha(Q^\alpha)$ is a set $Q^{\alpha+1}(P^{\alpha+1})$.*

From properties 3, 5, 1, 2, and the fact that $E_1 - E_2 = E_1 \,.\, CE_2$, we obtain

Property 7. *In a metric space the difference of two sets P^α or two sets Q^α is both a $P^{\alpha+1}$ and a $Q^{\alpha+1}$ $(\alpha < \Omega)$.*

Property 8 (Lusin). *If $3 \leqslant \alpha < \Omega$ then every set P^α contained in a metric space is the sum of a countable aggregate of disjoint sets $E_1, E_2, \ldots$ where E_n is a Q^{ξ_n}, $\xi_n < \alpha$ for $n = 1, 2, \ldots$.*

Proof. Let E be a P^α where $3 \leqslant \alpha < \Omega$. By property 1, $E = T_1 + T_2 + \ldots$, where T_n is a set Q^{η_n} and $2 \leqslant \eta_n < \alpha$ for $n = 1, 2, \ldots$. Put $S_n = T_1 + T_2 + \ldots + T_n$ and let $\xi_n = \max(\eta_1, \eta_2, \ldots, \eta_n)$; since $2 \leqslant \eta_n < \alpha$, we have $2 \leqslant \xi_n < \alpha$ for $n = 1, 2, \ldots$. Hence S_n is a set Q^{ξ_n} for $n = 1, 2, \ldots$ by properties 1 and 4.

Let $R_1 = S_1$ and $R_{n+1} = S_{n+1} - S_n$ for $n = 1, 2, \ldots$; then $R_{n+1} = S_{n+1}CS_n$ for $n = 1, 2, \ldots$. But CS_n is a set P^{ξ_n} by property 3; we may therefore write[94] $CS_n = T_{n,1} + T_{n,2} + \ldots$, where $T_{n,k}$ is a set $Q^{\xi_{n,k}}$ and $\xi_{n,k} < \xi_n$ for $k = 1, 2, \ldots$. Let $\zeta_{n,k} = \max(\xi_{n,1}, \xi_{n,2}, \ldots, \xi_{n,k})$; then $\zeta_{n,k} < \xi_n$ for $k = 1, 2, \ldots$, and so, from properties 1 and 4, the set

$$S_{n,k} = T_{n,1} + T_{n,2} + \ldots + T_{n,k} \text{ is a } Q^{\zeta_{n,k}} \text{ for all } k.$$

Put $R_{n,1} = S_{n,1}$ and $R_{n,k} = S_{n,k} - S_{n,k-1}$ for $k = 2, 3, \ldots$; since $\zeta_{n,k-1} \leqslant \zeta_{n,k}$ for $k = 2, 3, \ldots$ we conclude, from property 7, that $R_{n,k}$ is a set $Q^{\zeta_{n,k}+1}$ for $k = 1, 2, \ldots$ and so, since $\zeta_{n,k} < \xi_n$ for all k and $\zeta_{n,k} + 1 \leqslant \xi_n$, a set Q^{ξ_n}. Since S_{n+1} is a $Q^{\xi_{n+1}}$ and since $\xi_n \leqslant \xi_{n+1}$ we conclude, from properties 1 and 2, that the set $S_{n+1} \cdot R_{n,k}$ is a $Q^{\xi_{n+1}}$ for $k = 1, 2, \ldots$. But the set E is obviously the sum of disjoint sets as given by the relation

$$E = S_1 + \sum_{n=1}^{\infty} \sum_{k=1}^{\infty} S_{n+1} \cdot R_{n,k},$$

where S_1 is a Q^{ξ_1} and $S_{n+1} \cdot R_{n,k}$ is a $Q^{\xi_{n+1}}$ for $n = 1, 2, \ldots, k = 1, 2, \ldots$ and where $\xi_n < \alpha$ for $n = 1, 2, \ldots$. This establishes property 8.

A family of sets is called a *field* if the sum and difference of any two sets of the family also belong to the family.

Denote by R^α, $1 \leqslant \alpha < \Omega$, all sets which are both a P^α and a Q^α. It follows from properties 2 and 3 that for a given α $(1 \leqslant \alpha < \Omega)$ the family of all sets R^α is a field.

If $\xi < \alpha < \Omega$ then a set Q^ξ is, by property 1, a Q^α and, by property 5, a $P^{\xi+1}$; therefore it is a P^α. Hence, every set Q^ξ is both a P^α and a Q^α and so an R^α for $\xi < \alpha < \Omega$. Thus every P^α is the sum of a countable aggregate of sets R^α, i.e., a set R_σ^α. Passing to complements we conclude that every Q^α is the intersection of a countable aggregate of sets R^α, i.e., a set R_δ^α. Since every R^α is a set P^α it follows from property 2 that R_σ^α is a $P^\alpha (\alpha < \Omega)$. Similarly, since R^α is a Q^α, it follows from property 2 that R_δ^α is a Q^α. Thus a set which is both an R_σ^α and an R_δ^α is a set $R^\alpha (\alpha < \Omega)$ and, as seen above, the converse is also true for $1 < \alpha < \Omega$.

THEOREM 86. *If, in a metric space, K is a set P^α, H a set Q^α, and $H \subset K$, then there exists a set E which is both a P^α and a Q^α, $1 < \alpha < \Omega$, such that $H \subset E \subset K$.*

We first prove the following

LEMMA. *If $\{K_n\}$ and $\{H_n\}$, $n = 1, 2, \ldots$, are two infinite sequences of sets such that*

$$K_1 \subset K_2 \subset \ldots, \qquad H_1 \supset H_2 \supset \ldots, \tag{114}$$

and

(115) $$K_1 + K_2 + \ldots \supset H_1 . H_2 . H_3 \ldots,$$

then

(116) $$K_1 . H_1 + K_2 . H_2 + \ldots = H_1(K_1 + H_2)(K_2 + H_3)(K_3 + H_4) \ldots.$$

Proof. Let $\{K_n\}$ and $\{H_n\}$ be two infinite sequences satisfying (114) and (115); put

(117) $$P = K_1 . H_1 + K_2 . H_2 + \ldots$$

and

(118) $$Q = H_1(K_1 + H_2)(K_2 + H_3) \ldots.$$

Let p be an element of the set P; by (117), there exists an index n such that $p \in K_n . H_n$ and so, by (114), $p \in K_j$ for $j \geqslant n$ and $p \in H_j$ for $j \leqslant n$. This gives $p \in H_1$ and $p \in K_j + H_{j+1}$ for $j = 1, 2, \ldots$ and so, from (118), $p \in Q$. Consequently $P \subset Q$.

Let q be an element of Q. Put

(119) $$H = H_1 . H_2 . H_3 . .$$

and consider two cases.

1. $q \in H$; there exists, by (115) and (119), an index n such that $q \in K_n$ and, since $q \in H_n$, we have $q \in K_n . H_n$ and so, by (117), $q \in P$.

2. $q \notin H$; hence there exists an index k such that $q \notin H_k$. Let m be the smallest of such indices k; we cannot have $m = 1$ since $q \in Q$ and so, by (118), $q \in H_1$. Hence $m = n + 1$ where n is a natural number. Thus $q \in H_{m-1}$, that is, $q \in H_n$. Since $q \in Q$ it follows from (118) that $q \in K_n$ and so, $q \in K_n . H_n$; consequently $q \in P$.

In either case $Q \subset P$ and so (117) and (118) give (116); this proves the lemma.

To prove the theorem let α denote an ordinal number in $1 < \alpha < \Omega$, K a set P^α, and H a set Q^α such that $H \subset K$. Then there exist infinite sequences $\{M_n\}$ and $\{N_n\}$ such that M_n is a Q^{ξ_n}, and N_n is a P^{η_n}, where $\xi_n < \alpha$ and $\eta_n < \alpha$ for $n = 1, 2, \ldots$, and such that

(120) $$K = M_1 + M_2 + \ldots, \quad H = N_1 . N_2 . N_3 . \ldots$$

Put

(121) $$K_n = M_1 + M_2 + \ldots + M_n, \quad H_n = N_1 . N_2 \ldots N_n, \quad n = 1, 2, \ldots.$$

Let $\mu_n = \max(\xi_1, \xi_2, \ldots, \xi_n)$, $\nu_n = \max(\eta_1, \eta_2, \ldots, \eta_n)$; then K_n is a Q^{μ_n} and H_n a P^{ν_n} by property 4, where $\mu_n < \alpha$ and $\nu_n < \alpha$ for $n = 1, 2, \ldots$. From (121) we obtain relations (114) and from (120) and (121)

(122) $$K = K_1 + K_2 + \ldots, \quad H = H_1 . H_2 . H_3 . \ldots$$

Since $H \subset K$, (122) gives (115). Hence, by the above lemma, we have (116). Next define a set P by relation (117); from (116)

$$(123) \qquad P = H_1(K_1 + H_2)(K_2 + H_3)(K_3 + H_4)\ldots.$$

It follows from (117) and (122) that $P \subset K$ and from (122) and (123) that $H \subset P$; hence $H \subset P \subset K$.

Since K_n is a Q^{μ_n} and H_n is a P^{ν_n}, $\mu_n < \alpha$, $\nu_n < \alpha$, it follows from properties 1 and 7 that K_n is a P^α, H_n is a P^α and therefore, by property 4, $K_n . H_n$ is a P^α; hence, by (117) and property 2, the set P is a P^α. Similarly we find that K_n is a Q^α, H_n is a Q^α, and so, by property 4, $K_n + H_{n+1}$ is a Q^α; thus, by (123) and property 2, the set P is a Q^α. Hence the set $E = P$ is both a P^α and a Q^α and, since $H \subset P \subset K$, the set P is the required set. Theorem 86 is therefore proved.

COROLLARY. *If α is an ordinal number such that $1 < \alpha < \Omega$ and M and N are disjoint sets Q^α contained in a metric space then there exist sets S and T each of which is both a P^α and a Q^α and such that*

$$M \subset S, \quad N \subset T, \quad \textit{and } S . T = 0.$$

Proof. Let M and N be two disjoint sets Q^α; then $M \subset CN$ where CN is a P^α by property 3. Hence, by Theorem 86, there exists a set E which is both P^α and a Q^α and such that $M \subset E \subset CN$. From property 3 the set CE is both a Q^α and a P^α; hence the sets $S = E$ and $T = CE$ are the required sets and the corollary is thus established.[95]

70. Sets which are locally P^α and Q^α. A set E is said to be *locally a* $P^\alpha(Q^\alpha)$, $1 < \alpha < \Omega$, *at the point* p if there exists an open set U such that $p \in U$ and the set $E . U$ is a $P^\alpha(Q^\alpha)$. A set which is locally a $P^\alpha(Q^\alpha)$ at every one of its points is said to be locally a $P^\alpha(Q^\alpha)$.

THEOREM 87. *A set contained in a metric space which is locally a $P^\alpha(Q^\alpha)$ is a $P^\alpha(Q^\alpha)$.*

We first prove two lemmas.

LEMMA 1.[96] *Let X denote any ordered set and suppose that every element $\xi \in X$ is associated with an open set U^ξ and a closed set E^ξ both contained in a metric space M and such that, for $\nu \in X$, we have*

$$E^\nu \subset U^\nu - \sum_{\xi<\nu} U^\xi;$$

then the set

$$H = \sum_{\xi \in X} E^\xi$$

is an F_σ.

Proof. For $\xi \in X$ and n a natural number denote by E_n^ξ the set of all elements $p \in E^\xi$ for which $S(p, 1/n) \subset U^\xi$. It can easily be shown that E_n^ξ is closed and $E^\xi = \sum_{n=1}^{\infty} E_n^\xi$. Put

$$H_n = \sum_{\xi \in X} E_n^\xi; \tag{124}$$

then $H = \sum_{n=1}^{\infty} H_n$. To prove the lemma it will be sufficient to show that the sets H_n $(n = 1, 2, \ldots)$ are closed.

Hence let $a \in H'_n$. We note that if $p \in E_n^\xi$, $q \in E_n^\eta$, and $\rho(p, q) < 1/n$, then $\xi = \eta$. For, from the definition of the sets E_n^ξ and E_n^η, we have $S(p, 1/n) \subset U^\xi$, $S(q, 1/n) \subset U^\eta$; if $\xi < \eta$ we have $E^\eta \subset U^\eta - U^\xi$ and so $E^\eta \,.\, U^\xi = 0$. Since $q \in E_n^\eta$ and $S(p, 1/n) \subset U^\xi$ it follows that $q \notin S(p, 1/n)$ and so $\rho(p, q) \geqslant 1/n$ contrary to assumption. Similarly we cannot have $\eta < \xi$; consequently $\xi = \eta$.

It follows from the above that elements p of the set H_n for which $\rho(a, p) < 1/2n$ belong to the same term in the sum (124) and therefore to the same closed set, say $E_n^{\xi_0}$. Hence if $a \in H'_n$, then $a \in E_n^{\xi_0}$ and so $a \in H_n$. The set H_n is therefore closed and Lemma 1 is proved.

LEMMA 2. *Let* $\{U^\xi\}$, $\xi < \phi$, *denote a transfinite sequence* (of type ϕ)[97] *of open sets and* $\{E^\xi\}$, $\xi < \phi$, *a transfinite sequence of sets* $P^\alpha(Q^\alpha)$, *all the sets being contained in a metric space* M, *where*

$$E^\nu \subset U^\nu - \sum_{\xi<\nu} U^\xi, \qquad \nu < \phi.$$

Then

$$H = \sum_{\nu<\phi} E^\nu$$

is also a set $P^\alpha(Q^\alpha)$.

Proof. We first prove the result for sets P^2. If E^ν is a P^2, i.e., an F_σ, then $E^\nu = \sum_{k=1}^{\infty} E_k^\nu$, where $E_k^\nu (k = 1, 2, \ldots)$ is a closed set. From the hypothesis of Lemma 2,

$$E_k^\nu \subset U^\nu - \sum_{\xi<\nu} U^\xi \qquad \text{for } \nu < \phi;$$

hence, by Lemma 1, the set

$$\sum_{\nu<\phi} E_k^\nu, \qquad k = 1, 2, \ldots,$$

is an F_σ. So the set

$$\sum_{\nu<\phi} E^\nu = \sum_{\nu<\phi} \sum_{k=1}^{\infty} E_k^\nu = \sum_{k=1}^{\infty} \sum_{\nu<\phi} E_k^\nu$$

is also an F_σ, i.e., a P^2.

Let $E^\nu (\nu < \phi)$ be a set Q^2. Then the set

$$T^\nu = (U^\nu - \sum_{\xi<\nu} U^\xi) - E^\nu$$

is a P^2 and so, since Lemma 2 holds for sets P^2, it follows that the set

$\sum_{\nu<\phi} E^\nu$ is a P^2. It remains to show that

$$\sum_{\nu<\phi} E^\nu = \sum_{\nu<\phi} U^\nu - \sum_{\nu<\phi} T^\nu. \tag{125}$$

Suppose that $p \in \sum_{\nu<\phi} E^\nu$; then there exists an ordinal number ν such that $p \in E^\nu$. From the definition of the sets T^ν, $p \notin T^\nu$; but by hypothesis $p \in U^\nu - \sum_{\xi<\nu} U^\xi$; hence $p \notin U^\xi$ for $\xi < \nu$ and so certainly $p \notin T^\xi$ for $\xi < \nu$. But from $p \in U^\nu$ it follows that

$$p \notin U^\eta - \sum_{\xi<\eta} U^\xi \text{ for } \eta > \nu;$$

this gives $p \notin T^\eta$ for $\nu < \eta < \phi$. Hence $p \notin \sum_{\eta<\phi} T^\eta$ and so

$$p \in U^\nu - \sum_{\eta<\phi} T^\eta.$$

The left-hand side of (125) is therefore contained in the right-hand side.

On the other hand if

$$p \in \sum_{\nu<\phi} U^\nu - \sum_{\nu<\phi} T^\nu$$

then $p \in \sum_{\nu<\phi} U^\nu$ and so there exists a least ordinal number $\mu < \phi$ such that $p \in U^\mu$. Hence

$$p \in U^\eta - \sum_{\xi<\eta} U^\xi$$

and, since $p \notin T^\mu$, we must have $p \in E^\mu$. This gives (2).

Since the set $\sum_{\nu<\phi} U^\nu$ is open and the set $\sum_{\nu<\phi} T^\nu$ is a P^2 it follows from (125) that the set $\sum_{\nu<\phi} E^\nu$ is a Q^2. This proves Lemma 2 for $\alpha = 2$.

Next let β be an ordinal number such that $2 < \beta < \Omega$ and suppose that Lemma 2 holds for all ordinal numbers α where $1 < \alpha < \beta$ (this is true for $\beta = 3$).

Let $\{E^\xi\}$, $\xi < \phi$, denote a transfinite sequence of sets P^β such that

$$E^\nu \subset U^\nu - \sum_{\xi<\nu} U^\xi \qquad \text{for } \nu < \phi.$$

Since $\beta < \Omega$ the set of all ordinal numbers α, where $1 < \alpha < \beta$, is at most countable; so there exists an infinite sequence $\alpha_1, \alpha_2, \ldots$ in which each of the terms is repeated a countable number of times. A set E which is a P^β may obviously be expressed in the form $E = E_1 + E_2 + \ldots$, where E_n is a P^{α_n} (perhaps the null set) for $n = 1, 2, \ldots$; we may therefore write

$$E^\xi = \sum_{n=1}^{\infty} E_n^\xi \qquad \text{for } \xi < \phi,$$

where E_n^ξ is a P^{α_n} for $\xi < \phi$, and $n = 1, 2, \ldots$. Since Lemma 2 is assumed to hold for $\alpha < \beta$ the set $\sum_{\xi<\phi} E_n^\xi$ is a P^{α_n} and so the set

$$\sum_{\xi<\phi} E^\xi = \sum_{\xi<\phi} \sum_{n=1}^{\infty} E_n^\xi = \sum_{n=1}^{\infty} \sum_{\xi<\phi} E_n^\xi$$

is a set P^β. Hence Lemma 2 holds for sets P^β.

Finally, let $\{E^\xi\}$, $\xi < \phi$, denote a transfinite sequence of sets Q^β such that

$$E^\nu \subset U^\nu - \sum_{\xi<\nu} U^\xi \qquad \text{for } \nu < \phi.$$

Proceeding as in the case of sets Q^2 it can be proved that Lemma 2 holds for sets Q^β. Hence Lemma 2 holds for $\alpha = \beta$. It follows by transfinite induction that Lemma 2 holds for all ordinal numbers α, where $2 \leqslant \alpha < \Omega$.

Proof of Theorem 87. Let α be an ordinal number such that $1 < \alpha < \Omega$ and let E be a set contained in a metric space M which is locally a $P^\alpha(Q^\alpha)$. Let $\{p_\xi\}$, $\xi < \phi$, be a transfinite sequence of ordinal type ϕ consisting of all the elements of E. Thus for every ordinal number $\nu < \phi$ there exists an open set U^ν such that $p_\nu \in U^\nu$ and the set $E \,.\, U^\nu$ is a $P^\alpha(Q^\alpha)$. The set

$$E \,.\, U^\nu - \sum_{\xi<\nu} U^\xi$$

is therefore a $P^\alpha(Q^\alpha)$ also and so, by Lemma 2, since

$$E \,.\, U^\nu - \sum_{\xi<\nu} U^\xi \subset U^\nu - \sum_{\xi<\nu} U^\xi,$$

the set

$$H = \sum_{\nu<\phi} (E \,.\, U^\nu - \sum_{\xi<\nu} U^\xi)$$

is a $P^\alpha(Q^\alpha)$. But $H = E$; for if $p \in E$ then, for some $\nu < \phi$, $p = p_\nu \in U^\nu$. Let μ denote the least ordinal number less than or equal to ν such that $p \in U^\mu$; then $p \in E \,.\, U^\mu - \sum_{\xi<\mu} U^\xi$ and so $p \in H$. Hence $E \subset H$ and from the definition of H we have $H \subset E$. Consequently $H = E$. The set E is therefore a $P^\alpha(Q^\alpha)$ and this proves Theorem 87.[98]

In connection with Theorem 87 we note that it was proved by Zarankiewicz[99] that in complete spaces (which are discussed in Chapter VII) a set which is both an F_σ and a G_δ contains a point at which it is closed. This result is not true in some metric spaces. There exists, for instance, a linear set which is both a P^α and a Q^α for every ordinal α, where $2 < \alpha < \Omega$, but which is neither a P^ξ nor a Q^ξ ($\xi < \alpha$) at any one of its points.

71. Sets locally of the first category.

A set is called of the first category if it is the sum of a countable aggregate of nowhere-dense sets. A set which is not of the first category is said to be of the second category.

LEMMA. *Let* $\{U^\xi\}$, $\xi < \phi$, *denote a transfinite sequence (of type* ϕ*) of open sets contained in a metric space* M *and* $\{E^\xi\}$, $\xi < \phi$, *a transfinite sequence of sets which are nowhere-dense on* M *where*

$$E^\nu \subset U^\nu - \sum_{\xi<\nu} U^\xi \qquad \textit{for } \nu < \phi;$$

then the set

$$H = \sum_{\nu<\phi} E^\nu$$

is nowhere-dense.

Proof. Suppose that H is not nowhere-dense. Then there exists an open sphere S such that $S \subset \bar{H}$. Hence S contains points of H and, since $H \subset \sum_{\nu<\phi} U^\nu$, there exist ordinal numbers $\nu < \phi$ such that $S \,.\, U^\nu \neq 0$; let β be the smallest of all such numbers ν. Then $S \,.\, U^\beta \neq 0$ and $S \,.\, U^\xi = 0$ for $\xi < \beta$. The set $S \,.\, U^\beta$ is open, non-empty, and so contains a sphere, say S_1. Since $S_1 \subset S$ we have $S_1 \subset \bar{H}$; the set $S_1 \,.\, H$ is thus everywhere-dense on S_1. But from $S_1 \subset S \,.\, U^\beta \subset U^\beta$ and $S_1 \,.\, U^\xi = 0$ for $\xi < \beta$ we have $S_1 \,.\, E^\xi = 0$ for $\xi < \beta$; from $E^\nu \,.\, U^\beta = 0$ for $\nu > \beta$ we have $S_1 \,.\, E^\xi = 0$ for $\xi > \beta$. Consequently $S_1 \,.\, H = S_1 \sum_{\nu<\phi} E^\nu = S_1 \,.\, E^\beta \subset E^\beta$ and so, since E^β is nowhere-dense, $S_1 \,.\, H$ is nowhere-dense; this is impossible. The lemma is therefore proved.

THEOREM 88. *In a metric space a set which is locally of the first category is of the first category.*[100]

Proof. Let E be a set contained in a metric space M and locally of the first category in M. Let $\{p_\xi\}$, $\xi < \phi$, denote a transfinite sequence of type ϕ consisting of all the elements of E. Corresponding to every ordinal number $\nu < \phi$ there exists an open set U^ν such that $p_\nu \in U^\nu$ and the set $E \,.\, U^\nu$ is of the first category; hence the set

$$E^\nu = E \,.\, U^\nu - \sum_{\xi<\nu} U^\xi$$

is of the first category for $\nu < \phi$. Consequently $E^\nu = \sum_{n=1}^{\infty} E_n^\nu$, where E_n^ν $(n = 1, 2, \ldots)$ is a nowhere-dense set. We thus have

$$E_n^\nu \subset E^\nu \subset U^\nu - \sum_{\xi<\nu} U^\xi \quad \text{for } \nu < \phi,\, n = 1, 2, \ldots$$

and so, by the preceding Lemma, the sets $\sum_{\nu<\phi} E_n^\nu (n = 1, 2, \ldots)$ are nowhere-dense. The set

$$\sum_{\nu<\phi} E^\nu = \sum_{\nu<\phi} \sum_{n=1}^{\infty} E_n^\nu = \sum_{n=1}^{\infty} \sum_{\nu<\phi} E_n^\nu$$

is therefore of the first category. But, as shown in the proof of Theorem 87, the set $E = \sum_{\nu<\phi} E^\nu$; hence E is of the first category.

COROLLARY 1. *For every set E contained in a metric space the set E_1 of all points of E at which E is of the first category is itself of the first category.*

Proof. Since $E_1 \subset E$ and E is of the first category at $p \in E_1$ it follows that E_1 is of the first category at p. The set E_1 is therefore locally of the first category and so is a set of the first category by Theorem 88.

COROLLARY 2. *If a set E contained in a metric space is of the second category then there exists an open sphere K such that E is of the second category at every point of K.*

Proof. Let E_1 be the set of all points of E at which E is of the first category. Then E_1 is of the first category by Lemma 1. The set $E - E_1$ cannot be nowhere-dense (for then the set $E = E_1 + (E - E_1)$ would be of the first category) and so there exists an open sphere $S \subset \overline{E - E_1}$. Let q be any point of S, and S_1 a sphere such that $q \in S_1 \subset S$. Since $q \in S_1 \subset S \subset \overline{E - E_1}$ there exists a point $p \in (E - E_1)S_1 \subset E \,.\, S_1$. If the set $E \,.\, S_1$ were of the first category then the set E would be of the first category at p; this is impossible since $p \in E - E_1$. The set $E \,.\, S_1$ is therefore of the second category. Thus E is of the second category in every sphere containing q and contained in S; hence E is of the second category at q. Since q is any point of S, Corollary 2 is established.

It can be proved with the aid of the axiom of choice that a set contained in a separable metric space which is of the second category at every point of an open sphere is the sum of two sets each with the same property.[101] With the aid of the continuum hypothesis it can be proved that such a set is the sum of a countable aggregate of sets each with the same property.[102] It is impossible however to prove, even with the aid of the continuum hypothesis, that a set contained in any metric space which is everywhere of the second category is the sum of two sets each of the second category everywhere.

A set which is of the first category on every perfect set is said to be *always of the first category.* It can be proved with the aid of the axiom of choice that there exist non-countable linear sets which are always of the first category. Without the continuum hypothesis we cannot prove the existence of linear sets of cardinal $\mathbf{c}$ which are always of the first category. Sets which are always of the first category are of some importance in the theory of functions of a real variable and an extensive literature has been developed in connection with these sets.

72. Oscillation of a function. Suppose that we have a function $f(p)$ defined in a set E contained in a metric space M with metric ρ and taking on values in the same or another metric space M_1 with metric ρ_1. Let p_0 denote an element of $\bar{E}$ (if $p_0 \notin E, f(p_0)$ may not be defined). For every $\epsilon > 0$ denote by $\omega(p_0, \epsilon)$ the upper bound of all the numbers $\rho_1(f(p), f(q))$, where p and q are any two points of the set $E \,.\, S(p_0, \epsilon)$. Clearly

$$\omega(p_0, \epsilon') \leqslant \omega(p_0, \epsilon) \qquad \text{for } \epsilon' < \epsilon;$$

hence

$$\omega(p_0) = \lim_{\epsilon \to 0} \omega(p_0, \epsilon) = \lim_{n \to \infty} \omega(p_0, 1/n) \tag{126}$$

exists and is a non-negative number finite or infinite (the latter in case $\omega(p_0, \epsilon) = +\infty$ for $\epsilon > 0$). This limit is called *the oscillation of the function f in the set E at the point p_0.*

THEOREM 89. *Let the function $f(p)$ be defined in a set E contained in a metric space; then the set E_1 of all the points of $\bar{E}$ at which the oscillation of f in E is $\geqslant \alpha$ (α a real number, finite or infinite) is closed.*

Proof. Let $f(p)$ denote a function defined in a set E, $\omega(p)$ the oscillation of f in E at $p \in \bar{E}$ and let α be a given real number. Denote by P the set of all points of $\bar{E}$ for which $\omega(p) \geqslant \alpha$. Let p_0 be a limit point of P and ϵ an arbitrary positive number. Since $p_0 \in P'$ there exists a point $p \in P$ such that $p \in S(p_0, \epsilon)$ and, since $S(p_0, \epsilon)$ is open, there exists a number $\eta > 0$ such that $S(p, \eta) \subset S(p_0, \epsilon)$. But $p \in P$; hence $\omega(p) \geqslant \alpha$ and so certainly $\omega(p, \eta) \geqslant \alpha$. It follows from the definition of the number $\omega(p, \eta)$ that there exist points p_1 and p_2 of the set $E \,.\, S(p, \eta)$ such that

$$\rho_1(f(p_1), f(p_2)) \geqslant \alpha - \epsilon; \tag{127}$$

since $S(p, \eta) \subset S(p_0, \epsilon)$, the points p_1 and p_2 belong to the set $E \,.\, S(p_0, \epsilon)$. Therefore (127) gives

$$\omega(p_0, \epsilon) \geqslant \alpha - \epsilon$$

and, since ϵ is arbitrary, (126) gives $\omega(p_0) \geqslant \alpha$, that is, $p_0 \in P$. The set P is therefore closed and Theorem 89 is established.

COROLLARY. *If $f(p)$ is a function defined in a set E contained in a metric space then the set of all points of the set $\bar{E}$ at which the oscillation of f in E is zero is a set G_δ.*

For let P denote the set of all points of $\bar{E}$ at which $\omega(p) = 0$ and P_n, $n = 1, 2, \ldots$, the sets of all elements of $\bar{E}$ at which $\omega(p) \geqslant 1/n$. The sets P_n are closed by Theorem 89; the set $S = P_1 + P_2 + \ldots$ is therefore an F_σ and so the set CS is a G_δ. Obviously $P = \bar{E} - S = \bar{E} \,.\, CS$. Since $\bar{E}$ is closed it is a G_δ by Theorem 82; hence P is the intersection of sets G_δ and so is a G_δ.

THEOREM 90. *A function $f(p)$ defined in a set E contained in a metric space is continuous in E at a point p_0 if and only if the oscillation of f in E at p_0 is equal to zero.*

Proof. Suppose that $f(p)$ is continuous in E at the point $p_0 \in E$. Let $\epsilon > 0$ be given. The sphere $S(f(p_0), \epsilon)$ is open and contains $f(p_0)$ and so, since $f(p)$ is continuous at p_0, there exists an open set U containing p_0 and such that $p \in U \,.\, E$ implies that $f(p) \in S(f(p_0), \epsilon)$. Since $p_0 \in U$ and U is open, there exists a number $\eta > 0$ such that $S(p_0, \eta) \subset U$. Thus, for $p \in E \,.\, S(p_0, \eta) \subset E \,.\, U$, we have $f(p) \in S(f(p_0), \epsilon)$ and so $\rho_1(f(p_0), f(p)) < \epsilon$. Hence, if $p_1 \in E \,.\, S(p_0, \eta)$ and $p_2 \in E \,.\, S(p_0, \eta)$, then $\rho_1(f(p_1), f(p_2)) \leqslant \rho_1(f(p_0), f(p_1)) + \rho_1(f(p_0), f(p_2)) < 2\epsilon$. Consequently $\omega(p_0, \eta) < 2\epsilon$ and so $\omega(p_0) < 2\epsilon$; since ϵ is arbitrary we have $\omega(p_0) = 0$. The condition of the theorem is therefore necessary.

Next suppose that for a given element $p_0 \in E$, $\omega(p_0) = 0$. Let V be an open set containing $f(p_0)$. There exists a number $\epsilon > 0$ such that $S(f(p_0), \epsilon) \subset V$. Since $\omega(p_0) = 0$ there exists a number $\eta > 0$ such that $\omega(p_0, \eta) < \epsilon$. It follows from the definition of the number $\omega(p_0, \eta)$ that, if $p \in U . E$, where $U = S(p_0, \eta)$, then $\rho_1(f(p_0), f(p)) \leqslant \omega(p_0, \eta) < \epsilon$ and so $f(p) \in S(f(p_0), \epsilon) \subset V$. But this proves that the function f is continuous in E at p_0. The sufficiency of the condition is thus established and Theorem 90 is proved.

Theorem 90 and the Corollary to Theorem 89 give

THEOREM 91. *If $f(p)$ is a function defined in a closed set E contained in a metric space then the set of all points at which $f(p)$ is continuous in E is a set G_δ.*

Example. Show that a given set Q of real numbers is the set of all points of continuity of some function of a real variable if and only if the set Q is a G_δ.

Proof. Because of Theorem 91 it is obviously sufficient to show that every linear set G_δ is the set of all points of continuity of some function of a real variable.

Let Q be a given linear set G_δ. Therefore $Q = G_1 . G_2 . G_3 \ldots$ where each G_n is an open set. It may be assumed that G_1 is the set of all real numbers and $G_n \supset G_{n+1}$. We therefore write

$$G_1 = Q + (G_1 - G_2) + (G_2 - G_3) + \ldots + (G_n - G_{n+1}) + \ldots,$$

all the terms on the right being disjoint sets.

We next define a function $f(x)$ of the real variable x. Put $f(x) = 0$ for $x \in Q$; for $x \in G_n - G_{n+1}$ and for a given n put $f(x) = 1/n$ if x is rational and $f(x) = -1/n$ if x is irrational.

The proof that Q is the set of all points of continuity of the function $f(x)$ defined above is left to the reader.[103]

CHAPTER VII

COMPLETE SPACES

73. Complete spaces. An infinite sequence $p_1, p_2, \ldots$ of points of a metric space M with distance function ρ is said to satisfy *Cauchy's condition* or to be *a Cauchy sequence* if for every positive number ϵ there exists an index μ such that

$$(1) \qquad \rho(p_{n+k}, p_n) < \epsilon \qquad \text{for } n > \mu \text{ and } k = 1, 2, \ldots.$$

If an infinite sequence $p_1, p_2, \ldots$ of points of a metric space M possesses a limit, i.e. if

$$\lim_{n\to\infty} p_n = p_0 \in M,$$

then

$$\lim_{n\to\infty} \rho(p_n, p_0) = 0$$

and so, for $\epsilon > 0$, there exists an index μ such that $\rho(p_n, p_0) < \epsilon/2$ for $n > \mu$. Thus for any natural number k, $\rho(p_{n+k}, p_0) < \epsilon/2$ for $n > \mu$ and therefore, by the triangle law, $\rho(p_{n+k}, p_n) \leqslant \epsilon$; but this is the inequality (1). Consequently, every infinite sequence of points of a metric space which has a limit in this space satisfies Cauchy's condition. The converse is not necessarily true. The open interval (0, 1) is a metric space M in which the sequence $\{1/n\}$ $(n = 1, 2, \ldots)$ satisfies Cauchy's condition but it does not possess a limit in M.

A space M is called *complete* if every infinite sequence of points of M which satisfies Cauchy's condition has a limit in M.

There exist complete spaces of any cardinal; for a metric space in which the distance between any two distinct points is equal to 1 is complete.

It is evident that *an infinite sequence of points of a complete space M possesses a limit in M if and only if it satisfies Cauchy's condition.*

It is well known that the linear space is complete. But, as seen above, the open interval is not a complete space although it is a homeomorphic image of the linear space. Hence *completeness is not a topological property of space*; it may be thought of as a metric property since it depends on the type of metric in the space.

*It is easy to prove that a *metric space which is the sum of a finite number of complete spaces is complete*; however, a metric space which is the sum of a countable aggregate of complete spaces (e.g., finite spaces) may not be complete. Thus completeness of a space is a *finitely additive* property but not a countably additive one. However, completeness of space is an absolute multiplicative property, i.e., *the intersection of any aggregate of complete spaces* (where the metrics are identical in the common parts) *is a complete space.**

THEOREM 92. *In order that a metric space M be complete it is necessary and sufficient that every totally bounded set contained in M be compact in M.*

Proof. Let M be a complete space with metric ρ and E a totally bounded set contained in M; let E_1 be an infinite subset of E. Then E_1 is also totally bounded and is therefore contained in the sum of a finite number of spheres of radii < 1. Consequently, at least one of these spheres, say S_1, contains infinitely many points of E_1. The set $E_1 . S_1$ is therefore infinite and, since it is totally bounded, we conclude that there exists a sphere S_2 of radius $< 1/2$ such that the set $E_1 . S_1 . S_2$ is infinite. Continuing this argument we obtain an infinite sequence $S_1, S_2, \ldots$ of spheres such that S_n has radius $< 1/n$ and the set $E_1 . S_1 . S_2 \ldots S_n$ is infinite (for $n = 1, 2, \ldots$). Let p_n denote a point of the set $E_1 . S_1 . S_2 \ldots S_n$; then $\rho(p_{n+k}, p_n) < 2/n$ for $k = 1, 2, \ldots$ (since $p_{n+k} \in S_n$ for $k = 1, 2, \ldots$). Consequently the sequence $p_1, p_2, \ldots$ satisfies Cauchy's condition. Since M is complete

$$\lim_{n \to \infty} p_n = p$$

exists in M (obviously $p \in \bar{S}_n$ for $n = 1, 2, \ldots$). But the sphere S_n contains infinitely many points of E_1; hence p is a limit point of E_1, that is, $E'_1 \neq 0$. The set E is therefore compact and the condition of the theorem is necessary.

Suppose now that the metric space M is not complete. Then there exists an infinite Cauchy sequence $p_1, p_2, \ldots$ of points of M which does not possess a limit in M. Thus for every $\epsilon > 0$ there exists a natural number μ such that $\rho(p_{n+k}, p_n) < \epsilon$ for $n > \mu$ and $k = 1, 2, \ldots$. This implies that the set $E = \{p_1, p_2, \ldots\}$ is contained in the sum of $\mu + 1$ spheres $S(p_i, \epsilon)$, $i = 1, 2, \ldots, \mu + 1$. The set E is therefore totally bounded but it is not compact since a limit point of an infinite Cauchy sequence would be obviously the limit of that sequence whereas the above sequence does not possess a limit in M. The condition of the theorem is therefore sufficient and Theorem 92 is proved.

COROLLARY 1. *A metric space which is compact-in-itself is complete.*

For if a metric space M is compact-in-itself then every subset of M is compact and so, by Theorem 92, the space M is complete.

The converse of Corollary 1 need not be true. For example, an infinite metric space in which the distance between two distinct elements is equal to 1 is complete but not compact.

COROLLARY 2. *A metric space is compact-in-itself if and only if it is complete and totally bounded.*

Proof. The necessity of the condition follows from Corollary 1 and Theorem 67, the sufficiency from Theorem 92.

We remark further that Niemytzki and Tychonoff[1] have proved that a metric space M is compact-in-itself if and only if every homeomorphic image of M is complete.

In connection with Theorem 92 we note that, in a complete space, a bounded set need not be compact. For example, the infinite space in which the distance between two distinct points is 1 is complete and bounded but it is not compact.

In m-dimensional Euclidean space ($m = 1, 2, \ldots$) bounded sets are compact (the Bolzano-Weierstrass theorem) and conversely. However, to prove that every compact linear set is bounded it is necessary to employ the axiom of choice.

A metric space M is said to satisfy *the condition of Ascola* if the intersection of every descending sequence of closed spheres contained in M with radii tending to zero consists of only one point.

THEOREM 93. *A metric space is complete if and only if it satisfies the condition of Ascola.*

Proof. The necessity of the condition follows from Theorem 92. Assume that the metric space M satisfies the condition of Ascola. Let $p_1, p_2, \ldots$ be an infinite sequence of points of M which satisfies Cauchy's condition. Then for every natural number k there exists a least natural number n_k such that

$$\rho(p_{n_k}, p_n) < 1/2^k \qquad \text{for } n \geqslant n_k. \tag{2}$$

Put

$$S_k = \bar{S}(p_{n_k}, 1/2^{k-1}) \qquad \text{for } k = 1, 2, \ldots.$$

If $q \in S_{k+1}$ then $\rho(q, p_{n_{k+1}}) \leqslant 1/2^k$ and so

$$\rho(q, p_{n_k}) \leqslant \rho(q, p_{n_{k+1}}) + \rho(p_{n_{k+1}}, p_{n_k}) < 1/2^k + 1/2^k = 1/2^{k-1}$$

by (2). Consequently $q \in S_k$ and, since q is any element of S_{k+1}, we have $S_{k+1} \subset S_k$. Since M satisfies the condition of Ascola there exists a point $p \in M$ such that $p \in S_k$ for $k = 1, 2, \ldots$. Hence $\rho(p, p_{n_k}) \leqslant 1/2^{k-1}$ and so, for $n \geqslant n_k$, (2) gives

$$\rho(p, p_n) \leqslant \rho(p, p_{n_k}) + \rho(p_{n_k}, p_n) < 1/2^{k-1} + 1/2^k < 1/2^{k-2}$$

$$\text{for } k = 1, 2, \ldots.$$

Consequently $\lim_{n\to\infty} p_n = p$. M is therefore complete and Theorem 93 is established.

Note that in a complete space the intersection of a descending sequence of closed spheres may be empty. For example, let $P = \{p_1, p_2, \ldots\}$ denote a countable set in which ρ is defined by the relation $\rho(p_k, p_l) = 1 + 1/(k + l)$

for $k \neq l$. Evidently P is a complete metric space. Let S_n denote a closed sphere with centre p_n and radius $1 + 1/2n$ for $n = 1, 2, \ldots$. If $p_k \in S_n$ then $\rho(p_k, p_n) \leqslant 1 + 1/2n$; but $\rho(p_k, p_n) = 1 + 1/(k + n)$. Hence, from $1 + 1/(k + n) \leqslant 1 + 1/2n$, we obtain $k \geqslant n$; this gives $S_n = \{p_n, p_{n+1}, \ldots\}$ and so $S_1 \supset S_2 \supset S_3 \ldots$. Clearly $S_1 . S_2 . S_3 \ldots = 0$.

THEOREM 94. *The intersection of an infinite descending sequence of non-empty closed sets with diameters tending to zero and contained in a complete space consists of only one point.*[2]

Proof. Let $E_1 \supset E_2 \supset \ldots$ denote an infinite descending sequence of non empty closed sets contained in a complete space M where

$$\lim_{n\to\infty} \delta(E_n) = 0.$$

Since $E_n \neq 0$ for $n = 1, 2, \ldots$ there exists, for every natural n, a point $p_n \in E_n$. Now $E_1 \supset E_2 \supset \ldots$, $p_{n+k} \in E_n$ for all natural numbers n and k, and so $\rho(p_{n+k}, p_n) \leqslant \delta(E_n)$; since $\lim_{n\to\infty} \delta(E_n) = 0$ there exists, for every $\epsilon > 0$, an index μ such that $\delta(E_n) < \epsilon$ for $n > \mu$; this gives the inequality (1). Consequently, since M is complete, we have

$$\lim_{n\to\infty} p_n = p \in M.$$

But $p_{n+k} \in E_n$ for $k = 1, 2, \ldots$ and E_n is closed; hence $p \in E_n$ for $n = 1, 2, \ldots$ and so $p \in E_1 . E_2 . E_3 \ldots$. If also $q \in E_1 . E_2 \ldots$ then $\rho(p, q) \leqslant \delta(E_n)$ for $n = 1, 2, \ldots$ and, since $\lim_{n\to\infty} \delta(E_n) = 0$, we have $\rho(p, q) = 0$ or $p = q$. Theorem 94 is therefore established.

We note here that in a complete space the intersection of a descending sequence of non-empty closed sets may be empty. For instance, in the countable space $M = \{a_1, a_2, \ldots\}$ where $\rho(a_k, a_l) = 1$ for $k \neq l$, the sequence $\{E_n\}$, where $E_n = \{a_n, a_{n+1} \ldots\}$, is such a sequence.

THEOREM 95. *Hilbert space is complete.*

Proof. Let $p_1, p_2, \ldots$ denote an infinite Cauchy sequence of points in the Hilbert space H. Let $p_n = (x_1^{(n)}, x_2^{(n)}, \ldots)$ and let ϵ denote a positive number; then there exists an index μ for which inequality (1) holds. Since $p_n \in H$ for $n = 1, 2, \ldots$ we have, for all natural numbers n and k,

$$\rho(p_{n+k}, p_n) = \left(\sum_{i=1}^{\infty} (x_i^{(n+k)} - x_i^{(n)})^2\right)^{\frac{1}{2}} \tag{3}$$

and so from (1)

$$|x_i^{(n+k)} - x_i^{(n)}| < \epsilon \qquad \text{for } n > \mu,\ k = 1, 2, \ldots,\ i = 1, 2, \ldots.$$

Consequently each of the sequences $x_i^{(1)}, x_i^{(2)}, \ldots$ $(i = 1, 2, \ldots)$ is convergent. Put

$$(4) \qquad x_i = \lim_{n\to\infty} x_i^{(n)} \qquad \text{for } i = 1, 2, \ldots,$$

and

$$(5) \qquad p = (x_1, x_2, \ldots).$$

From (3) and (1) we obtain

$$\left(\sum_{i=1}^{m} (x_i^{(n+k)} - x_i^{(n)})^2\right)^{\frac{1}{2}} < \epsilon \qquad \text{for } n > \mu, k = 1, 2, \ldots, m = 1, 2, \ldots;$$

so, from (4), as $k \to \infty$,

$$(6) \qquad \left(\sum_{i=1}^{m} (x_i - x_i^{(n)})^2\right)^{\frac{1}{2}} \leqslant \epsilon \qquad \text{for } n > \mu, m = 1, 2, \ldots.$$

Hence the series $\sum_{i=1}^{\infty} (x_i - x_i^{(n)})^2$ is convergent for $n > \mu$. But the series $\sum_{i=1}^{\infty} (x_i^{(n)})^2$ is also convergent for $n = 1, 2, \ldots$; hence, from the inequality[3]

$$x_i^2 \leqslant 2(x_i - x_i^{(n)})^2 + 2(x_i^{(n)})^2,$$

we may conclude that the series $\sum_{i=1}^{\infty} x_i^2$ is convergent and therefore $p \in H$.

Letting m tend to infinity in (6) we obtain

$$(7) \qquad \rho(p, p^{(n)}) \leqslant \epsilon \qquad \text{for } n > \mu;$$

since ϵ is arbitrary, (7) gives

$$\lim_{n\to\infty} \rho(p, p^{(n)}) = 0 \text{ or } \lim_{n\to\infty} p_n = p.$$

Hence every Cauchy sequence contained in H possesses a limit in H. Consequently H is complete. This proves Theorem 95.

Obviously, *a closed subset of a complete space is complete.* Since n-dimensional Euclidean space ($n = 1, 2, \ldots$) is isometric with a closed subset of Hilbert space it follows from Theorem 95 that *Euclidean space of any number of dimensions is complete.* This follows also from that fact that the metric product of two, and therefore of any finite number, of complete spaces is a complete space.

Examples

1. Prove that Fréchet's space E_ω is complete.

Proof. Let $p_n = (x_1^{(n)}, x_2^{(n)}, \ldots)$, $n = 1, 2, \ldots$, be an infinite sequence of points of E_ω which satisfies Cauchy's condition; then there exists, for $\epsilon > 0$ and every natural number i, an index μ (from the definition of distance in E_ω, §59) such that

$$\frac{|x_i^{(n+k)} - x_i^{(n)}|}{i!(1 + |x_i^{(n+k)} - x_i^{(n)}|)} < \epsilon \qquad \text{for } n > \mu, k = 1, 2, \ldots.$$

Thus, for $\epsilon < 1/2$, we have $|x_i^{(n+k)} - x_i^{(n)}| < 2i!\epsilon$, for $n > \mu$, $k = 1, 2, \ldots$; hence there exists a finite limit

$$x_i = \lim_{n\to\infty} x_i^{(n)} \qquad \text{for } i = 1, 2, \ldots.$$

Setting $p = (x_1, x_2, \ldots)$ we have

$$\lim_{\to\infty} p^{(n)} = p \qquad \text{in } E_\omega.$$

2. Show that the space (C) of § **67** is complete.

Proof. If the sequence f_n $(n = 1, 2, \ldots)$ of functions belonging to (C) satisfies Cauchy's condition then it is uniformly convergent in the interval $0 \leqslant x \leqslant 1$ to a continuous function which therefore belongs to (C). This is a consequence of the definition of distance in (C).

3. Show that the 0-dimensional Baire space is complete.

Proof. The 0-dimensional Baire space is a closed subset of the space E_ω (§ **60**); since the latter is complete (example 1) the first is also complete.

4. Show that a metric product of two non-empty metric spaces is complete if and only if each of the factor spaces is complete.

Proof. Let A and B be two non-empty complete metric spaces with distance functions ρ_1 and ρ_2 respectively; let (a_n, b_n), $n = 1, 2, \ldots$, be an infinite Cauchy sequence of points belonging to the metric space $A \times B$. Given $\epsilon > 0$, there exists an index μ such that $\rho((a_{n+k}, b_{n+k}), (a_n, b_n)) < \epsilon$ for $n > \mu$ and $k = 1, 2, \ldots$. Hence, from the definition of distance in the space $A \times B$ (§ **60**), we have $\rho_1(a_{n+k}, a_n) < \epsilon$ and $\rho_2(b_{n+k}, b_n) < \epsilon$ for $n > \mu$, $k = 1, 2, \ldots$; thus the sequences $a_1, a_2, \ldots$ and $b_1, b_2, \ldots$ satisfy Cauchy's condition and, since A and B are complete, there exist limits

$$a = \lim_{n\to\infty} a_n \qquad \text{in } A$$

and

$$b = \lim_{n\to\infty} b_n \qquad \text{in } B.$$

Hence

$$(a, b) = \lim_{n\to\infty} (a_n, b_n) \qquad \text{in } A \times B.$$

Consequently $A \times B$ is complete and the condition is proved sufficient.

Next let $A \times B$ denote a complete space and let $a_1, a_2, \ldots$ be an infinite Cauchy sequence in A. Let b_0 be any point of B and ϵ a positive number. Then there exists a natural number μ such that $\rho_1(a_{n+k}, a_n) < \epsilon$ for $n > \mu$, $k = 1, 2, \ldots$; so $\rho((a_{n+k}, b_0), (a_n, b_0)) = \rho_1(a_{n+k}, a_n) < \epsilon$ for $n > \mu$ and $k = 1, 2, \ldots$. Therefore the infinite sequence (a_n, b_0), $n = 1, 2, \ldots$, of points of $A \times B$ satisfies Cauchy's condition and since $A \times B$ is complete the sequence has a limit

$$(a, b_0) = \lim_{n\to\infty} (a_n, b_0) \in A \times B.$$

This gives

$$\lim_{\to\infty} a_n = a \in A.$$

Thus every infinite Cauchy sequence contained in A has a limit in A and so A is complete. Similarly we prove that B is complete. The condition is therefore necessary.

74. The complete space containing a given metric space.

THEOREM 96. *Every metric space is a subset of a complete space.*

Proof. Let M denote a given metric space with distance function ρ. Divide all infinite sequences of points of M which satisfy Cauchy's condition into classes assigning two sequences $p_1, p_2, \ldots$ and $q_1, q_2, \ldots$ to the same class if and only if

$$\lim_{n\to\infty} \rho(p_n, q_n) = 0.$$

Let K denote the aggregate of all such classes. Let k_1 and k_2 be two different elements of the aggregate K. Let $p_1, p_2, \ldots$ be any sequence belonging to the class k_1 and $q_1, q_2, \ldots$ any sequence belonging to the class k_2. Then there exists for every positive ϵ an index μ such that

$$\rho(p_{n+k}, p_n) < \epsilon \text{ and } \rho(q_{n+k}, q_n) < \epsilon \qquad \text{for } n > \mu,\ k = 1, 2, \ldots;$$

but from the triangle law

$$\rho(p_{n+k}, q_{n+k}) \leqslant \rho(p_{n+k}, p_n) + \rho(p_n, q_n) + \rho(q_n, q_{n+k})$$

and

$$\rho(p_n, q_n) \leqslant \rho(p_n, p_{n+k}) + \rho(p_{n+k}, q_{n+k}) + \rho(q_{n+k}, q_n).$$

Hence

$$-2\epsilon < \rho(p_{n+k}, q_{n+k}) - \rho(p_n, q_n) < 2\epsilon \qquad \text{for } n > \mu,\ k = 1, 2, \ldots;$$

since ϵ is arbitrary the sequence $\rho(p_n, q_n)$, $n = 1, 2, \ldots$, is convergent and since its terms are $\geqslant 0$, $\lim_{n\to\infty} \rho(p_n, q_n)$ is a finite real number $\geqslant 0$. Moreover, this limit cannot equal 0 for then the sequences $p_1, p_2, \ldots$ and $q_1, q_2, \ldots$ would be in the same class contrary to the assumption that k_1 and k_2 are different elements of the aggregate K. Consequently

$$\lim_{n\to\infty} \rho(p_n, q_n) > 0.$$

It is evident that this limit depends only on the classes k_1 and k_2 and not on the sequences $\{p_n\}$ and $\{q_n\}$ which are members of k_1 and k_2. For let p'_1, $p'_2, \ldots$ be any other sequence of the class k_1 and $q'_1, q'_2, \ldots$ any other sequence of the class k_2; then $\lim_{n\to\infty} \rho(p_n, p'_n) = 0$ and $\lim_{n\to\infty} \rho(q_n, q'_n) = 0$. But

$$\rho(p'_n, q'_n) \leqslant \rho(p'_n, p_n) + \rho(p_n, q_n) + \rho(q_n, q'_n),$$
$$\rho(p_n, q_n) \leqslant \rho(p_n, p'_n) + \rho(p'_n, q'_n) + \rho(q'_n, q_n);$$

so

$$-\rho(p'_n, p_n) - \rho(q_n, q'_n) \leqslant \rho(p'_n, q'_n) - \rho(p_n, q_n) \leqslant \rho(p'_n, p_n) + \rho(q'_n, q_n).$$

Therefore

$$\lim_{n\to\infty} \rho(p'_n, q'_n) = \lim_{n\to\infty} \rho(p_n, q_n)$$

since the limit on the right-hand side was shown to exist and

$$\lim_{n\to\infty} \rho(p'_n, p_n) = \lim_{n\to\infty} \rho(q'_n, q_n) = 0.$$

it follows that $\lim_{n\to\infty} \rho(p_n, q_n)$ depends on k_1 and k_2 only and we denote it by $r(k_1, k_2)$.

The function r defined in the set $K \times K$ satisfies all the distance axioms. It is only necessary to prove the triangle law. Let k_1, k_2, k_3, be three elements of the aggregate K, $\{p'_n\}$, $\{p''_n\}$, and $\{p'''_n\}$ three sequences belonging to the classes k_1, k_2, and k_3. From the definition of the function r we obtain

$$r(k_1, k_2) = \lim_{n\to\infty} \rho(p'_n, p''_n), r(k_2, k_3) = \lim_{n\to\infty} \rho(p''_n, p'''_n),$$

and

$$r(k_1, k_3) = \lim_{n\to\infty} \rho(p'_n, p'''_n).$$

But

$$\rho(p'_n, p'''_n) \leqslant \rho(p'_n, p''_n) + \rho(p'''_n, p''_n) \qquad \text{for } n = 1, 2, \ldots;$$

hence

$$r(k_1, k_3) \leqslant r(k_1, k_2) + r(k_2, k_3).$$

The function r may therefore be employed as the distance function in K. We next show that K is complete.

Let $k_1, k_2, \ldots$ denote an infinite Cauchy sequence of points of K; then for a given $\epsilon > 0$ there exists a natural number μ such that

$$r(k_m, k_n) < \epsilon \qquad \text{for } m > \mu, n > \mu. \tag{8}$$

Let $p_1^{(m)}, p_2^{(m)}, \ldots$ be one of the sequences of the class k_m. For m and n any two natural numbers $> \mu$, we have

$$\lim_{s\to\infty} \rho(p_s^{(m)}, p_s^{(\mu+1)}) = r(k_m, k_{\mu+1}) < \epsilon$$

and

$$\lim_{s\to\infty} \rho(p_s^{(\mu+1)}, p_s^{(n)}) = r(k_{\mu+1}, k_n) < \epsilon.$$

Consequently there exists a natural number σ such that

$$(9) \qquad \rho(p_s^{(m)}, p_s^{(\mu+1)}) < \epsilon \qquad \text{for } s > \sigma,$$

$$\rho(p_s^{(\mu+1)}, p_s^{(n)}) < \epsilon \qquad \text{for } s > \sigma.$$

Since the sequence $p_1^{\mu+1}, p_2^{\mu+1}, \ldots$ satisfies Cauchy's condition there exists for ϵ given above, a natural number $\nu > \sigma$ such that

$$(10) \qquad \rho(p_i^{(\mu+1)}, p_j^{(\mu+1)}) < \epsilon \qquad \text{for } i > \nu, j > \nu.$$

Let i and j be two natural numbers $> \nu$; since $\nu > \sigma$, (9) gives

$$(11) \qquad \rho(p_i^{(m)}, p_i^{(\mu+1)}) < \epsilon \text{ and } \rho(p_j^{(\mu+1)}, p_j^{(n)}) < \epsilon.$$

Then from the triangle law, (10), and (11),

$$\rho(p_i^{(m)}, p_j^{(n)}) \leqslant \rho(p_i^{(m)}, p_i^{(\mu+1)}) + \rho(p_i^{(\mu+1)}, p_j^{(\mu+1)}) + \rho(p_j^{(\mu+1)}, p_j^{(n)}) < 3\epsilon.$$

Consequently for every number $\epsilon > 0$ there exist indices μ and ν such that

$$(12) \qquad \rho(p_i^{(m)}, p_j^{(n)}) < 3\epsilon \qquad \text{for } m > \mu, n > \mu, i > \nu, j > \nu.$$

It follows from (12) that

$$\rho(p_m^{(m)}, p_n^{(n)}) < 3\epsilon \qquad \text{for } m > \mu + \nu \text{ and } n > \mu + \nu,$$

that is, the sequence $p_1^{(1)}, p_2^{(2)}, p_3^{(3)}, \ldots$ satisfies Cauchy's condition and therefore is a member of some element k of K. We then have

$$(13) \qquad r(k_m, k) = \lim_{n\to\infty} \rho(p_n^{(m)}, p_n^{(n)}).$$

Let ϵ be a given positive number and μ and ν two natural numbers for which (12) holds; let m be a natural number $> \mu$. Then for $n > \mu + \nu$ we have

$$\rho(p_n^{(m)}, p_n^{(n)}) < 3\epsilon$$

and so, from (13),

$$r(k_m, k) \leqslant 3\epsilon, \qquad m > \mu;$$

this gives

$$\lim_{m\to\infty} r(k_m, k) = 0, \text{ that is, } \lim_{m\to\infty} k_m = k.$$

Thus every Cauchy sequence of elements of K has a limit in K. Hence the space K is complete.

For every element p of the space M, denote by p' the sequence $p' = (p, p, \ldots)$; then p' is an element of K. It follows from the definition of the distance function r that

$$r(p', q') = \rho(p, q) \qquad \text{for } p \in M, q \in M.$$

The subset of K consisting of all the elements p' for $p \in M$ is therefore isometric with M. This proves Theorem 96.

The elements $(p, p, \ldots)$, where $p \in M$, are dense on K. For if $k \in K$, and $k = (p_1, p_2, \ldots)$ then, since $p_1, p_2, \ldots$ is a Cauchy sequence, there exists, for $\epsilon > 0$, an index μ such that $\rho(p_m, p_n) < \epsilon$ for $m > \mu$ and $n > \mu$. But

$$r(p'_m, k) = \lim_{n\to\infty} \rho(p_m, p_n);$$

hence $r(p'_m, k) \leqslant \epsilon$ for $m > \mu$ and so

$$\lim_{n\to\infty} r(p'_m, k) = 0, \text{ that is, } k = \lim_{n\to\infty} p'_m.$$

Thus every element of K is the limit of a sequence of elements $p' = (p, p, \ldots)$, where $p \in M$.

If, in particular, the space M is the set of all rational numbers then the above proof gives the Cantor theory of irrational numbers. It follows at once from the theory of irrational numbers that *the linear space* (i.e. the set of all real numbers) *is complete*.

75. Absolutely closed spaces. Complete topological spaces. A metric space M is called *absolutely closed* if it is closed in every space in which it can be embedded.

THEOREM 97. *A metric space is absolutely closed if and only if it is complete.*

Proof. Assume M is absolutely closed. There exists, by Theorem 96, a complete space P such that $M \subset P$ and, since M is absolutely closed, it is closed in P. Let $p_1, p_2, \ldots$ be a Cauchy sequence of points of M; since P is complete this sequence has a limit $p \in P$. But M is closed in P; hence $p \in M$. The space M is therefore complete.

Suppose conversely that M is complete and that it can be embedded in a metric space M_1. It may be supposed that $M \subset M_1$. Let p denote a limit point of M which belongs to M_1. By Theorem 53, there exists a sequence $p_1, p_2, \ldots$ of points of M such that

$$\lim_{n\to\infty} p_n = p;$$

hence $p_1, p_2, \ldots$ is a Cauchy sequence (§ 73) and so has a limit in M. This limit must be p since a sequence cannot have two different limits; consequently $p \in M$. The space M is therefore closed in M_1; this proves the theorem. From Theorems 67, 92, and 96 we obtain the following corollaries:

COROLLARY 1. *A metric space M is totally bounded if and only if there exists a complete space which contains M and in which M is compact.*

COROLLARY 2. *A metric space M is an absolute G_δ if and only if there exists a complete space K containing M and such that M is a G_δ in K.*

Proof. The necessity of the condition follows from Theorem 96. Suppose that the metric space M is a G_δ in the complete space $K \supset M$ and let M_1 be any metric space $\supset M$. Since M is a G_δ in K we may write $M = K - (F_1 + F_2 + \ldots)$, where $F_n (n = 1, 2, \ldots)$ is a closed set $\subset K$. Since K is complete it is absolutely closed by Theorem 97; hence it is closed in the space $M_1 + K$ in which a metric may be established by the method given in § **47**, with the retention of the metrics in M_1 and K. The sets F_n $(n = 1, 2, \ldots)$ are therefore closed in $M_1 + K$. Moreover, the sets F_n and the set K are closed in the space $M_1 \subset M_1 + K$. Consequently the set

$$M = K(M_1 - F_1)(M_1 - F_2)(M_1 - F_3) \ldots$$

is a G_δ with respect to M_1. This proves Corollary 2.

Examples

1. Prove that every infinite subset of a set E contained in a metric space contains an infinite Cauchy sequence of different points if and only if the set E is totally bounded.[4]

2. Prove that a metric space is separable if and only if it is homeomorphic with a totally bounded metric space.[5]

3. Prove that every metric space is a (1, 1) continuous image of a complete space.

Proof. Let M denote a metric space with distance function ρ. Let ρ_1 be a new distance in M such that $\rho_1 = 1$ for any two different points of M; let the new metric space thus obtained be denoted by M_1. The space M_1 is obviously complete and, putting $f(p) = p$, we obtain a (1, 1) continuous mapping of M_1 on M.

4. Let E be the set of all points (x,y), $0 \leqslant x \leqslant 1$, $0 \leqslant y \leqslant 1$. Show that if $\rho((x_1, y), (x_2, y)) = |x_1 - x_2|$ for $0 \leqslant x_1 \leqslant 1$, $0 \leqslant x_2 \leqslant 1$, $0 \leqslant y \leqslant 1$, and $\rho((x_1, y_1), (x_2, y_2)) = 1$ for $0 \leqslant x_1 \leqslant 1$, $0 \leqslant x_2 \leqslant 1$, $0 \leqslant y_1 \leqslant 1$, $0 \leqslant y_2 \leqslant 1$, $y_1 \neq y_2$, then E is complete.

A metric space is called *topologically complete* if it possesses a homeomorphic image which is complete.

We state without the proofs (which are rather complicated) the following two theorems.

A metric space M is topologically complete[6] *if and only if there exists a complete metric space which is a G_δ and contains M;* (or else, *if and only if M is an absolute G_δ*, i.e., a G_δ in every metric space containing M).

A metric space M is topologically complete if and only if with every point $p \in M$ and every natural number n can be associated an open set $U_n(p)$ containing

p and satisfying the following condition;[7] *if $p_1, p_2, \ldots$ is an infinite sequence of points of M such that*

$$U_1(p_1) \,.\, U_2(p_2) \ldots U_n(p_n) \neq 0, \qquad n = 1, 2, \ldots,$$

then the infinite sequence of sets $U_n(p_n)$ tends to a point of M (i.e., for every $\epsilon > 0$ there exists an index μ such that $U_n(p) \subset S(p, 1/n)$ for $n > \mu$).

From these theorems it may be deduced that *every inner mapping* (i.e., a continuous mapping which maps open sets into open sets) *transforms a topologically complete space into a topologically complete space.*[8]

76. The category of a complete space.

THEOREM 98. *A (non-empty) complete space is of the second category on itself.*

Proof. Let T denote a non-empty complete space, and E a subset of T which is the first category on T. It will be sufficient to show that $T - E \neq 0$.

Since E is of the first category on T, we may write $E = E_1 + E_2 + \ldots$ where the sets E_n $(n = 1, 2, \ldots)$ are nowhere-dense on T.

Since $T \neq 0$ there exists a point $p \in T$. Put $U = S(p, 1)$; since E_1 is nowhere-dense on T there exists an open set $V \subset U$ such that $V \neq 0$ and $E_1 \,.\, V = 0$. Since $V \neq 0$ and is open there exists a point $p_1 \in V$ and a positive number $r_1 < 1$ such that $U_1 = S(p_1, r_1) \subset \bar{U}_1 \subset V$ (§**48**). Now $p_1 \in U_1$, U_1 is open, and E_2 is nowhere-dense on T; hence there exists an open set $V_1 \subset U_1$ such that $V_1 \neq 0$ and $V_1 \,.\, E_2 = 0$. Again, since $V_1 \neq 0$ and is open there exists an element $p_2 \in V_1$ and a positive number $r_2 < \frac{1}{2}$ such that $U_2 = S(p_2, r_2) \subset \bar{U}_2 \subset V$. Continuing this argument we obtain an infinite sequence $p_1, p_2, p_3, \ldots$ of points of the space T and an infinite sequence $U_1, U_2, \ldots$ of open sets contained in T such that $U_n \supset \bar{U}_{n+1}$, $p_n \in U_n$, and $U_n \,.\, E_n = 0$, where $\delta(U_n) < 2r_n < 2/n$ for $n = 1, 2, \ldots.$ Consequently $\rho(p_{n+k}, p_n) < 2/n$ for $n = 1, 2, \ldots$, $k = 1, 2, \ldots$; since T is complete it follows that there exists a point $p_0 \in T$ such that

$$\lim_{n \to \infty} p_n = p_0.$$

But $p_{n+k} \in U_{n+k} \subset \bar{U}_{n+1}$ for $k = 1, 2, \ldots$, and $\bar{U}_{n+1}$ is closed; hence $p_0 \in \bar{U}_{n+1} \subset U_n$ for $n = 1, 2, \ldots$ and so, since $U_n \,.\, E_n = 0$, we have $p_0 \notin E_n$ for $n = 1, 2, \ldots.$ Consequently $p_0 \notin E$ and since $p_0 \in T$ we have $p_0 \in T - E$; hence $T - E \neq 0$. Theorem 98 is therefore proved.

If T denotes a linear space then T is complete (§ **74**) and so satisfies Theorem 98. In a linear space every set consisting of a single element is nowhere-dense; hence every countable set, in particular, the set E of all rational numbers, is of the first category. If the set $T - E$ were of the first category on T then the set $T = E + (T - E)$ would be of the first category on T contrary to Theorem 98. We thus have

COROLLARY 1. *The set of all irrational numbers is of the second category on the set of all real numbers.*

Assume that the set N of all irrational numbers is an F_σ in the set of all real numbers; then $N = E_1 + E_2 + \ldots$, where each E_n is closed. Let n be a given natural number and δ any open interval. Then there exists a rational number $r \in \delta$ and, since $r \notin N$, $r \notin E_n$. Since E_n is closed there exists an open interval $d \subset \delta$ such that $r \in d$ and $E_n \cdot d = 0$. The set E_n is therefore nowhere-dense on the set of real numbers for $n = 1, 2, \ldots$; this implies that the set $N = E_1 + E_2 + \ldots$ is of the first category on the straight line contrary to Corollary 1. Consequently N is not an F_σ. Thus

COROLLARY 2. *The set of all irrational numbers is not an F_σ.*

On the other hand, since it is the complement of the set of all rational numbers the set of all irrational numbers is a G_δ. This gives

COROLLARY 3. *There exists in the set of all real numbers a set G_δ which is not an F_σ.*

Also, the set of all rational numbers is not a G_δ (since its complement is not an F_σ). We thus obtain

COROLLARY 4. *There exists a set of real numbers which is an F_σ but not a G_δ.*

77. Continuity extended to a set G_δ. Let $f(p)$ denote a function defined and continuous in a set E contained in a metric space M with distance function ρ and taking on values in a complete metric space N with distance function ρ_1. Denote by T the set of all those elements of $\bar{E}$ at which the oscillation of f in E is equal to zero. Then T is a G_δ by the corollary to Theorem 89; since f is continuous in E it follows from Theorem 90 that $E \subset T$. Let p_0 be a given element of the set $T - E$; since $T \subset \bar{E}$, p_0 is a limit point of E and so there exists an infinite sequence $p_1, p_2, \ldots$ of points of E such that

$$\lim_{n \to \infty} p_n = p_0.$$

Let ϵ denote an arbitrary positive number. Since $\omega(p_0) = 0$ there exists, according to the definition of oscillation (§ **70**), a number $r > 0$ such that for any two points p and q of the set $E \cdot S(p_0, r)$ we have the inequality

$$\rho_1(f(p), f(q)) < \epsilon.$$

But, since $\lim\limits_{n \to \infty} p_n = p_0$, there exists an index μ such that $p_n \in S(p_0, r)$ for $n > \mu$. Consequently

$$\rho_1(f(p_{n+k}), f(p_n)) < \epsilon \qquad \text{for } n > \mu,\ k = 1, 2, \ldots;$$

since N is complete, the sequence $\{f(p_n)\}$ has a limit, say b, in N.

The point b of the set N depends solely on the point p_0 and not on the sequence $p_1, p_2, \ldots$ of points of E which has p_0 as its limit. For let $q_1, q_2, \ldots$ be an infinite sequence of points of E such that

$$\lim_{n\to\infty} q_n = p_0.$$

As shown above, the sequence $\{f(q_n)\}$ has a limit in the set N; denote this limit by c. Since $\lim_{n\to\infty} p_n = p_0$ and $\lim_{n\to\infty} q_n = p_0$ the sequence $p_1, q_1, p_2, q_2, \ldots$ also has the limit p_0 (§ **35**); hence the sequence $f(p_1), f(q_1), f(p_2), f(q_2), \ldots$ has a limit which we denote by g. But the sequences $f(p_1), f(p_2), \ldots$ and $f(q_1), f(q_2), \ldots$ are subsequences of a sequence with limit g; they must therefore themselves have the limit g. Consequently $b = g = c$.

This proves that to every point p_0 of the set $T - E$ corresponds a uniquely determined point $b = \phi(p_0)$ such that if $p_n \in E$ and $\lim_{n\to\infty} p_n = p_0$ then

$$\lim_{n\to\infty} f(p_n) = \phi(p_0).$$

The function $\phi(p)$ is thus defined in the set $T - E$. Put $\phi(p) = f(p)$ for $p \in E$; this defines the function $\phi(p)$ in the whole set T. We next show that the function $\phi(p)$ is continuous in T.

Let $p_0 \in T$ and $\epsilon > 0$ be given. Denote by $\omega(p_0, \epsilon)$ the upper bound of the set G of all numbers $\rho_1(f(p), f(q))$ where p and q are any two points of the set $E \,.\, S(p_0, \epsilon)$ and by $\omega_1(p_0, \epsilon)$ the upper bound of the set H of all numbers $\rho_1(\phi(p), \phi(q))$ where p and q are any two points of the set $T \,.\, S(p_0, \epsilon)$.

Let h be any number of the set H; then there exist elements p and q of the set $T \,.\, S(p_0, \epsilon)$ such that

$$h = \rho_1(\phi(p), \phi(q)). \tag{14}$$

Since $p \in T \subset \bar{E}$ there exists an infinite sequence $p_1, p_2, \ldots$ of points of E such that

$$\lim_{n\to\infty} p_n = p;$$

since $p \in S(p_0, \epsilon)$, $p_n \in S(p_0, \epsilon)$ for $n > \mu$. Similarly, since $q \in T \,.\, S(p_0, \epsilon)$, there exists an infinite sequence $q_1, q_2, \ldots$ of points of E such that

$$\lim_{\to\infty} q_n = q \text{ and } q_n \in S(p_0, \epsilon) \qquad \text{for } n > \nu.$$

Thus for $n > \mu + \nu$ the numbers $\rho_1(f(p_n), f(q_n))$ belong to the set G; therefore

$$\rho_1(f(p_n), f(q_n)) \leqslant \omega(p_0, \epsilon) \qquad \text{for } n > \mu + \nu. \tag{15}$$

But, since $p_n \in E$, $q_n \in E$, $\lim_{n\to\infty} p_n = p$, $\lim_{n\to\infty} q_n = q$, it follows that

$$\lim_{n\to\infty} f(p_n) = \phi(p) \text{ and } \lim_{n\to\infty} f(q_n) = \phi(q). \tag{16}$$

From (15), (16), and the continuity of the function ρ_1 (§ **49**), we obtain

$$\rho_1(\phi(p), \phi(q)) \leqslant \omega(p_0, \epsilon)$$

and so, from (14), $h \leqslant \omega(p_0, \epsilon)$. Since h is any number of the set H it follows that the upper bound of this set, that is, the number $\omega_1(p_0, \epsilon)$, is $\leqslant \omega(p_0, \epsilon)$. Hence

$$(17) \qquad \omega_1(p_0) = \lim_{\epsilon \to 0} \omega_1(p_0, \epsilon) \leqslant \lim_{\epsilon \to 0} \omega(p_0, \epsilon) = \omega(p_0),$$

where $\omega(p_0)$ is the oscillation of the function f in E at p_0 (§ **70**) and $\omega_1(p_0)$ the oscillation of the function ϕ in T at p_0. But $\omega(p_0) = 0$ by hypothesis; hence (17) gives $\omega_1(p_0) = 0$. Consequently, by Theorem 90, the function ϕ is continuous in T at the point p_0.

We have thus proved

THEOREM 99. *Let M be a metric space and N a complete space. If $f(p)$ is a function defined and continuous in a set $E \subset M$ and assuming values in N then there exists a function $\phi(p)$ defined and continuous in a certain set T such that T is a G_δ and such that $E \subset T \subset \bar{E}$ and $\phi(p) = f(p)$ for each point p of the set E.*

The set in the theorem could be chosen to be the set T of all elements of $\bar{E}$ at which the oscillation of the given function in E is zero. It follows readily that if a function $f(p)$ defined and continuous in a set E can be extended with the retention of continuity over a set S such that $E \subset S \subset \bar{E}$, then $S \subset T$. Such an extension of a function continuous in E over any set $S \subset T$ is achieved in only one way.

Note that the condition that the space N be complete cannot be omitted from Theorem 99. In fact, let M denote the linear space and N the metric space consisting of all rational numbers; let E be a subset of M consisting of all rational numbers. The function $f(x) = x$ is continuous in the set $E \subset M$ and assumes values in the space N. If the function $\phi(x)$ were defined and continuous in a set T such that $E \subset T \subset M$ and if $\phi(x) \in N$ where $\phi(x) = f(x)$ for $x \in E$, then $T = E$. For if $x_0 \in T - E$ then x_0 is irrational and since $\phi(x) = x$ for all rational x and the function ϕ is continuous in T we have $\phi(x_0) = x_0$ contrary to the assumption that $\phi(x_0) \in N$. But, by Corollary 2 to Theorem 98, the set E is not a G_δ in M; hence Theorem 99 does not hold.

78. Lavrentieff's theorem. Let E and F be two homeomorphic sets. Then there exists a function f defined and continuous in E such that $E\, h_f\, F$, and a function g, the inverse of f, continuous in F such that $F\, h_g\, E$.

By Theorem 99, there exists a function $\phi(p)$ defined and continuous in a certain set T such that $E \subset T \subset \bar{E}$, and $\phi(p) = f(p)$ for $p \in E$. Similarly, there exists a function $\psi(q)$ defined and continuous in a set H such that $F \subset H \subset \bar{F}$, and $\psi(q) = g(q)$, for $q \in F$.

Let M denote the set of all elements p of T for which $\phi(p) \in H$, and denote by N the set of all elements q of H for which $\psi(q) \in T$. Then $M\, h_\phi\, N$. For, since the function ϕ is continuous in T, it is continuous in M. Similarly, the function ψ is continuous in the set $N \subset H$. To prove the $M\, h_\phi\, N$, it is sufficient to show that the function ψ (considered in the set N) is inverse to the function ϕ (in the set M) or, in other words, that the relation

$$p \in M,\ \phi(p) = q \tag{18}$$

is equivalent to the relation

$$q \in N,\ \psi(q) = p. \tag{19}$$

Because of symmetry, it is sufficient to show that (18) implies (19).

Suppose that relations (18) are satisfied for a given p. Since $p \in M$, we have $\phi(p) \in H$ by the definition of M; so $q \in H$ by (18). Since $p \in M \subset T \subset \bar{E}$, there exists an infinite sequence $\{p_n\}$ such that $p_n \in E$ for $n = 1, 2, \ldots$, and $p_n \to p$; since $p \in T$, $E \subset T$, and the function ϕ is continuous in T, it follows that $\phi(p_n) \to \phi(p)$. But, since $p_n \in E$, $\phi(p_n) = f(p_n)$ and, since $E\, h_f\, F$, and the function g is the inverse of f in E we find, on setting $f(p_n) = q_n$ that $q_n \in F \subset H$, and $p_n = g(q_n)$. Consequently, according to the definition of the function ψ, we have $p_n = \psi(q_n)$, and, since $p_n \to p$, we have $\psi(q_n) \to p$. But $q_n \in H$, $q \in H$, and ψ is continuous in H; so $\psi(q_n) \to \psi(q)$. This, because of the relations $\psi(q_n) \to p$, gives $\psi(q) = p$. Since $q \in H$, and $p \in T$, it follows from the definition of the set N that $q \in N$. Hence relations (19) are established.

We have therefore proved that $M\, h_\phi\, N$. From $E\, h_f\, F$, $E \subset T$, $F \subset H$, $\phi(p) = f(p)$ for $p \in E$, and from the definition of the set M, we have $E \subset M$. Similarly, from $F\, h_g\, E$, $\psi(q) = g(q)$ for $q \in F$, and from the definition of the set N, we get $F \subset N$. It will be shown that M and N are sets G_δ. Because of the symmetry of the relations, it is sufficient to prove that one of them, say M, is a G_δ. To that end, we first prove the following

LEMMA. *If a function $\phi(p)$ is continuous in a set T which is a G_δ, and if V is an open set, then the set S of all points p of T for which $\phi(p) \in V$ is a G_δ.*

Proof. Let p be a point of the set S. Now ϕ is continuous in T, V is open, and $\phi(p) \in V$; so there exists an open set $U(p)$ such that

$$p \in U(p) \text{ and } \phi(q) \in V \qquad \text{when } q \in T\,.\,U(p).$$

Denote by U the sum of all the sets $U(p)$ where p ranges over all the points of S. Then U is open and $S = T\,.\,U$. But T is a G_δ; hence the set S is a G_δ.

COROLLARY. *If a function $\phi(p)$ is continuous in a set T which is a G_δ, and if the set H is a G_δ, then the set M of all points of T for which $\phi(p) \in H$ is a G_δ.*

Proof. Since the set H is a G_δ, we may write $H = V_1 . V_2 . V_3 \ldots$, where each V_n is an open set. Denote by S_n the set of all points p of T for which $\phi(p) \in V_n$; the set S_n is a G_δ by the above Lemma. It follows that $M = S_1 . S_2 . S_3 \ldots$. For if $p \in M$, then $\phi(p) \in H \subset V_n$; this gives $p \in S_n$ for $n = 1, 2, \ldots$; then $\phi(p) \in V_n$ for $n = 1, 2, \ldots$, and so $\phi(p) \in H$ and $p \in M$. Since S_n is a G_δ, the set M is a G_δ.

Collecting the results obtained in this section we may state

THEOREM 100 (Lavrentieff).[9] *If E and T are sets contained in complete spaces and $E \, h_f \, T$, then there exist sets M and N, each a G_δ, such that $M \, h_\phi \, N$, where $E \subset M \subset \bar{E}$, $T \subset N \subset \bar{T}$, and $\phi(p) = f(p)$ for $p \in E$.*

In other words, *a homeomorphism between two sets contained in complete spaces can always be extended to two sets G_δ which contain the corresponding sets and are contained in their closures.*

Furthermore, it can be shown that the preceding extension of the homeomorphism between the sets E and T to the sets M and N is the best possible.[10]

79. Conclusions from Lavrentieff's theorem. Let E denote a set G_δ contained in a complete space; then $E = E_1 . E_2 . E_3 \ldots$, where each E_n is an open set. Let T denote a homeomorphic image of E; assume T is also contained in a complete space. Consider the sets M and N which satisfy the conditions of Theorem 100. Put $Q_n = M . E_n$, $n = 1, 2, \ldots$. Since E_n is open, the set $M - E_n = M . CE_n$ is closed in M and so, under the homeomorphism between M and N, is mapped into a set closed in N; it may therefore be written in the form $N . F_n$, where F_n is closed. But, from $M \, h_\phi \, N$, we have $(M - E_n) \, h_\phi \, N . F_n$; hence $M . E_n \, h_\phi \, (N - F_n)$, that is, $Q_n \, h_\phi \, N . U_n$, where $U_n = CF_n$ is open. Since $E \subset M$, $E = E_1 . E_2 . E_3 \ldots$, and $Q_n = M . E_n$, we have $E = M . E = Q_1 . Q_2 . Q_3 \ldots$, while the relation $Q_n \, h_\phi \, N . U_n$ gives (since $M \, h_\phi \, N$, and $Q_n \subset M$, for $n = 1, 2, \ldots$) $Q_1 . Q_2 . Q_3 \ldots h_\phi \, N . U_1 . U_2 . U_3 \ldots$, i.e., $\phi(E) = T = N . U_1 . U_2 . U_3 \ldots$. Since the set N is a G_δ and the sets U_n are open, it follows that T is a G_δ. We have thus proved

THEOREM 101.[11] *In a complete space a homeomorphic image of a set G_δ is a G_δ.*

Note that *in a complete space a homeomorphic image of an F_σ (or even of a closed set) may not be an F_σ.* In fact, let $Z = X \times Y$ denote the metric product of the Baire 0-dimensional space and the straight line; Z is a complete space since it is the product of two complete spaces (§ **73**). The set Z_1 consisting of the points $(x, 0)$, where $x \in X$, is closed in Z and is isometric with the 0-dimensional Baire space X; it is therefore (§ **60**) homeomorphic with the set of all irrational numbers and so with the set Z_2 of all points $(0, y)$ of the space Z, where y is irrational. If the set Z_2 were an F_σ (in Z), it would be an F_σ in the set Y, i.e., in the straight line; this is impossible since the set of all irrational numbers is not an F_σ in the line (§ 76). Thus in the complete space Z, the homeomorphic image Z_2 of the closed set Z_1 is not an F_σ.

A family **F** of sets contained in a complete space is said to be a *topological invariant* if every homeomorphic image in the space of a set of **F** also belongs to **F**.

THEOREM 102. *If a family* **F** *of sets contained in a complete space is a topological invariant, then the family of sets which are sums of a countable aggregate of sets belonging to* **F** *is a topological invariant.*

For let $E = E_1 + E_2 + E_3 + \ldots$, where $E_n \in \mathbf{F}$, and suppose that $E\, h_f\, T$. From $E_n \subset E$ we get $E_n\, h_f\, T_n$, where T_n is a certain subset of T; so $(E_1 + E_2 + \ldots)\, h_f\, (T_1 + T_2 + \ldots)$, i.e., $E\, h_f\, (T_1 + T_2 + \ldots)$. But, from $E_n \in \mathbf{F}$ and $E_n\, h_f\, T_n$, we have $T_n \in \mathbf{F}$ since **F** is a topological invariant. Theorem 102 is therefore proved.

THEOREM 103. *If a family* **F** *of sets contained in a complete space is a topological invariant, and if the intersection of a member of* **F** *with a* G_δ *belongs to* **F**, *then the family of all intersections of countable aggregates of sets belonging to* **F** *is a topological invariant.*

Proof. Let $E = E_1 . E_2 . E_3 \ldots$, where $E_n \in \mathbf{F}$, and assume that $E\, h_f\, T$. There exist, by Theorem 100, two sets M and N, each a G_δ, such that $M\, h_\phi\, N$, $E \subset M$, $T \subset N$, and $\phi(p) = f(p)$ for $p \in E$. It follows that the sets $M . E_n$ $n = 1, 2, \ldots$, belong to **F** and so their homeomorphic images $T_n = \phi(M . E_n)$ belong to **F**. Since $E = M . E_1 . M . E_2 \ldots$, and ϕ is (1, 1) in M, we have $T = \phi(E) = \phi(M . E_1) . \phi(M . E_2) \ldots$, i.e., T is the intersection of a countable aggregate of sets belonging to **F** as required.

THEOREM 104. *If a family* **F** *of sets contained in a complete space is a topological invariant and if the intersection of a member of* **F** *with a set* G_δ *and the sum of a member of* **F** *and a set* F_σ *always belong to* **F**, *then the family of the complements of all sets belonging to* **F** *is a topological invariant.*

Proof. Suppose that $E = CX$, where $X \in \mathbf{F}$, and assume that $E\, h_f\, T$. Let M and N be the sets of Theorem 100 and ϕ the homeomorphic mapping of M on N with the usual properties. It follows that $M . X \in \mathbf{F}$ and $Q = \phi(M . X) \in \mathbf{F}$. Since $E = CX \subset M$, we have $E = M - X$ and so $T = \phi(E) = \phi(M - M . X) = \phi(M) - \phi(M . X) = N - Q$. Therefore $CT = Q - CN$; since $Q \in \mathbf{F}$ and CN is an F_σ, it follows that $CT \in \mathbf{F}$. This proves the theorem.

THEOREM 105. *If a family* **F** *of sets contained in a complete space is a topological invariant, and if the intersection of a member of* **F** *with a set* G_δ *belongs to* **F**, *then the family of the differences of all pairs of sets belonging to* **F** *is a topological invariant.*

Proof. Suppose that $E_1 \in \mathbf{F}$, $E_2 \in \mathbf{F}$, $E = E_1 - E_2$, and $E\,h\,T$. Let M, N be the sets of Theorem 100 and ϕ the homeomorphism from M to N. Since $E = E_1 - E_2$, and $E \subset M$, we have $E = M \,.\, E_1 - M \,.\, E_2$, where $M \,.\, E_1$ and $M \,.\, E_2$ belong to $\mathbf{F}$ by hypothesis; hence the sets T_1 and T_2, where $M \,.\, E_1\, h_\phi\, T_1$ and $M \,.\, E_2\, h_\phi\, T_2$, belong to $\mathbf{F}$. But $(M \,.\, E_1 - M \,.\, E_2)\, h_\phi\, (T_1 - T_2)$; hence $\phi(M \,.\, E_1 - M \,.\, E_2) = \phi(E) = T = T_1 - T_2$, where $T_1 \in \mathbf{F}$ and $T_2 \in \mathbf{F}$. Theorem 105 is therefore proved.

80. Topological invariance of sets P^α and Q^α. From the fact that sets G_δ are sets Q^2 (§ 69) and from properties 5, 4, and 1 of the sets P^α and Q^α it follows that, for $\alpha \geqslant 3$, the intersection of a set P^α with a G_δ is a P^α. Similarly, since an F_σ is a set P^2, it follows from properties 1 and 2 of the sets P^α and Q^α that, for $\alpha \geqslant 2$, the sum of a P^α and an F_σ is a P^α.

THEOREM 106. *In a complete space the family of Hausdorff sets P^α, for $\alpha \geqslant 3$, and the family of sets Q^α, for $\alpha \geqslant 2$, are each a topological invariant.*

Proof. For sets Q^2, i.e., for sets G_δ, the result follows from Theorem 101.

Let β be an ordinal number such that $3 \leqslant \beta < \Omega$ and suppose that Theorem 106 holds for every ordinal α such that $2 \leqslant \alpha < \beta$ (this is certainly true for $\beta = 3$). Let E denote a given set P^β; then $E = E_1 + E_2 + \dots$, where E_n is a set Q^{ξ_n}, and $\xi_n < \beta$, for $n = 1, 2, \dots$, and where, since $\beta \geqslant 3$, we may suppose from property 1 of sets Q^α, that $\xi_n \geqslant 2$, for $n = 1, 2, \dots$. Let T be a set such that $E\,h_f\,T$. Then $E_n\,h_f\,T_n$, where $T_n \subset T$, and $T = T_1 + T_2 + \dots$. Since E_n is a set Q^{ξ_n}, where $2 \leqslant \xi_n < \beta$, and the theorem is assumed to hold for numbers α such that $2 \leqslant \alpha < \beta$, we conclude from $E_n\,h_f\,T_n$ that T_n is a set Q^{ξ_n}. Since $T = T_1 + T_2 + \dots$, it follows that T is a set P^β.

The family of sets P^β is therefore a topological invariant. However, since $\beta \geqslant 3$ and from the remarks made at the beginning of this section, it follows that the intersection of a set P^β with a set G_δ and the sum of a set P^β and an F_σ is a P^β. Consequently, the family of sets P^β satisfies the conditions of Theorem 104 and therefore the family of the complements of sets P^β, i.e., the family of sets Q^β, is a topological invariant. Theorem 106 is thus proved by transfinite induction.

This theorem does not hold in the case of sets P^1, P^2, and Q^1. That it is not true for sets P^1 follows from the fact that in the complete linear space consisting of all real numbers $\geqslant 0$ and the number -1 the set $\{-1\}$ is open but its homeomorphic image $\{1\}$ is not open.

Next let X denote the 0-dimensional Baire space, Y the set of all real numbers; put $Z = X \times Y$. The set Z is a complete space. The set of points $(x, 0)$, where $x \in X$, is closed in Z and, because it is isometric with the 0-dimensional Baire space, it is homeomorphic with the set of all irrational numbers (§ 60); this last set is isometric with the set of points $(0, y)$ of the space Z, where y is an irrational number. But this last set is not an F_σ on the line and so certainly not an F_σ in the set Z. Thus a homeomorphic image of a set Q^1 (and so of a set P^2) may not be a P^2 (and so not a Q^1). Hence Theorem 106 does not hold for sets P^2 and Q^1.

From Theorem 106 we obtain

COROLLARY 1. *The family of those sets of a complete space which has the property that each member set is both a P^α and a Q^α, is a topological invariant for $\alpha \geqslant 3$.*

Corollary 1 is not true for $\alpha = 2$ because, as proved above, in a complete space a homeomorphic image of a set Q^1 (and so of a set which is both a P^2 and a Q^2) need not be a P^2. It is clear that Corollary 1 is not true for $\alpha = 1$.

Sets $P^{\alpha+1}$, which are not sets $P^{\xi+1}$ for any $\xi < \alpha$, are identical with Lebesgue's sets O of class α; similarly, sets $Q^{\alpha+1}$ which are not sets $Q^{\xi+1}$ for any $\xi < \alpha$, are Lebesgue's sets F of class α. (Lebesgue's definition of sets O and F of class α is different from but equivalent to the above.)

COROLLARY 2. *The families of Lebesgue sets O and F of class α in a complete space are each a topological invariant for $2 \leqslant \alpha < \Omega$.*

81. Borel sets: their topological invariance. Denote by $\mathbf{B}$ the family of all sets P^α and Q^α for $1 \leqslant \alpha < \Omega$. Sets which belong to this family are called *measurable in the Borel sense* or simply *Borel sets.* It follows from the properties of the sets P^α and Q^α (§ 67) that the family $\mathbf{B}$ satisfies the following conditions:

1′. *Every closed set belongs to* $\mathbf{B}$;
2′. *The sum of a countable aggregate of sets belonging to* $\mathbf{B}$ *belongs to* $\mathbf{B}$;
3′. *The intersection of a countable aggregate of sets belonging to* $\mathbf{B}$ *belongs to* $\mathbf{B}$.

Condition 1′ follows from the definition of the sets Q^1 and the family $\mathbf{B}$. Let now $E = E_1 + E_2 + \ldots$, where each E_n belongs to $\mathbf{B}$. We may suppose, by property 5 (§ 69), that E_n is a set Q^{ξ_n}, where ξ_n is some ordinal number $< \Omega$. Since, for every infinite sequence of ordinals ξ_n which are less than Ω, there exists an ordinal $\alpha < \Omega$ such that $\xi_n < \alpha$, it follows that E is a P^α and so belongs to $\mathbf{B}$. Thus the family $\mathbf{B}$ satisfies condition 2′. Similarly it may be proved that the family $\mathbf{B}$ satisfies condition 3′.

We next show that the family of all sets P^α and Q^α $(1 \leqslant \alpha < \Omega)$ is the *smallest* family which satisfies conditions 1′, 2′, and 3′, that is, if a family $\mathbf{B}^*$ satisfies conditions 1′, 2′, and 3′, then $\mathbf{B} \subset \mathbf{B}^*$.

Let $\mathbf{B}^*$ denote any family of sets contained in the metric space considered and satisfying conditions 1′, 2′, and 3′. The sets Q^1 belong to $\mathbf{B}^*$ by condition 1′. It is evident from conditions 1′ and 2′ that the sets F_σ, i.e., the sets P^2 belong to $\mathbf{B}^*$; from this and property 1 (§ 69), it follows that the sets P^1 belong to $\mathbf{B}^*$.

Let α be an ordinal number, $1 < \alpha < \Omega$, and suppose that all sets P^ξ and Q^ξ belong to $\mathbf{B}^*$ for $1 \leqslant \xi < \alpha$. Let E be a set P^α; since $\alpha > 1$, we may write $E = E_1 + E_2 + \dots$, where E_n is a set Q^{ξ_n}, and $\xi_n < \alpha$; by hypothesis, the sets E_n, $n = 1, 2, \dots$, belong to $\mathbf{B}^*$, and, so by condition 2′, the set E belongs to $\mathbf{B}^*$. Similarly, if E is a set Q^α, we may write $E = E_1 . E_2 \dots$, where E_n is a set P^{ξ_n} and $\xi_n < \alpha$; by hypothesis, the sets E_n belong to $\mathbf{B}^*$ and so, by condition 3′, the set E belongs to $\mathbf{B}^*$. Consequently, by transfinite induction, all sets P^α and Q^α belong to $\mathbf{B}^*$. Thus $\mathbf{B} \subset \mathbf{B}^*$. We have thus proved that

The family of all Borel sets is the smallest family which satisfies conditions 1′, 2′, *and* 3′.

Borel sets may therefore be defined without the aid of ordinal numbers or transfinite induction as sets belonging to the smallest (or, if we like, every) family $\mathbf{B}^*$ of sets satisfying conditions 1′, 2′, and 3′.

It follows from properties 3 and 7 (§ **69**) that the family $\mathbf{B}$ of all Borel sets also satisfies the following conditions:

4′. *The complement of a set belonging to* $\mathbf{B}$ *belongs to* $\mathbf{B}$.

5′. *The difference of two sets belonging to* $\mathbf{B}$ *belongs to* $\mathbf{B}$.

The definition of Borel sets and Theorem 106 lead to

THEOREM 107. *In a complete space, a homeomorphic image of a Borel set is a Borel set.*[12]

We prove the following further property of Borel sets:

If a family $\mathbf{F}$ *of sets satisfies the following three conditions*

1″. *Every open set belongs to* $\mathbf{F}$;

2″. *The sum of a countable aggregate of disjoint sets belonging to* $\mathbf{F}$ *belongs to* $\mathbf{F}$;

3″. *The intersection of a countable aggregate of sets belonging to* $\mathbf{F}$ *belongs to* $\mathbf{F}$; *then every Borel set belongs to* $\mathbf{F}$.

Proof. It is clear from conditions 1″ and 3″ that every set G_δ (i.e., every set Q^2) belongs to $\mathbf{F}$; hence, by properties 1 and 5 of sets P^α and Q^α (§ **69**), the sets Q^1 and P^1 belong to $\mathbf{F}$; from condition 2″ and property 8, it follows that sets P^3 belong to $\mathbf{F}$, and so, by property 1, the sets P^2 also belong to $\mathbf{F}$.

Let α denote a given ordinal number such that $3 < \alpha < \Omega$, and suppose that all sets P^ξ and Q^ξ belong to $\mathbf{F}$, where $\xi < \alpha$. By condition 2″ and property 8 every set P^α belongs to $\mathbf{F}$; by condition 3″ and the definition of sets Q^α every set Q^α belongs to $\mathbf{F}$. We have thus proved by transfinite induction that all sets P^α and Q^α, $1 \leqslant \alpha < \Omega$, belong to $\mathbf{F}$.

As a result of this property it follows that *the family of all Borel sets is the smallest family* $\mathbf{F}$ *which satisfies conditions* 1″, 2″, *and* 3″.

The Borel character of a set (or its class) depends on the space in which the set is contained. The space itself is both a P^1 and a Q^1. It would be possible to speak of sets (or metric spaces) which are absolutely Borel sets or absolutely sets P^α or Q^α, i.e., sets which are Borel sets P^α or Q^α in every metric space which contains them.

It is readily seen that there exists no metric space which is absolutely open; for no metric space M is open in the metric product $M \times R_1$, where R_1 is the linear space. On the other hand, Theorem 97, Corollary 2 to Theorem 97, and the fact that a closed subset of a complete space is complete lead to the result that *a set is absolutely a* P^α *for* $2 \leqslant \alpha < \Omega$, *or absolutely a* Q^α *for* $1 \leqslant \alpha < \Omega$, *if and only if it is a* P^α *or a* Q^α *in some complete space in which it is contained.* From this it follows readily that *a set is absolutely a Borel set if and only if it is a Borel set in some complete space in which it is contained.*

82. Analytic sets; defining systems. We consider next a generalization of Borel sets. To arrive at this generalization in a natural way, we consider first sets $F_{\sigma\delta}$, i.e., intersections of a countable aggregate of sets F_σ. Let E be a given set $F_{\sigma\delta}$. Since every set F_σ is the sum of a countable aggregate of closed sets, we may write

$$E = \prod_{k=1}^{\infty} (E_1^k + E_2^k + E_3^k + \dots) = \prod_{k=1}^{\infty} \sum_{n=1}^{\infty} E_n^k, \tag{20}$$

where E_n^k $(k = 1, 2, \dots; n = 1, 2, \dots)$ is a closed set. But (20) is the same as

$$E = \sum E_{n_1}^{1} \,.\, E_{n_2}^{2} \,.\, E_{n_3}^{3} \dots, \tag{21}$$

where the summation ranges over all infinite sequences $n_1, n_2, n_3, \dots$ of natural numbers.

For every finite sequence of indices $n_1, n_2, \dots, n_k$ which we shall denote briefly by $n(k)$, put

$$E_{n_1}^{1} \,.\, E_{n_2}^{2} \dots E_{n_k}^{k} = E_{n_1, n_2, \dots, n_k} = E_{n(k)}; \tag{22}$$

these are obviously closed sets (and may be empty).

On account of (22), (21) may be written in the form

$$E = \sum E_{n(1)} \,.\, E_{n(2)} \,.\, E_{n(3)} \dots, \tag{23}$$

where the summation extends over all infinite sequences of natural numbers.

Every set E of the form (23), where $E_{n(k)}$ is closed, is called an *analytic* set, briefly a set (A); it is also referred to as the *nucleus of the defining system* $[E_{n(k)}]$. The defining system $[E_{n(k)}]$ is known if, corresponding to every finite combination $n(k)$ of indices, the corresponding set $E_{n(k)}$ is known. We have therefore proved that every set $F_{\sigma\delta}$ is a set (A).

It is evident from the definition that the analytic character of a set will depend on the space in which the set is contained. A space M is always an analytic set relative to itself (since it is the nucleus of the defining system consisting of the closed sets each equal to M), but it is not necessarily analytic in a space in which it is contained. It is possible to consider sets which are absolutely analytic, i.e., analytic in every metric space in which they are contained. It is readily seen that *a set* (or a metric space) *is absolutely analytic if and only if it is analytic in some complete space in which it is contained.*

*A different definition of analytic sets is given by Menger.[13] He calls a metric space M *analytic* if there exists a defining system $[E_{n(k)}]$ consisting of open sets of the space M such that: (1) for every infinite sequence $n(k)$, $k = 1, 2, \ldots$, of natural numbers for which $E_{n(k)} \neq 0$, the set $\Pi_{k=1}^{\infty} E_{n(k)}$ consists of only one element; (2) for every point $p \in M$ there exists an infinite sequence $n(k)$, $k = 1, 2, \ldots$, of natural numbers such that the set $\Pi_{k=1}^{\infty} E_{n(k)}$ consists only of the point p. It can be shown that the Fréchet space D_{ω} is analytic in the Menger sense.

However, the Menger definition of analytic spaces seems to be too general. In fact, as pointed out by Marczewski, there exist $2^{\mathfrak{c}}$ linear sets which are analytic spaces in the Menger sense. For it can be proved that if E is any subset of the interval $(0, 1)$, E_1 the complement of E with respect to this interval, and E_2 a translation of E_1 (along the line) through unit length, then the set $E + E_2$ is analytic in the Menger sense.

Menger[14] has proved that the nuclei of defining systems $[E_{n(k)}]$, where $E_{n(k)}$ are finite and closed m-dimensional parallelotopes in Euclidean m-dimensional space R_m, and where for each natural k there exists only a finite number of non-empty sets $E_{n(k)}$, are identical with the closed and bounded subsets of R_m.*

83. The operation A. Lusin's sieve. Suppose that to every finite combination $n(k)$ of indices, there is assigned a certain set $E_{n(k)}$ (not necessarily contained in a metric space). We then say that a defining system $[E_{n(k)}]$ of sets is given and the set

$$E = \sum E_{n(1)} \cdot E_{n(2)} \cdot E_{n(3)} \cdots,$$

where the summation extends over all infinite sequences $n(k)$, $k = 1, 2, \ldots$, of natural numbers, is said to be the *nucleus of the system.*

If all the sets $E_{n(k)}$ belong to a certain family $\mathbf{F}$ of sets, then the nucleus of the defining system $[E_{n(k)}]$ is called the *result of the operation A* on the sets of the family $\mathbf{F}$.

*It can be proved[15] that if $\mathbf{F}$ is a ring (i.e., the sum and intersection of any finite number of sets of $\mathbf{F}$ belong to $\mathbf{F}$) then the family of all sets of the form

$$\sum E_{\alpha(1)} \cdot E_{\alpha(2)} \cdot E_{\alpha(3)} \cdots,$$

where the summation extends over all infinite sequences $\alpha(k)$, $k = 1, 2, \ldots$, consisting of the numbers 0 and 1 and where $E_{\alpha(k)} \in \mathbf{F}$, is the same as the family $\mathbf{F}_{\delta}$, that is, the family of all intersections of countable aggregates of sets belonging to the family $\mathbf{F}$.

Denote by W the set of all rational numbers between 0 and 1. If with every number w of the set W is associated a set E_w, we then have a *sieve* $[E_w]$. A set *sifted through a sieve* $[E_w]$ is the set of all elements p for which there exists a certain infinite decreasing sequence (dependent on p) of numbers of the set W, say

$$w_1 > w_2 > w_3 > \dots,$$

such that

$$p \in E_{w_n}, \qquad n = 1, 2, 3, \dots.$$

It will be shown that a set sifted through a sieve is the result of the operation A on the sets constituting the sieve.

Denote by

$$r_1, r_2, r_3, \dots \tag{24}$$

the infinite sequence consisting of all different numbers of the set W. Put

$$E_n = E_{r_n}, \qquad n = 1, 2, \dots.$$

Let k be a given natural number and suppose that we have already defined all sets $E_{n(k)}$, where $n(k)$ is any combination of k natural numbers and where the sets $E_{n(k)}$ belong to the sieve $[E_w]$. Let $n(k+1)$ denote a combination of $k+1$ natural numbers. Then, by hypothesis, $E_{n(k)} = E_{r_s}$, where s is a certain natural number. Put $E_{n(k+1)} = E_{r_q}$, where r_q is the n_{k+1}th term of the sequence (24) satisfying the inequality $r_q < r_s$.

The sets $E_{n(k)}$ are thus defined by induction. We shall show that the set P sifted through the sieve $[E_w]$ is the nucleus Q of the system $[E_{n(k)}]$.

Suppose that $p \in P$. Then there exists an infinite sequence of indices $m_1, m_2, m_3, \dots$ such that

$$r_{m_1} > r_{m_2} > r_{m_3} > \dots, \tag{25}$$

and

$$p \in E_{r_{m_i}}, \qquad i = 1, 2, \dots \tag{26}$$

Put $n_1 = m_1$; then from the definition of the sets E_n we obtain $E_{r_{m_1}} = E_{n_1}$. From $r_{m_2} < r_{m_1}$ and the definition of the sets E_{n_1, n_2}, we have

$$E_{r_{m_2}} = E_{n_1, n_2},$$

for some natural number n_2. Furthermore, since $r_{m_3} < r_{m_2}$, there exists a natural number n_3 such that

$$E_{r_{m_3}} = E_{n_1, n_2, n_3}.$$

Continuing this argument, we obtain from (26) an infinite sequence $n_1, n_2, n_3, \dots$ of natural numbers such that $p \in E_{n(k)}$ for $k = 1, 2, \dots$, and so $p \in Q$.

Suppose now that $p \in Q$. Then there exists an infinite sequence $n(k)$, $k = 1, 2, \dots$, of natural numbers such that $p \in E_{n(k)}$ for $k = 1, 2, \dots$. It follows readily from the definition of the sets $E_{n(k)}$ that there exists an infinite sequence (25) of numbers of the set W such that

$$E_{n(k)} = E_{r_{m_k}} \qquad \text{for } k = 1, 2, \ldots$$

and so, from (26) and the definition of the set P, $p \in P$. Consequently $P = Q$; this proves the theorem.

Moreover, it can be shown[16] that if $\mathbf{F}$ is a family of sets such that $\mathbf{F} = \mathbf{F}_d$, where $\mathbf{F}_d$ is the family of all intersections of a finite number of sets of $\mathbf{F}$, then a result of the operation A on sets of the family $\mathbf{F}$ may be considered to be a set sifted through the sieve $[E_w]$, where E_w (for $w \in W$) is a set of the family $\mathbf{F}$.

However, not every result of the operation A carried out on sets of a family $\mathbf{F}$ can be obtained as a set sifted through a sieve $[E_w]$, where E_w is a set of the family $\mathbf{F}$. Thus for instance, in the case of a family $\mathbf{F}$ consisting of two different unit sets, say $\mathbf{F} = \{\{1\}, \{-1\}\}$, the null set is a result of the operation A carried out on the sets of the family $\mathbf{F}$ since it is the nucleus of the system $[E_{n(k)}]$, where $E_{n(k)} = \{(-1)^k\}$ for every finite sequence $n(k)$ of natural numbers. But no set sifted through a sieve $[E_w]$, where $E_w \in \mathbf{F}$ for rational w, is empty.*

84. Fundamental properties of the operation A.

THEOREM 108. *If each of the sets $E^{r(s)}$ is a result of the operation A on the sets of a family $\mathbf{F}$, then the nucleus of the system $[E^{r(s)}]$ is also a result of the operation A carried out on the sets of the family $\mathbf{F}$.*

Proof. Every natural number k can be expressed uniquely in the form

$$k = 2^{p_k - 1}(2q_k - 1),$$

where p_k and q_k are two natural numbers; put

$$\phi(k) = p_k, \ \psi(k) = q_k,$$

and, for natural p and q,

$$\nu(p, q) = 2^{p-1}(2q - 1).$$

Then

$$\psi(k) \leqslant k \qquad \text{for } k = 1, 2, \ldots; \tag{27}$$

$$\nu(\phi(k), \psi(k)) = k \qquad \text{for } k = 1, 2, \ldots; \tag{28}$$

$$\nu(i, \psi(k)) \leqslant k, \qquad k = 1, 2, \ldots; i = 1, 2, \ldots, \phi(k); \tag{29}$$

and

$$\phi(\nu(p, q)) = p, \ \psi(\nu(p, q)) = q, \ p = 1, 2, \ldots; \ q = 1, 2, \ldots. \tag{30}$$

Each of the sets $E^{r(s)}$ is by hypothesis a result of the operation A on the sets of the family **F**. Hence for each combination $r(s)$ of indices there exists a system

$$S^{r(s)} = [E^{r(s)}_{n(k)}]$$

of sets of the family **F** whose nucleus is the set $E^{r(s)}$. It follows from (27) and (29) that every finite sequence $n(k)$ of indices determines uniquely the natural numbers

$$\phi(n_i) \text{ and } \psi(n_{\nu(i,\psi(k))}), \qquad i = 1, 2, \ldots, k;$$

we may therefore define a system $[E_{n(k)}]$ by putting, for every finite sequence $n(k)$ of indices

$$(31) \qquad E_{n(k)} = E^{\phi(n_1),\phi(n_2),\ldots,\phi(n_{\psi(k)})}_{\psi(n_{\nu(1,\psi(k))}),\psi(n_{\nu(2,\psi(k))}),\ldots,\psi(n_{\nu(\phi(k),\psi(k))})}.$$

These sets belong to the family **F**.

It will be shown that the nucleus of the system $[E^{r(s)}]$ is also the nucleus of the system $[E_{n(k)}]$. Let x be a given element of the nucleus of the system $[E_{n(k)}]$. Then there exists an infinite sequence $n(k)$, $k = 1, 2, \ldots$ of indices such that

$$(32) \qquad x \in E_{n(k)}, \qquad k = 1, 2, 3, \ldots$$

Put

$$(33) \qquad r_s = \psi(n_s), \qquad s = 1, 2, \ldots$$

and let s be a given natural number. Put

$$(34) \qquad j_h = \psi(n_{\nu(h,s)}), \qquad h = 1, 2, \ldots.$$

Since, by (30), $\psi(\nu(h, s)) = s$ and $\phi(\nu(h, s)) = h$, we obtain

$$E_{n_1,n_2,\ldots,n_{\nu(h,s)}} = E^{r(s)}_{j(h)} \qquad \text{for } h = 1, 2, \ldots$$

Hence, from (32),

$$x \in E^{r(s)}_{j(h)} \qquad \text{for } h = 1, 2, \ldots$$

consequently x belongs to the nucleus $E^{r(s)}$ of the system $S^{r(s)}$. We thus have

$$(35) \qquad x \in E^{r(s)} \qquad \text{for } s = 1, 2, \ldots$$

and so x belongs to the nucleus of the system $[E^{r(s)}]$.

Suppose now that x belongs to the nucleus of the system $[E^{r(s)}]$. Then there exists an infinite sequence $r_1, r_2, r_3, \ldots$ of natural numbers such that (35) holds. But $E^{r(s)}$ is the nucleus of the system $[E^{r(s)}_{n(k)}]$; thus for every natural s there exists an infinite sequence of indices $m_1{}^{(s)}, m_2{}^{(s)}, m_3{}^{(s)}, \ldots$ or, briefly, $m^{(s)}(k)$ such that

$$(36) \qquad x \in E^{r(s)}_{m^{(s)}(k)}, \qquad k = 1, 2, \ldots; s = 1, 2, \ldots.$$

Put

$$(37) \qquad n_h = \nu(r_h, m^{(\psi(h))}_{\phi(h)}), \qquad h = 1, 2, \ldots$$

From (37) and (30) we obtain

$$\phi(n_h) = r_h,\ \psi(n_h) = m_{\phi(h)}^{(\psi(h))}, \qquad h = 1, 2, \ldots,$$

and therefore, for $h = \nu(i, \psi(k))$, we have

$$\psi(n_{\nu(i,\psi(k))}) = m_i^{(\psi(k))}, \quad i = 1, 2, \ldots; k = 1, 2, \ldots.$$

Hence from (31) and (37)

$$E_{n(k)} = E^{r(\psi(k))}_{m^{(\psi(k))}(\phi(k))}, \qquad k = 1, 2, \ldots;$$

so, from (36),

$$x \in E_{n(k)}, \qquad k = 1, 2, \ldots.$$

Consequently x belongs to the nucleus of the system $[E_{n(k)}]$.

It follows that the systems $[E^{r(s)}]$ and $[E_{n(k)}]$ have the same nucleus. But the sets $E_{n(k)}$ belong to the family $\mathbf{F}$ by (31); Theorem 108 is therefore proved.

COROLLARY. *The sum and intersection of a countable aggregate of sets which belong to a family* $\mathbf{F}$ *are results of the operation* A *on the sets of* E.

Proof. If $E = T_1 + T_2 + T_3 + \ldots$, where $T_n \in \mathbf{F}$ for $n = 1, 2, \ldots$, then. for every finite combination $n(k)$ of indices, put

$$E_{n(k)} = T_{n_1}.$$

It is obvious that E is the nucleus of the system $[E_{n(k)}]$. If $E = T_1 \,.\, T_2 \,.\, T_3 \ldots$, put

$$E_{n(k)} = T_k$$

for every combination $n(k)$ of indices. Again, it is clear that E is the nucleus of the system $[E_{n(k)}]$.

Let $\mathbf{F}$ be a given family of sets and let $A(\mathbf{F})$ denote the family of all sets which are results of the operation A on the sets of $\mathbf{F}$. Then Theorem 108 may be expressed by the relation

$$(38) \qquad A(A(\mathbf{F})) = A(\mathbf{F}).$$

If we denote by $S(\mathbf{F})$ and $P(\mathbf{F})$ the families of all sets which are sums and intersections of a countable aggregate of sets which belong to $\mathbf{F}$, then the corollary to Theorem 108 can be expressed in the form

$$(39) \qquad S(\mathbf{F}) \subset A(\mathbf{F}) \text{ and } P(\mathbf{F}) \subset A(\mathbf{F}).$$

Relations (38) and (39) hold for every family $\mathbf{F}$ of sets.

These relations give

$$(40) \qquad \mathbf{F} \subset A(\mathbf{F})$$

for every family $\mathbf{F}$ of sets.

THEOREM 109. *If a family* **F** *of sets contained in a complete space is a topological invariant and if the intersection of every member of* **F** *with a* G_δ *belongs to* **F**, *then the family* $A(\mathbf{F})$ *is a topological invariant.*

Proof. Let E be a set of the family $A(\mathbf{F})$. Then

$$(41) \qquad E = \sum E_{n(1)} \cdot E_{n(2)} \ldots E_{n(k)} \ldots,$$

where the sets $E_{n(k)}$ belong to **F** and the summation extends over all infinite sequences $n(k)$, $k = 1, 2, \ldots$, of indices. Let T be a homeomorphic image of E. By Theorem 100, there exist two sets P and Q, each a G_δ, and a function f defined in P such that $E \subset P$, $T \subset Q$, $P\, h_f\, Q$, and $f(E) = T$.

Put $P \cdot E_{n(k)} = Y_{n(k)}$; these sets belong to **F** since they are intersections of a G_δ with sets which belong to **F**. It follows that

$$E = \cdot \sum Y_{n(1)} \cdot Y_{n(2)} \cdot Y_{n(3)} \ldots$$

and so, since f is (1, 1) in $E \subset P$,

$$T = f(E) = \sum f(Y_{n(1)}) \cdot f(Y_{n(2)}) \cdot f(Y_{n(3)}) \ldots;$$

since the sets $f(Y_{n(k)})$ belong to **F**, T belongs to the family $A(\mathbf{F})$. This establishes the theorem.

85. Every Borel set an analytic set. Let **C** denote the family of all closed sets contained in a metric space; it follows from the definitions of analytic sets (§ **82**) and of the operation $A(\mathbf{F})$ (§ **83**) that $A(\mathbf{C})$ is the family of all analytic sets. From (38) we obtain

$$A(A(\mathbf{C})) = A(\mathbf{C});$$

this gives

THEOREM 110. *A result of the operation A carried out on analytic sets of a given metric space is an analytic set.*

From (39) and (38) we get

$$S(A(\mathbf{C})) \subset A(A(\mathbf{C})) = A(\mathbf{C}) \text{ and } P(A(\mathbf{C})) \subset A(\mathbf{C});$$

hence the sum and the intersection of a countable aggregate of analytic sets are analytic sets.

Finally, from (40),

$$\mathbf{C} \subset A(\mathbf{C}),$$

i.e., *every closed set is an analytic set.*

The family **A** of all analytic sets is therefore one of the families **B*** which satisfy the condition 1′, 2′, and 3′ of § **81**, and, since every one of such families **B*** contains the family **B** of all Borel sets (§ **81**), we have

THEOREM 111. *Every Borel set contained in a metric space is an analytic set.*

86. Regular defining system. A defining system $[E_{n(k)}]$ is called *regular* if the sets $E_{n(k)}$ are closed and satisfy the conditions

$$\delta(E_{n(k)}) < 1/k, \tag{42}$$

$$E_{n(k+1)} \subset E_{n(k)}, \tag{43}$$

$$E_{n(k)} \neq 0 \tag{44}$$

for every finite combination of indices $n(k+1)$.

THEOREM 112. *Every non-empty analytic set contained in a separable complete space is the nucleus of a certain regular system.*

Proof. Let E denote a given non-empty analytic set contained in a separable complete space M. Then E is the nucleus of some system $[F_{n(k)}]$, where the sets $F_{n(k)}$ are closed.

Let k be a given natural number. Since M is separable it is the sum of a countable aggregate of sets whose diameters are $\leqslant 1/k$; moreover, it may be supposed that these sets are closed (replacing them by their closures if necessary). We may therefore write

$$M = \sum_{n=1}^{\infty} M_n^{(k)}, \tag{45}$$

where $M_n^{(k)}$ is closed and

$$\delta(M_n^{(k)}) \leqslant 1/k, \qquad k = 1, 2, \ldots \tag{46}$$

Put

$$E_n = M_n^{(2)}, \qquad n = 1, 2, \ldots, \tag{47}$$

and, for all n_1 and n_2,

$$E_{n_1, n_2} = E_{n_1}; \tag{48}$$

finally, for $k > 1$ and for every finite combination $n_1, n_2, \ldots, n_{2k}$ of $2k$ natural numbers, put

$$E_{n_1, n_2, \ldots, n_{2k-1}} = E_{n_1, n_2, \ldots, n_{2k}} = F_{n_2, n_4, \ldots, n_{2k-2}} \cdot M_{n_{2k-1}}^{(2k)}. \tag{49}$$

It follows from (47), (48), (49), and (46) that condition (42) is satisfied for every finite combination of indices $n_1, n_2, \ldots, n_k$; also the sets $E_{n(k)}$ are closed. It will be shown that the set E is the nucleus of the system $[E_{n(k)}]$.

Suppose that $x \in E$. Since E is the nucleus of the system $[F_{n(k)}]$ there exists an infinite sequence $m(k)$, $k = 1, 2, \ldots$, of indices such that

$$x \in F_{m(k)}, \qquad k = 1, 2, \ldots. \tag{50}$$

Since $x \in M$ it follows from (45) that there exists for every natural number k an index i_k such that

$$x \in M_{i_k}^{(2k)}, \qquad k = 1, 2, \ldots. \tag{51}$$

Denote by $n_1, n_2, n_3, \ldots$ the successive terms of the infinite sequence

$$i_1, m_1, i_2, m_2, i_3, m_3, \ldots;$$

then (51), (47), (48), and (49), give

$$(52) \qquad x \in E_{n(k)}, \qquad k = 1, 2, \ldots;$$

hence x belongs to the nucleus of the system $[E_{n(k)}]$. Suppose, on the other hand, that x is an element of the nucleus of the system $[E_{n(k)}]$. Then there exists an infinite sequence $n_1, n_2, n_3, \ldots$ of indices for which (52) holds. Therefore, from (49),

$$(53) \qquad x \in F_{n_2, n_4, \ldots, n_{2k-2}}, \qquad k = 2, 3, 4, \ldots.$$

Consequently x belongs to the nucleus of the system $[F_{n(k)}]$ and so to the set E.

It follows that the set E is the nucleus of the system $[E_{n(k)}]$. We have therefore shown thus far that every analytic set E is the nucleus of a certain system $[E_{n(k)}]$, where the sets $E_{n(k)}$ are closed and satisfy condition (42).

If, for every finite combination $n(k)$ of indices, we put

$$(54) \qquad X_{n(k)} = E_{n(1)} \cdot E_{n(2)} \ldots E_{n(k)},$$

then the sets $X_{n(k)}$ will be closed,

$$(55) \qquad \delta(X_{n(k)}) < 1/k \qquad \text{by (42)},$$

and

$$(56) \qquad X_{n(k+1)} \subset X_{n(k)}.$$

The set E is clearly the nucleus of the system $[X_{n(k)}]$.

If the set E is not empty there exists an element x_0 of E. Corresponding to a given finite combination $r(s)$ of indices, put

$$(57) \qquad X^{r(s)} = \sum X_{r(s),n(1)} \cdot X_{r(s),n(2)} \cdot X_{r(s),n(3)} \cdots$$

where the summation ranges over all infinite sequences $n_1, n_2, \ldots$ of natural numbers. If, for a given set $r(s)$ of indices, the set (57) is not empty, denote one of its elements by $x_{r(s)}$; on account of (56) and (57) this will be an element of E.

We next define, for every finite combination $r(s)$ of indices, sets $Y_{r(s)}$ as follows:

If $X^{r(s)} \neq 0$, then $Y_{r(s)} = X_{r(s)}$.

If $X^{r(s)} = 0$ and $X^{r(1)} = 0$, then $Y_{r(s)} = \{x_0\}$.

If $X^{r(s)} = 0$ and $X^{r(1)} \neq 0$, and if $p + 1$ is the smallest index such that $X^{r(p+1)} = 0$ (i.e., $0 < p < s$, and $X^{r(p)} \neq 0$), put $Y_{r(s)} = \{x_{r(p)}\}$.

It follows from (57), (56), and (55) that the sets $Y_{r(s)}$ satisfy the conditions

$$\delta(Y_{n(k)}) \leqslant 1/k,$$

$$Y_{n(k+1)} \subset Y_{n(k)},$$

and $$Y_{n(k)} \neq 0$$

for every finite combination $n(k+1)$. Furthermore, it follows readily from the definition of these sets, from (57) and (55) and from the fact that E is the nucleus of the system $[X_{n(k)}]$, that E is the nucleus of the system $[Y_{n(k)}]$. Theorem 112 is therefore proved.

*Concerning linear analytic sets we state here, without proof, the following theorem:[17]

A bounded linear set is analytic if and only if it is the nucleus of a defining system $[D_{n(k)}]$ *such that*

(i) *each set* $D_{n(k)}$ *is a closed interval*

$$l2^{1-k} \leqslant x \leqslant (l+1)2^{1-k},$$

where l *is some integer dependent on the sequence* $n(k)$;

(ii) *for every finite sequence* $n(k+1)$ *of natural numbers*

$$D_{n(k+1)} \subset D_{n(k)};$$

(iii) *for every natural number* k *there exists only a finite number of different sets* $D_{n(k)}$.*

87. Condition for a set to be analytic.

THEOREM 113. *A non-empty set* E *contained in a complete separable space is analytic if and only if it is a continuous image of the set of all irrational numbers.*

Proof. Let E be a given non-empty analytic set. Then, by Theorem 112, it is the nucleus of some regular system $[E_{n(k)}]$, where $E_{n(k)}$ are closed sets satisfying conditions (42), (43), and (44). Let x denote a given irrational number, and

$$x = [x] + \frac{1}{n_1 +} \, \frac{1}{n_2 +} \cdots \tag{58}$$

the development of x as a continued fraction. Put

$$F(x) = E_{n(1)} \cdot E_{n(2)} \cdot E_{n(3)} \cdot \ldots. \tag{59}$$

The set $F(x)$ consists of only one element of the space M by (42), (43), (44), and Theorem 94; denote this element by $f(x)$. The function $f(x)$ is thus defined for every irrational x and, as seen from its definition, it assumes values in the set E. On the other hand, it is evident that every point of E is one of the values of the function $f(x)$ for an irrational x. For if $p \in E$, there exists an infinite sequence $n(k)$, $k = 1, 2, \ldots$, of indices, such that $p \in E_{n(k)}$ for $k = 1, 2, \ldots$ and so, if x is an irrational number defined by (58), we conclude that $p \in F(x)$; since $F(x)$ consists of only one element, it follows that $p = f(x)$. Consequently, E is the set of all values of the function $f(x)$ for x irrational.

We next show that $f(x)$ is continuous in the set of irrational numbers. Let

$$x_0 = [x_0] + \frac{1}{n_1^0 +} \frac{1}{n_2^0 +} \frac{1}{n_3^0 +} \cdots \tag{60}$$

denote a given irrational number, and ϵ a given positive number. Let k be a natural number such that

$$1/k < \epsilon. \tag{61}$$

It follows from the properties of continued fractions that, corresponding to the numbers x_0 and k, there exists a positive number η such that every irrational number x which satisfies the inequality

$$|x - x_0| < \eta \tag{62}$$

may be expressed as the continued fraction (58) such that

$$n_i = n_i^0, \qquad i = 1, 2, \ldots, k;$$

therefore

$$E_{n(k)} = E_{n^0(k)}. \tag{63}$$

But, from the definition of $f(x)$ and from (58) and (59), we have

$$f(x) \in E_{n(k)}, \; f(x_0) \in E_{n^0(k)};$$

hence, from (63), (42), and (61), we obtain

$$\rho(f(x), f(x_0)) < \epsilon. \tag{64}$$

Thus for every irrational number x_0 and every positive number ϵ there exists a positive number η such that the inequality (62) implies the inequality (64); hence the function $f(x)$ is continuous in the set of all irrational numbers. Consequently, the condition of Theorem 113 is necessary.

Let $f(x)$ be a function of a real variable defined and continuous in the set of all irrational numbers and assuming values in a complete separable metric space M. It will be shown that the set of all values of $f(x)$ for irrational x (a set which is obviously not empty) is an analytic set.

Since the sum of a countable aggregate of analytic sets is an analytic set (§ **85**) it is sufficient to show that the set E of all values of $f(x)$ for irrational x in the interval $(k, k + 1)$ or in the interval $(0, 1)$ is an analytic set.

Let $n_1, n_2, \ldots, n_k$ denote any finite sequence of natural numbers. Denote by $X_{n(k)}$ the set of all irrational numbers x in the interval $(0, 1)$ whose developments as continued fractions have

$$\frac{1}{n_1 +} \frac{1}{n_2 +} \cdots \frac{1}{n_{k-1} +} \frac{1}{n_k}$$

for their kth convergents, and put

$$E_{n(k)} = f(X_{n(k)}); \tag{65}$$

these will be closed sets (not necessarily bounded). It follows that E is the nucleus of the system $[E_{n(k)}]$.

For if $p \in E$, then by the definition of the set E there exists an irrational x in the interval (0, 1) such that $f(x) = p$; let

$$x = \frac{1}{n_1 +} \frac{1}{n_2 +} \frac{1}{n_3 +} \cdots$$

be the development of x as a continued fraction. Then, from the definition of the sets $X_{n(k)}$,

$$x \in X_{n(k)}, \qquad k = 1, 2, \ldots$$

and so, from (65)

$$f(x) \in E_{n(k)}, \qquad k = 1, 2, \ldots;$$

hence the point $p = f(x)$ belongs to the nucleus of the system $[E_{n(k)}]$.

On the other hand, let p denote an element of the nucleus of the system $[E_{n(k)}]$. Then there exists an infinite sequence $n_1^0, n_2^0, n_3^0, \ldots$ of indices, such that

$$p \in E_{n^0(k)}, \qquad k = 1, 2, \ldots. \tag{66}$$

Put

$$x_0 = \frac{1}{n_1^0 +} \frac{1}{n_2^0 +} \frac{1}{n_3^0 +} \cdots; \tag{67}$$

this will be an irrational number of the interval (0, 1). Let ϵ denote a given positive number. Since $f(x)$ is continuous in the set of irrational numbers, there exists a positive number η, dependent on x_0 and ϵ, and such that the inequality

$$|x - x_0| < \eta \tag{68}$$

implies that

$$\rho(f(x), f(x_0)) < \epsilon \tag{69}$$

for irrational x of the interval (0, 1).

Moreover, it follows from the properties of continued fractions that, corresponding to every x_0 and η, there exists a natural number k such that every irrational x whose kth convergent is the same as that of (67), i.e., every number of the set $X_{n0(k)}$, satisfies the inequalities (68) and (69). Hence

$$\delta(f(X_{n^0(k)})) \leqslant \epsilon;$$

from (65),

$$\delta(E_{n^0(k)}) < \epsilon. \tag{70}$$

But from (67) and from the definition of the sets (65) we have

$$f(x_0) \in E_{n^0(k)}$$

and so, from (66) and (70),

$$\rho(p, f(x_0)) < \epsilon.$$

Since ϵ is arbitrary, this gives $p = f(x_0)$, i.e., $p \in E$. This proves that E is the nucleus of the system $[E_{n(k)}]$, where the sets $E_{n(k)}$ are closed; consequently E is an analytic set. The condition of Theorem 113 is thus seen to be sufficient.

We remark in connection with this theorem that *a linear set is analytic if and only if it is the set of all values of a real function* (defined in the set of all real numbers) *which is continuous on the right.*[18]

Linear analytic sets can be obtained from closed plane sets as follows: Let F be a closed plane set bounded on every line parallel to the y-axis; on each such line which meets F we take a point of F with greatest ordinate; the set so obtained is then projected on the y-axis. It can be shown that this projection is always an analytic set and that, by a suitable choice of the closed plane set F, this projection coincides with any given analytic linear set.[19]

88. Continuous images of analytic sets. Topological invariance of their complements.

THEOREM 114. *A continuous image of an analytic set contained in a separable complete space is an analytic set.*

Proof. Let E denote an analytic set and T its continuous image. Then there exists a function $f(p)$ defined and continuous in E such that $T = f(E)$. But, by Theorem 113, there exists a function $\phi(x)$, defined and continuous in the set N of all irrational numbers, such that $E = \phi(N)$. Put $\psi(x) = f(\phi(x))$ for irrational x; the function $\psi(x)$ is continuous in N and $T = \psi(N)$. Hence, by Theorem 113, T is an analytic set.

Theorems 111 and 114 lead to the following

COROLLARY. *A continuous image of a Borel set contained in a separable complete space is an analytic set.*

Hurewicz[20] has shown that all analytic sets are continuous images of every linear analytic set which is not an F_σ. It can also be shown[21] that of two analytic linear sets, neither of which contains an interval, at least one is a continuous image of the other except in the case when one of the sets is countable and not closed and the other is non-countable, closed, and bounded.

Furthermore, a linear set E is analytic if and only if there exists a real continuous function $f(x)$ such that E is the set of all values of $f'(x)$, the derivative of $f(x)$, for all real values of x for which $f(x)$ has a finite derivative.

However, Theorem 114 and the Corollary to it need not hold in non-separable complete spaces.

In fact, let M be the set of all real numbers and let the distance between two numbers x and $y \neq x$ be defined as follows: If $0 \leqslant x \leqslant 1$ and $0 \leqslant y \leqslant 1$ then $\rho(x, y) = |x - y|$; otherwise $\rho(x, y) = 1$. It is clear that M is complete under this metric and the set of all real numbers in the interval $[1, 2]$ is closed. It is obvious that every subset of the interval $[0, 1]$ of cardinal **c** is a continuous and $(1, 1)$ image of the isolated set of all numbers in the interval $[1, 2]$. There exist, however, as will be shown later, non-analytic sets in the interval $[0, 1]$ and hence in the space M. Thus in a complete space a continuous $(1, 1)$ image of a closed set need not be analytic.

We have, however,

THEOREM 115. *In a complete space (not necessarily separable) a homeomorphic image of an analytic set is an analytic set.*

Proof. Let E denote an analytic set contained in a complete space M and let E be the nucleus of the defining system $[E_{n(k)}]$ consisting of the closed sets of M; let $T \subset M$ be a homeomorphic image of E. There exist, by Theorem 100, sets P and Q, each a G_δ and contained in M, such that $E \subset P$, $T \subset Q$, $P \; h_\phi \; Q$, and $E \; h_\phi \; T$. For every finite sequence $n(k)$ of indices, put

$$\phi(P \, . \, E_{n(k)}) = T_{n(k)};$$

since the set $P \, . \, E_{n(k)}$ is a G_δ, the set $T_{n(k)}$ is a G_δ by Theorem 101. Thus, by Theorems 111 and 108, the nucleus of the defining system $[T_{n(k)}]$ is an analytic set. Since $E \subset P$, we have $E = \sum P \, . \, E_{n(1)} \, . \, P \, . \, E_{n(2)} \, . \, . \, . \,$, and so

$$T = \phi(E) = \sum \phi(P \, . \, E_{n(1)}) \, . \, \phi(P \, . \, E_{n(2)}) \, . \, . \, . = \sum T_{n(1)} \, . \, T_{n(2)} \, . \, . \, . \, .$$

The set T is therefore the nucleus of the system $[T_{n(k)}]$ and so an analytic set. This proves Theorem 115.

Now let **A** denote the family of all analytic sets contained in a given complete space. The sets F_σ and G_δ, being Borel sets, belong to the family **A**. Furthermore, since the sum and intersection of two analytic sets are analytic sets (§ **85**), it follows from Theorem 115 that the family **A** satisfies the conditions of Theorem 104. Consequently the family of the complements of all sets of the family **A** is a topological invariant. We thus have

THEOREM 116.[22] *A homeomorphic image of the complement of an analytic set contained in a complete space is the complement of an analytic set.*

We note, however, that a continuous image (even $(1, 1)$ and continuous in one direction[23]) of the complement of an analytic linear set need not be the complement of an analytic set.

89. A new condition for a set to be analytic.

THEOREM 117. *If $f(x)$ is a function of a real variable x and if $f(x)$ takes on values in a metric space, then the set of all values of x at which the function is semi-continuous is at most countable.*[24]

Proof. Let x_0 denote a real number at which $f(x)$ is continuous on the left only. Since $f(x)$ is not continuous on the right at x_0, there exists a positive rational number u such that the inequality

$$\rho(f(x), f(x_0)) < 2u \qquad \text{for } x_0 < x < x_0 + \eta \tag{71}$$

is not satisfied for any positive η. On the other hand, since $f(x)$ is continuous on the left at x_0, there exists a rational number $v < x_0$ such that

$$\rho(f(x), f(x_0)) < u \qquad \text{for } v < x < x_0. \tag{72}$$

The numbers u and v may be chosen to be the first terms of a certain infinite sequence of rational numbers such that u does not satisfy (71) for any positive η and v (after u has been selected) satisfies (72). In this manner, a pair (u, v) of rational numbers corresponds to every element of the set E of all real numbers for which the function $f(x)$ is continuous on the left only. Different pairs correspond to different elements of E.

To prove this, suppose that the same pair is assigned to x_0 and to the number $x_1 > x_0$. We then have (72) and

$$\rho(f(x), f(x_1)) < u \text{ for } v < x < x_1, \text{ where } v < x_1. \tag{73}$$

Since $v < x_0 < x_1$, we may assume $x = x_0$ in (73); this gives $\rho(f(x_0), f(x_1)) < u$ and so, from (73),

$$\rho(f(x), f(x_0)) \leqslant \rho(f(x), f(x_1)) + \rho(f(x_1), f(x_0)) < 2u$$

for $v < x < x_1$ and, certainly, for $x_0 < x < x_1$. If we put $\eta = x_1 - x_0$, we have $\eta > 0$ and (71) satisfied, contrary to the definition of the number u. Hence to different elements of E correspond different pairs (u, v) of rational numbers and, since the set of all pairs of rational numbers is countable, the set E is at most countable. Similarly, it could be proved that the set of all values of x at which a function f is continuous on the right only is at most countable. Theorem 117 is therefore proved.

Theorem 117 will obviously remain true if the function $f(x)$ is defined only in a subset of the set of all real numbers.

THEOREM 118. *A non-empty set E contained in a separable complete space is analytic if and only if it is the set of values of a function of a real variable which is continuous on the left in the set of all real numbers.*[25]

Proof. Let E denote a given non-empty analytic set contained in a complete and separable space. Then, by Theorem 112, E is the nucleus of a certain regular system $[E_{n(k)}]$, where the sets $E_{n(k)}$ are closed and satisfy conditions (42), (43), and (44).

Let x denote a given real number. Corresponding to every real number x there exists a definite infinite sequence $n_1, n_2, \ldots$ of natural numbers such that

(74) $$x = [x] + 2^{-n_1} + 2^{-n_1-n_2} + 2^{-n_1-n_2-n_3} + \ldots.$$

Put
$$F(x) = E_{n(1)} \cdot E_{n(2)} \cdot E_{n(3)} \cdot \ldots.$$

As in the proof of Theorem 113, we show that $F(x)$ consists of one element only and, if this element be denoted by $f(x)$, then E is the set of all values of the function $f(x)$ for x real.

We now prove that the function $f(x)$ is continuous on the left in the set of all real numbers. Let

(75) $$x_0 = [x_0] + 2^{-n_1^0} + 2^{-n_1^0-n_2^0} + 2^{-n_1^0-n_2^0-n_3^0} + \ldots$$

denote a given real number and ϵ an arbitrary positive number. Let k be a natural number such that

(76) $$1/k < \epsilon;$$

put

(77) $$x_1 = [x_0] + 2^{-n_1^0} + 2^{-n_1^0-n_2^0} + \ldots + 2^{-n_1^0-n_2^0-\ldots-n_k^0}.$$

Then, from (75), $x_1 < x_0$.

Let x be a real number such that

(78) $$x_1 < x < x_0$$

and let (74) be the development of x as an infinite binary fraction. It follows from (74), (75), (77), and (78) that

$$n_i = n_i^0, \qquad i = 1, 2, \ldots, k:$$

consequently

(79) $$E_{n(k)} = E_{n^0(k)}.$$

But

$$f(x) \in E_{n(k)}, \; f(x_0) \in E_{n^0(k)};$$

therefore, from (79), (42), and (76),

(80) $$\rho(f(x), f(x_0)) < \epsilon.$$

Thus, corresponding to every real number x_0 and every positive number ϵ, there exists a number $x_1 < x_0$ such that (78) implies (80); this proves that the function $f(x)$ is continuous on the left in the set of all real numbers. The condition of Theorem 118 is therefore necessary.

Let now $f(x)$ denote a function of a real variable, continuous on the left for every real x, and taking on values in a separable complete space. Denote by E the set of all values of $f(x)$ for x real; it will be shown that E is an analytic set.

Let M be the set of all values of x for which f is continuous on both sides, and N the set of all remaining values of x, that is, of those for which f is continuous on the left only. The set N is at most countable by Theorem **117**; it is therefore a set F_σ and so M is a set G_δ, that is, a Borel set. The function f is continuous in M; hence $f(M)$ is a continuous image of a Borel set and so, by the Corollary to Theorem **114**, an analytic set. The set $f(N)$ is at most countable; it is therefore an F_σ and so an analytic set. The set $E = f(M + N) = f(M) + f(N)$ is therefore the sum of two analytic sets and hence an analytic set. The condition of the theorem is therefore sufficient and Theorem **118** is proved.

90. Generalized sieves. Let P denote a given plane set and W a given property of linear sets. Denote by $\Gamma_W(P)$ the set of all real numbers a such that the linear set of points in which the line $x = a$ meets P has the property W. For a given family $\mathbf{F}$ of sets, denote by F_Γ the family of all sets $\Gamma_W(P)$ where $P \in \mathbf{F}$. The operation Γ_W with respect to a family of sets (and depending on a property W of linear sets) is called *the operation of the generalized sieve.*

In particular, if W is the property of "being a non-empty set," the operation Γ_W is identical with the operation of projection on the x-axis.

Furthermore, if $\mathbf{F}$ be the family of all closed plane sets and W the property "consisting of only one element," then the family F_{Γ_W} is the family of all linear sets $F_{\sigma\delta}$.[26]

If W is the property of "consisting of more than one point" then F_{Γ_W} is the family of all linear sets F_σ.[27]

If W is the property of "being an infinite set" then F_{Γ_W} is the family of all linear sets $F_{\sigma\delta}$.[28]

If W is the property of "being a non-countable set" then F_{Γ_W} is the family of all linear analytic sets.[29]

If W is the property of "being a set ordered according to increasing ordinates but not well-ordered"[30] then F_{Γ_W} is the family of all linear analytic sets.

If W denotes the property of "containing at least one point with an irrational ordinate" F_{Γ_W} is the family of all linear analytic sets.[31]

We shall prove the last statement. Let F denote a given closed plane set. Removing from F all points on the lines $y = b$, where b is a rational number, we obtain all points of F with irrational ordinates. The sum S of the above lines is obviously an F_σ; hence the set $E = F - S$ is a G_δ. The set $\Gamma_W(F)$ is the projection of the set E on the x-axis, that is, the projection of a set G_δ and so, by the corollary to Theorem **114**, an analytic set.

On the other hand, let E denote a given analytic linear set. Then by Theorem **113** there exists a function $x = f(y)$ defined and continuous in the set N of irrational numbers such that $f(N) = E$. Let T be the set of points (x, y) such that $y \in N$ and $x = f(y)$. Since f is continuous in N the set $\bar{T} - T$ is contained in the sum of all the lines $y = b$, where b is a rational number.

Consequently $\Gamma_W(\bar{T}) = \Gamma_W(T)$. It is evident that $\Gamma_W(T) = E$; hence $E = \Gamma_W(\bar{T})$ and, since $\bar{T}$ is closed, $E \in F_{\Gamma_W}$. This proves the theorem.

*So far the family **F** consisted of plane closed sets. Now let **F** denote the family of all plane Borel sets.

It can be shown (although the proof is not easy) that if W is the property of "consisting of one element" or of "being countable" then F_{Γ_W} is the family of all complements (with respect to the line) of analytic linear sets.[32] A similar result is obtained when W is the property of "being a non-empty closed set."[33] It follows from this that if E is a plane Borel set which is met by every parallel to the y-axis in a closed set then the projection of E on the x-axis is a Borel set.[34]

Finally, let **F** denote the family of all plane analytic sets. It can be shown that if W denotes the property of "consisting only of one element" then F_{Γ_W} is the family of all linear sets each of which is the difference of two analytic sets.[35] And if W denotes the property of "consisting of more than one element" or of "being infinite" or of "being non-countable" then F_{Γ_W} is the family of all analytic linear sets.[36]

We remark further that sieves were also considered in metric spaces (where the plane was replaced by the parallelotope contained in the metric space).[37]*

91. The power of an analytic set.

LEMMA. *Let E be a given set contained in a separable metric space and let S be an open sphere such that the set $E \cdot S$ is not countable; then there exist open spheres S_0 and S_1 of arbitrarily small radii such that $S_0 \subset S$, $S_1 \subset S$, $\bar{S}_0 \cdot \bar{S}_1 = 0$, and the sets $E \cdot S_0$, $E \cdot S_1$ are non-countable.*

Proof. The non-countable set $E \cdot S$ contains, by Theorem 69, a non-countable subset of elements of condensation; let p_0 and p_1 be two of them. Since $p_0 \in S$ and $p_1 \in S$, and since S is open, we have, for r_0 and r_1 sufficiently small, $S_0 = S(p_0, r_0) \subset S$, and $S_1 = S(p_1, r_1) \subset S$, where it may be supposed that $r_0 + r_1 < \rho(p_0, p_1)$; this results in $\bar{S}_0 \cdot \bar{S}_1 = 0$. Furthermore, since S_0 and S_1 are open sets containing p_0 and p_1 respectively, and since p_0 and p_1 are elements of condensation of $E \cdot S$, we conclude that the sets $E \cdot S_0$ and $E \cdot S_1$ are non-countable. The lemma is therefore proved.

THEOREM 119. *Every non-countable analytic set contained in a separable complete space contains a non-empty perfect subset.*[38]

Proof. Let E denote a non-countable analytic set contained in a separable complete space; then it is the nucleus of a regular system $[E_{n(k)}]$.

For every finite combination $r(s)$ of indices, put

$$(81)\qquad E^{r(s)} = \sum E_{r(1)} \cdot E_{r(2)} \cdots E_{r(s)} \cdot E_{r(s),n(1)} \cdot E_{r(s),n(2)} \cdot E_{r(s),n(3)} \cdots,$$

where the summation ranges over all infinite sequences $n(k)$, $k = 1, 2, \ldots$, of natural numbers.

It follows from (81) that

$$E = E^1 + E^2 + E^3 + \dots, \tag{82}$$

and

$$E^{r(s)} = E^{r(s),1} + E^{r(s),2} + E^{r(s),3} + \dots \tag{83}$$

for every finite combination $r(s)$ of indices.

With each finite combination $\alpha(k)$ of the numbers 0 and 1 associate a sphere $S_{\alpha(k)}$ and a natural number $m_{\alpha(k)}$ subject to the following conditions:

$$\delta(S_{\alpha(k)}) < 1/k, \tag{84}$$

$$S_{\alpha(k)} \subset S_{\alpha(k-1)},\text{[39]} \tag{85}$$

$$\bar{S}_{\alpha(k-1),0} \cdot \bar{S}_{\alpha(k-1),1} = 0, \tag{86}$$

$$E^{m_{\alpha(1)}, m_{\alpha(2)}, \dots, m_{\alpha(k)}} \cdot S_{\alpha(k)} \tag{87}$$

is non-countable. We shall show that such a correlation is possible.

Since E is non-countable, it contains an element of condensation p. Put $S = S(p, 1)$; the set $E \cdot S$ is non-countable; we may therefore apply to it our lemma. Hence there exist spheres S_0 and S_1 such that $\bar{S}_0 \cdot \bar{S}_1 = 0$, $\delta(S_0) < 1$, $\delta(S_1) < 1$, and the sets $E \cdot S_0$ and $E \cdot S_1$ are non-countable. From (82)

$$E \cdot S_0 = E^1 \cdot S_0 + E^2 \cdot S_0 + E^3 \cdot S_0 + \dots$$

and so, since $E \cdot S_0$ is non-countable, there exists an index m_0 such that the set $E^{m_0} \cdot S_0$ is non-countable. Similarly we deduce the existence of m_1 such that the set $E^{m_1} \cdot S_1$ is non-countable.

Now let k be a given natural number and suppose that we have already defined all spheres $S_{\alpha(k)}$ and natural numbers $m_{\alpha(k)}$ (where $\alpha(k)$ is any combination of k numbers, each either 0 or 1) so as to have conditions (84), (85), (86), and (87) satisfied. Let $\alpha(k)$ be any combination of k numbers consisting of the numbers 0 and 1. It follows from (87) and our lemma that there exist two spheres $S_{\alpha(k),0}$ and $S_{\alpha(k),1}$ such that

$$S_{\alpha(k),0} \subset S_{\alpha(k)}, \quad S_{\alpha(k),1} \subset S_{\alpha(k)},$$

$$\bar{S}_{\alpha(k),0} \cdot \bar{S}_{\alpha(k),1} = 0,$$

$$\delta(S_{\alpha(k),0}) < 1/(k+1), \delta(S_{\alpha(k),1}) < 1/(k+1)$$

and the sets

$$E^{m_{\alpha(1)}, m_{\alpha(2)}, \dots, m_{\alpha(k)}} \cdot S_{\alpha(k),0}$$

and

$$E^{m_{\alpha(1)}, m_{\alpha(2)}, \dots, m_{\alpha(k)}} \cdot S_{\alpha(k),1}$$

are non-countable. From this last property and from (83) we deduce the existence of indices $m_{a(k),0}$ and $m_{a(k),1}$ such that the sets

$$E^{m_{a(1)},m_{a(2)},\ldots,m_{a(k),0}} \,.\, S_{a(k),0}$$

and

$$E^{m_{a(1)},m_{a(2)},\ldots,m_{a(k),1}} \,.\, S_{a(k),1}$$

are non-countable.

The spheres $S_{a(k)}$ and the integers $m_{a(k)}$ which satisfy conditions (84), (85), (86), and (87) are thus defined by induction for every finite combination $a(k)$ of indices consisting of the numbers 0 and 1.

Next denote by F_k, for every natural number k, the set

$$F_k = \sum \bar{E}^{m_{a(1)},m_{a(2)},\ldots,m_{a(k)}} \,.\, \bar{S}_{a(k)}, \tag{88}$$

where the summation ranges over all combinations of k numbers $a_1, a_2, \ldots, a_k$, each of which is 0 or 1. The sets F_k are closed and bounded (since they are sums of a finite number of closed and bounded sets); it follows readily from (83) and (85) that

$$F_{k+1} \subset F_k, \qquad k = 1, 2, \ldots. \tag{89}$$

and from (87) that

$$F_k \neq 0, \qquad k = 1, 2, \ldots. \tag{90}$$

The sequence $F_1, F_2, F_3, \ldots$ is therefore a descending sequence of closed sets; hence the set

$$F = F_1 \,.\, F_2 \,.\, F_3 \ldots \tag{91}$$

is closed and non-empty.

Let p denote a given element of the set F. From (91) $p \in F_1$ and from (88)

$$F_1 = \bar{E}^{m_0} \,.\, \bar{S}_0 + \bar{E}^{m_1} \,.\, \bar{S}_1,$$

where $\bar{S}_0 \,.\, \bar{S}_1 = 0$ by (86). Consequently we have either

$$p \in \bar{E}^{m_0} \,.\, \bar{S}_0 \text{ or } p \in \bar{E}^{m_1} \,.\, \bar{S}_1;$$

in the first case put $\beta_1 = 0$, in the second put $\beta_1 = 1$. Then

$$p \in \bar{E}^{\beta_1} \,.\, \bar{S}_{\beta_1}. \tag{92}$$

But (91) gives $p \in F_2$ and from (88)

$$F_2 = \bar{E}^{m_0,m_{0,0}} \,.\, \bar{S}_{0,0} + \bar{E}^{m_0,m_{0,1}} \,.\, \bar{S}_{0,1} + \bar{E}^{m_1,m_{1,0}} \,.\, \bar{S}_{1,0} + \bar{E}^{m_1,m_{1,1}} \,.\, \bar{S}_{1,1}$$

where, on account of (86) and (85), the terms of the sum F_2 are disjoint sets. Thus the element p of F_2 belongs to only one of the four terms and it is easily deduced from (92) and (85) that this term has the form $\bar{E}^{m_{\beta(1)},m_{\beta(2)}} \,.\, \bar{S}_{\beta(2)}$ ($\beta(2)$ being the abbreviation for the sequence β_1, β_2), where β_2 is one of the numbers 0 and 1.

Again from (91) we have $p \in F_3$ whence, arguing as before, we deduce from (88), (86), and (85) that, for a certain β_3 which is either 0 or 1, we have

$$p \in \bar{E}^{m_{\beta(1)}, m_{\beta(2)}, m_{\beta(3)}} \cdot \bar{S}_{\beta(3)}.$$

Continuing in this manner we obtain an infinite sequence

$$\beta_1, \beta_2, \beta_3, \ldots$$

whose terms are the numbers 0 and 1; further

$$p \in \bar{E}^{m_{\beta(1)}, m_{\beta(2)}, \ldots, m_{\beta(k)}} \cdot \bar{S}_{\beta(k)}, \qquad k = 1, 2, \ldots. \tag{93}$$

Let ϵ denote an arbitrary positive number. Denote by s a natural number such that

$$1/s < \epsilon, \tag{94}$$

and put $\gamma_{s+1} = 1 - \beta_{s+1}$; this will be either 0 or 1 and $\gamma_{s+1} \neq \beta_{s+1}$. Hence from (86)

$$\bar{S}_{\beta(s), \beta_{s+1}} \cdot \bar{S}_{\beta(s), \gamma_{s+1}} = 0. \tag{95}$$

Put

$$\gamma_i = \beta_i \text{ for } i \leqslant s, \gamma_i = 0 \text{ for } i \geqslant s + 2. \tag{96}$$

From (83), (84), (85), and Theorem 94 we conclude that the intersection

$$P = \prod_{k=1}^{\infty} \bar{E}^{m_{\gamma(1)}, m_{\gamma(2)}, \ldots, m_{\gamma(k)}} \cdot \bar{S}_{\gamma(k)} \tag{97}$$

consists of a single point which we shall denote by q. It follows from (97) and (88) that $q \in F_k$ for $k = 1, 2, \ldots$, and so, from (91), $q \in F$. From (93), (95), (96), (97), and the fact that $\gamma_{s+1} \neq \beta_{s+1}$, we find that $p \neq q$.

Finally, from (93), (96), and (97) we have

$$p \in \bar{S}_{\beta(s)} \text{ and } q \in \bar{S}_{\beta(s)};$$

so, from (84) and (94) (since $\delta(\bar{Q}) = \delta(Q)$ for every set Q),

$$\rho(p, q) < \epsilon.$$

Thus, corresponding to every element p of the set F and every positive number ϵ, there exists an element q of F different from p and such that $\rho(p, q) < \epsilon$. Hence p is a limit element of the set F. Consequently F is dense-in-itself and, being closed, is perfect.

It remains to show that $F \subset E$. If $p \in F$ then, from (93),

$$p \in \prod_{k=1}^{\infty} \bar{E}^{m_{\beta(1)}, m_{\beta(2)}, \ldots, m_{\beta(k)}} \cdot \bar{S}_{\beta(k)} \subset \prod_{k=1}^{\infty} \bar{E}^{m_{\beta(1)}, m_{\beta(2)}, \ldots, m_{\beta(k)}}.$$

Since $E_{r(s)}$ is closed for all s and, from (81), $E^{r(s)} \subset E_{r(s)}$, it follows that $\bar{E}^{r(s)} \subset E_{r(s)}$ for every finite combination $r(s)$ of indices; consequently

$$\prod_{k=1}^{\infty} \bar{E}^{m_{\beta(1)}, m_{\beta(2)}, \ldots, m_{\beta(k)}} \subset \prod_{k=1}^{\infty} E_{m_{\beta(1)}, m_{\beta(2)}, \ldots, m_{\beta(k)}} \subset E.$$

The set F is therefore a non-empty perfect subset of E; this proves the theorem.

Theorems 119 and 111 give

THEOREM 120. *A non-countable Borel set E contained in a separable complete space contains a non-empty perfect subset.*[40]

For every infinite sequence $a_1, a_2, \ldots$ of the numbers 0 and 1 put

$$P(a_1, a_2, \ldots) = \prod_{k=1}^{\infty} \bar{E}^{m_{a(1)}, m_{a(2)}, \ldots, m_{a(k)}} \,.\, \bar{S}_{a(k)}.$$

As shown above the set $P\ (a_1, a_2, \ldots)$ consists of only one element of E and to different infinite sequences of the numbers 0 and 1 correspond different elements of E. Hence the set E has cardinal $\geqslant \mathbf{c}$. Since a separable space has cardinal $\leqslant \mathbf{c}$ it follows that the cardinal of E is $\mathbf{c}$. Hence

COROLLARY 1. *A non-countable analytic set contained in a separable complete space has cardinal* **c**.

From Theorem 111 we obtain

COROLLARY 2. *A non-countable Borel set contained in a separable complete space has cardinal* **c**.[41]

92. Souslin's theorem. Two sets P and Q are said to be exclusive B if there exist two Borel sets M and N such that

$$P \subset M,\ Q \subset N, \text{ and } M \,.\, N = 0. \tag{98}$$

LEMMA. *If*

$$P = P_1 + P_2 + P_3 + \ldots, \quad Q = Q_1 + Q_2 + Q_3 + \ldots, \tag{99}$$

and if P and Q are not exclusive B, there exist indices p and q such that the sets P_p *and* Q_q *are not exclusive B.*

Proof. Suppose that the sets P_p and Q_q are exclusive B for all integers p and q. Then for every pair of integers p and q there exist Borel sets $M_{p,q}$ and $N_{p,q}$ such that

$$P_p \subset M_{p,q}, \quad Q_q \subset N_{p,q}, \quad \text{and } M_{p,q} \,.\, N_{p,q} = 0. \tag{100}$$

Put

$$M = \sum_{p=1}^{\infty} \prod_{q=1}^{\infty} M_{p,q}, \quad N = \sum_{q=1}^{\infty} \prod_{p=1}^{\infty} N_{p,q}. \tag{101}$$

It is clear that M and N are Borel sets and, from (99), (100), and (101), we may conclude that (98) is satisfied; this contradicts the hypothesis of the lemma which is therefore proved.

THEOREM 121. *Two disjoint analytic sets contained in a complete separable space are exclusive B.*

Proof.[42] Let E and T be two analytical sets and $[E_{n(k)}]$, $[T_{n(k)}]$ the regular systems defining these sets. For every finite set $r(s)$ of indices let the sets $E^{r(s)}$, $T^{r(s)}$ be defined by (81) of § **91**. We shall then have (82) and (83) and analogous relations for the sets T and $T^{r(s)}$.

Suppose that E and T are disjoint but not exclusive B. By the preceding lemma, there exist indices p_1 and q_1 such that the sets E^{p_1} and T^{q_1} are not exclusive B. Since

$$E^{p_1} = E^{p_1,1} + E^{p_1,2} + E^{p_1,3} + \dots,$$

$$T^{q_1} = T^{q_1,1} + T^{q_1,2} + T^{q_1,3} + \dots$$

and, since E^{p_1} and T^{q_1} are not exclusive B, we deduce as before the existence of indices p_2 and q_2 such that E^{p_1,p_2} and T^{q_1,q_2} are not exclusive B.

Continuing this argument we obtain two infinite sequences of indices $p_1, p_2, p_3, \dots$ and $q_1, q_2, q_3, \dots$ such that the sets

$$E^{p(k)} \text{ and } T^{q(k)} \qquad k = 1, 2, \dots,$$

are not exclusive B. From (81)

$$E^{p(k)} \subset E_{p(k)} \tag{102}$$

and

$$T^{q(k)} \subset T_{q(k)}. \tag{103}$$

If the closed (Borel) sets on the right of (102) and (103) are disjoint, then the sets on the left of (102) and (103) are exclusive B; this contradicts the previous conclusion. Consequently

$$P_k = E_{p(k)} \,.\, T_{q(k)} \neq 0, \qquad k = 1, 2, \dots. \tag{104}$$

Since the systems $[E_{n(k)}]$ and $[T_{n(k)}]$ are regular, we have (42), (43), and (44) of § **86** satisfied, i.e.,

$$\delta(E_{p(k)}) < 1/k, \quad \delta(T_{q(k)}) < 1/k,$$

$$E_{p(k)} \supset E_{p(k+1)},$$

and

$$T_{q(k)} \supset T_{q(k+1)}, \qquad k = 1, 2, \dots;$$

hence, from (104), $P_k \supset P_{k+1}$. Moreover, the sets P_k are closed and non-empty. By Theorem 94, the set $P_1 \,.\, P_2 \,.\, P_3 \ldots$ is not empty. Hence there exists an element $x \in P_k$ for $k = 1, 2, \ldots$, and so, from (104),

$$x \in E_{p(k)} \text{ and } x \in T_{q(k)} \qquad \text{for } k = 1, 2, \ldots;$$

Consequently x belongs to the nucleus of the system $[E_{n(k)}]$ and to the nucleus of the system $[T_{n(k)}]$; it is therefore a common point of the sets E and T contrary to the assumption that $E \,.\, T = 0$. Hence Theorem 121 has been proved.

Suppose that a set E and its complement $T = CE$ are each analytic sets. Since $E \,.\, T = 0$, we may apply Theorem 121 to the sets E and T. Hence there exist two Borel sets M and N such that $E \subset M$, $T \subset N$, and $M \,.\, N = 0$. Thus $M \,.\, T = 0$, i.e., $M \subset CT = E$; but $E \subset M$ and so $E = M$. Thus E is a Borel set. This proves that if the complement of an analytic set E is an analytic set, then E is a Borel set. On the other hand, if E is a Borel set then, by property 4 (§ **81**) of Borel sets, the complement of E is also a Borel set and so, by Theorem 111, the sets E and CE are analytic sets. This gives

THEOREM 122 (Souslin). *In order that a set E contained in a separable complete space be a Borel set it is necessary and sufficient that the set E and its complement be analytic sets.*

From Theorem 122 we obtain the

COROLLARY. *In order that an analytic set contained in a separable complete space be a Borel set, it is necessary and sufficient that its complement be an analytic set.*

We note that Theorem 121 may be generalized to

THEOREM 123. *If $P_1, P_2, P_3, \ldots$ is an infinite sequence of disjoint analytic sets there exist Borel sets $M_1, M_2, M_3, \ldots$ such that*

$$P_k \subset M_k, \qquad k = 1, 2, \ldots, \tag{105}$$

and

$$M_pM_q = 0, \qquad p \neq q. \tag{106}$$

Proof. Let p and q be two different natural numbers. Since the analytic sets P_p and P_q are disjoint there exist, by Theorem 121, Borel sets $M_{p,q}$ and $M_{q,p}$ such that

$$P_p \subset M_{p,q}, \quad P_q \subset M_{q,p} \tag{107}$$

and

$$M_{p,q} \,.\, M_{q,p} = 0. \tag{108}$$

Thus (107) and (108) hold for every pair of different natural numbers p and q.

For every natural k put

(109) $$M_k = \prod_{n \neq k} M_{k,n},$$

where the intersection extends over all natural numbers n different from k. The sets (109) are clearly Borel sets. From (107) and (109) we obtain (105) and from (108) and (109) we get (106) (since for $p \neq q$ (109) gives $M_p \subset M_{p,q}$ and $M_q \subset M_{q,p}$). Theorem 123 is therefore proved.

Note that with a certain modification in its statement the preceding theorem may be generalized to arbitrary spaces (not necessarily metric).[43] However, it does not hold in complete and locally separable spaces.

*In fact, let I denote the closed interval $0 \leqslant x \leqslant 1$ and $M = I \times I$ the combinatorial square. Put

$$\rho((x_1, y), (x_2, y)) = |x_1 - x_2|, \qquad x_1, x_2, y \in I,$$

and

$$\rho((x_1, y_1), (x_2, y_2)) = 1, \text{ for } x_1, x_2, y_1, y_2 \in I, y_1 \neq y_2.$$

It is easily verified that the function ρ satisfies the distance axioms and that the space M with metric ρ is complete and locally separable.

It can be shown[44] that for every ordinal number $\alpha < \Omega$ there exists a Borel set B_α of class α of real numbers. It follows from the axiom of choice that there exists a transfinite sequence $\{y_\alpha\}$, $\alpha < \Omega$, of type Ω consisting of different numbers of the interval I. Let E_α denote the set of all points (x, y_α) of the set M, where $x \in B_\alpha$; then E_α is a Borel set of class α. Put

$$E = \sum_{\alpha < \Omega} E_\alpha;$$

the set E is not a Borel set in M; for if it were a Borel set of class β in M, then the set of all points $(x, y_{\beta+1})$ of the set E (as the intersection of the set E with a set closed in M) would be a Borel set of class $\leqslant \beta$; this is impossible since this set is identical with the set $E_{\beta+1}$.

On the other hand, it is easy to see that the sets E and $M - E$ are analytic. In fact, for $\alpha < \Omega$ the set B_α, as a Borel set, is analytic; it is therefore the nucleus of a defining system $H^\alpha_{n(k)}$ composed of closed sets of the interval I. For every finite set $n(k)$ of natural numbers denote by $E^\alpha_{n(k)}$ the set of all points $(x, y_\alpha) \in M$, for which $x \in H^\alpha_{n(k)}$ and put

$$E_{n(k)} = \sum_{\alpha < \Omega} E^\alpha_{n(k)}.$$

The sets $E_{n(k)}$ are closed in M and the set E is the nucleus of the defining system $[E_{n(k)}]$; therefore E is an analytic set.

Let K denote the set of all points (x, y) of the space M, where $y \in \{y_\alpha\}$, $\alpha < \Omega$. Employing the sets $I - B_\alpha$ (which are Borel sets in I) instead of the sets B_a and arguing as above, we conclude that $K - E$ is analytic in M. But the set $M - K$ is closed in M; hence the set $(M - K) + (K - E)$ is analytic in M. Thus the set E and its complement are both analytic in M but E is not a Borel set in M. Consequently Souslin's theorem does not hold in M.[45*]

93. Biuniform and continuous images of the set of irrational numbers. We have shown in § 87 that a continuous image of the set of all irrational numbers is an analytic set. We now prove

THEOREM 124. *In a separable complete space a* (1, 1) *continuous image of the set of all irrational numbers is a Borel set.*

Proof. Since the set of all irrational numbers is homeomorphic with the set E of all irrational numbers in the interval (0, 1), it is sufficient to show that, if f is a function defined and continuous in E and taking on different values (belonging to a separable complete space) for different elements of E, then $T = f(E)$ is a Borel set.

For every finite combination $n_1, n_2, \ldots, n_k$ or, in our notation $n(k)$, of natural numbers denote by $E_{n(k)}$ the set of all irrational numbers in the interval (0, 1) whose kth convergent is

$$\frac{1}{n_1 +} \frac{1}{n_2 +} \cdots \frac{1}{n_{k-1} +} \frac{1}{n_k}.$$

Each of the sets $E_{n(k)}$ is homeomorphic with the set of all irrational numbers; hence, by Theorem 113, the sets

$$T_{n(k)} = f(E_{n(k)}) \tag{110}$$

are analytic.

Let k denote a given natural number. It follows from the definition of the sets $E_{n(k)}$ that, if $p(k)$ and $q(k)$ are two different sets of k natural numbers, then

$$E_{p(k)} \cdot E_{q(k)} = 0; \tag{111}$$

so, from (13), since f is (1, 1) in E,

$$T_{p(k)} \cdot T_{q(k)} = 0. \tag{112}$$

Furthermore,

$$E = \sum E_{n(k)} \tag{113}$$

where the summation ranges over all combinations of k natural numbers.

Finally, for every finite combination $n(k + 1)$ of natural numbers, we have

$$E_{n(k+1)} \subset E_{n(k)};$$

hence, from (110),

$$T_{n(k+1)} \subset T_{n(k)}. \tag{114}$$

With every finite combination $n(k)$ of natural numbers, we now associate a certain Borel set $M_{n(k)}$ subject to the following conditions:

(115) $$T_{n(k)} \subset M_{n(k)} \subset \bar{T}_{n(k)},$$

(116) $$M_{n(k)} \subset M_{n(k-1)},$$ [46]

(117) $$M_{p(k)} \cdot M_{q(k)} = 0$$

for different combinations $p(k)$ and $q(k)$ of k natural numbers ($k = 1, 2, \ldots$). We shall show that such a correlation is possible.

The sets $T_1, T_2, T_3, \ldots$ are disjoint by (112) and, since they are analytic, there exists (Theorem 123) for every natural n a Borel set M_n such that

$$T_n \subset M_n, \qquad n = 1, 2, \ldots$$

and

$$M_p \cdot M_q = 0, \qquad p \neq q;$$

moreover, it may be supposed that $M_n \subset \bar{T}_n$, for, if not, it would be sufficient to replace M_n by the set $M_n \cdot \bar{T}_n$, which is also a Borel set. Relations (115) and (117) are therefore true for $k = 1$.

Next let k be a given natural number and suppose that all sets $M_{n(k)}$ (where $n(k)$ is any combination of k natural numbers) are already defined so as to have conditions (115), (116), and (117) satisfied. Since the aggregate of all sets of $k + 1$ natural numbers is countable we deduce from (112), from the fact that the sets (110) are analytic, and from Theorem 121, the existence, for every combination of $k + 1$ natural numbers, of a Borel set $N_{n(k+1)}$ such that

(118) $$T_{n(k+1)} \subset N_{n(k+1)}$$

and

(119) $$N_{p(k+1)} \cdot N_{q(k+1)} = 0$$

for different combinations $p(k + 1)$ and $q(k + 1)$ of $k + 1$ integers. Put

(120) $$M_{n(k+1)} = M_{n(k)} \cdot \bar{T}_{n(k+1)} \cdot N_{n(k+1)};$$

these are Borel sets (being the intersection of three Borel sets); from (114), (115), (118), and (120), we find that

$$T_{n(k+1)} \subset M_{n(k+1)} \subset \bar{T}_{n(k+1)}.$$

Furthermore, (120) gives

$$M_{n(k+1)} \subset M_{n(k)}.$$

Finally, from (120) and (119), we get

$$M_{p(k+1)} \cdot M_{q(k+1)} = 0$$

for two different combinations of $k + 1$ natural numbers.

Hence (115), (116), and (117) remain true if we replace k by $k+1$. The Borel sets $M_{n(k)}$ which satisfy (115), (116), and (117) are thus defined by induction for every finite combination $n(k)$ of natural numbers.

Put

$$S_k = \sum M_{n(k)}, \tag{121}$$

where the summation ranges over all combinations $n(k)$ of k natural numbers. The sets (121) are Borel sets for all k since they are sums of a countable aggregate of Borel sets; so the set

$$S = S_1 . S_2 . S_3 \ldots \tag{122}$$

is a Borel set. It will be shown that $T = f(E) = S$.

From (113), (110), (115), and (121), we get $f(E) \subset S_k$ for $k = 1, 2, \ldots$ and so, from (122), $f(E) \subset S$. It is therefore sufficient to show that $S \subset F(E)$.

Let y denote an element of the set S. From (122), $y \in S_1$ and so, from (121), y is an element of the sum $M_1 + M_2 + \ldots$, i.e., $y \in M_{m_1}$ for some natural number m_1. Similarly, from (122), $y \in S_2$; hence we conclude, in virtue of (121), that $y \in M_{m'_1,m_2}$ for certain indices m'_1 and m_2. From (116), $M_{m'_1,m_2} \subset M_{m'_1}$; hence $y \in M_{m'_1}$ and, since $y \in M_{m_1}$, we conclude from (117) that $m'_1 = m_1$. Next starting with $y \in S_3$, we may deduce the existence of an index m_3 such that $y \in M_{m_1,m_2,m_3}$. Proceeding thus indefinitely we obtain an infinite sequence $m_1, m_2, m_3, \ldots$ of indices such that

$$v \in M_{m_1,m_2,\ldots,m_k}, \qquad k = 1, 2, \ldots. \tag{123}$$

Put

$$x = \frac{1}{m_1 +}\,\frac{1}{m_2 +}\,\frac{1}{m_3 +}\cdots; \tag{124}$$

this will be a number of the set E.

Let ϵ be a given positive number. Since the function f is continuous in E there exists, corresponding to ϵ, a number $\eta > 0$ such that

$$|x - x'| < \eta \tag{125}$$

implies that

$$\rho(f(x), f(x')) < \epsilon, \qquad \text{for all } x' \in E. \tag{126}$$

From (124) and the properties of continued fractions it follows that corresponding to the number η there exists an index k such that every number x' of E whose kth convergent is the same as the kth convergent of the number (124), i.e., every number of the set $E_{m_1,m_2,\ldots,m_k}$ satisfies (125) and therefore also (126). Thus, from (110), every number t of the set $T_{m_1,m_2,\ldots,m_k}$ satisfies the inequality

$$\rho(f(x), t) < \epsilon;$$

so every number t of the set $\bar{T}_{m_1,m_2,\ldots,m_k}$ satisfies the inequality

$$\rho(f(x), t) \leqslant \epsilon. \tag{127}$$

But, from (123) and (115), we have

$$y \in \bar{T}_{m_1,m_2,\ldots,m_k};$$

we may therefore put $t = y$ in (127), which gives

$$\rho(f(x), y) \leqslant \epsilon.$$

Since ϵ is arbitrary, it follows that $\rho(f(x), y) = 0$, i.e., $y = f(x)$. Consequently $S \subset f(E)$. Theorem 124 is therefore proved. We shall deduce from this theorem some important results later on.

94. The property of compact closed or compact open subsets of a metric space. Let E denote a closed and bounded (hence, compact) set of real numbers and $f(x)$ a function defined and continuous in E and assuming values in a metric space. Let y be an element of the set $T = f(E)$; among the numbers x of E for which $f(x) = y$, there is a greatest. For let $\phi(y)$ denote the upper bound of the set $P(y)$ of all numbers x belonging to E for which $f(x) = y$. Since the set $\{y\}$ is closed, it follows from the corollary to Theorem 24 that the set $P(y)$ is closed and so $P(y)$ contains its upper bound $\phi(y)$; this is a finite number since $P(y) \subset E$ and E is bounded. Since $\phi(y) \in P(y)$, we have $f(\phi(y)) = y$; so, from the definition of the number $\phi(y)$, it follows that it is the greatest number x of the set E for which $f(x) = y$.

Denote by X the set of all numbers $\phi(y)$ for which $y \in T$; then $X \subset E$, $f(X) = T$, and the function f is continuous and (1, 1) in X. The set T is therefore a continuous and (1, 1) image of the set X. It will be shown that the set X is a G_δ. Since E is closed, it will be sufficient to show that the set $E - X$ is an F_σ.

It follows from its definition that the set X is the set of all numbers x of E such that, for $x' \in E$ and $x' > x$, we have $f(x') \neq f(x)$. Thus, if $x \in E - X$, there exists a number $x' > x$ such that $x' \in E$ and $f(x') = f(x)$. Denote by F_n the set of all numbers x of E for which there exists a number x' of E such that $x' \geqslant x + 1/n$ and $f(x) = f(x')$; it is evident that $E - X = F_1 + F_2 + F_3 + \ldots.$ It remains to prove that the sets F_n, $n = 1, 2, \ldots$, are closed.

Let n denote a given natural number and x_0 a limit element of the set F_n. Then there exists an infinite sequence x_k, $k = 1, 2, \ldots$, of numbers of the set F_n such that $\lim_{k\to\infty} x_k = x_0$. Since $x_k \in F_n$ for $k = 1, 2, \ldots$, there exists for every natural number k a number x'_k of the set E such that $x'_k \geqslant x_k + 1/n$ and $f(x_k) = f(x'_k)$. The infinite sequence $\{x'_k\}$ is bounded; consequently, it contains a convergent subsequence x'_{k_j} $(j = 1, 2, \ldots)$. Put

$$\lim_{j\to\infty} x'_{k_j} = x'_0;$$

since $x'_{k_j} \in E$ for $j = 1, 2, \ldots$, and E is closed, x'_0 is an element of E. But, from $x'_k \geqslant x_k + 1/n, f(x_k) = f(x'_k)$, and the continuity of the function f in E, we have $x'_0 \geqslant x_0 + 1/n$ and $f(x_0) = f(x'_0)$, since

$$\lim_{k\to\infty} x_k = x_0 \text{ and } \lim_{j\to\infty} x'_{k_j} = x'_0;$$

hence $x_0 \in F_n$. The set F_n is therefore closed.

Hence if E is a closed and bounded set of real numbers and $f(x)$ a function continuous in E then the set $T = f(E)$ is a continuous and (1, 1) image of a certain set G_δ contained in E. Combining this result with Theorem 78 we obtain

THEOREM 125. *A closed and compact set of elements of a metric space is a continuous and* (1, 1) *image of a certain set G_δ of real numbers.*

Let G denote an open and compact set contained in a given metric space. Put $E = \bar{G}$; this will be a closed and compact set (§ **57**). Hence, by Theorem 125, there exists a set Γ of real numbers which is a G_δ, and a function f continuous and (1, 1) in Γ and such that $f(\Gamma) = E$. Denote by Γ_1 the set of all numbers x of Γ for which $f(x) \in G$; since $G \subset \bar{G} = E$, we have $f(\Gamma_1) = G$ and, since $\Gamma_1 \subset \Gamma$, the function f is continuous and (1, 1) in Γ_1. But the set Γ is a G_δ; we may, therefore, apply to it the lemma of § **78**, from which it follows that the set Γ_1 is a G_δ. We have thus proved

THEOREM 126. *An open and compact set of elements of a metric space is a continuous and* (1, 1) *image of a certain set G_δ of real numbers.*

95. The theorem of Mazurkiewicz about linear sets G_δ.

THEOREM 127. *Every set of real numbers which is a G_δ is the sum of two sets, one of which is the null set or a homeomorphic image of the set of all irrational numbers, and the other is an at-most-countable set.*

We first prove the following

LEMMA. *If U is an open set of real numbers containing a non-countable set N and η is a positive number, then there exists an infinite sequence of non-overlapping open intervals $D_1, D_2, D_3, \ldots$; each D_n has length $< \eta$, each $\bar{D}_n \subset U$, each $N . D_n$ is non-countable, and the set $N - (D_1 + D_2 + \ldots)$ is at most countable.*

Proof. Let U be an open set of real numbers containing a non-countable set N and let η be a given positive number. The set N contains an element of condensation x and, since $N \subset U$, we have $x \in U$. Since U is open, there exists a natural number k such that the closed interval $P_k = (x - 1/k, x + 1/k)$ is contained in U where it may be assumed that $2/k < \eta$. It is easily seen that the interval P_s differs from the sum of all the open intervals

$Q_n = (x + 1/(n + 1), x + 1/n)$ and $R_n = (x - 1/n, x - 1/(n + 1))$, where the summation ranges over all natural $n \geqslant s$, by a countable set of points. On the other hand, the interval P_s contains a non-countable set of elements of N since it contains x, an element of condensation of N, as an interior point. Hence we conclude that for every natural s there exists a natural number $n \geqslant s$ such that at least one of the intervals Q_n and R_n contains a non-countable subset of N. Consequently, infinitely many of the open intervals Q_n and R_n, for $n \geqslant k$, contain a non-countable subset of N; let these be the intervals $H_1, H_2, H_3, \ldots$. The set $N . \bar{P}_k - (H_1 + H_2 + \ldots)$ is at most countable.

Furthermore, since $\bar{P}_k \subset U$, $\delta(P_k) < \eta$, and $H_n \subset P_k$, for $n = 1, 2, \ldots$, it follows that $\bar{H}_n \subset U$, $\delta(H_n) < \eta$, and the sets $H_n . N$ are non-countable.

The set $U - \bar{P}_k$ is open and so may be divided into a countable aggregate of open intervals of length $< \eta$ by removing a countable set of points; let (a, b) denote one of the intervals thus obtained. Let $a_1, a_2, a_3, \ldots$ be an infinite decreasing sequence of numbers $< \frac{1}{2}(a + b)$ and approaching a; let $b_1, b_2, \ldots$ be an infinite increasing sequence of numbers $> \frac{1}{2}(a + b)$ and approaching b. The interval (a, b) differs from the sum of the intervals (a_1, b_1), (a_{n+1}, a_n), and (b_n, b_{n+1}), for $n = 1, 2, \ldots$, by a countable set of points; the closures of these intervals are contained in the open interval (a, b) and so also in U. Hence, except for a countable set of points, the set $U - \bar{P}_k$ can be divided into a countable aggregate of open intervals of length $< \eta$ whose closures are contained in U. Those intervals which contain a non-countable subset of N (if such intervals exist) may be denoted by $K_1, K_2, \ldots$. The set $N(U - \bar{P}_k) - (K_1 + K_2 + \ldots)$ is at most countable. The aggregate of all intervals $H_1, H_2, \ldots$ and $K_1, K_2, \ldots$ is countable; it may therefore be arranged as an infinite sequence $D_1, D_2, \ldots$. It is evident that the intervals D_n satisfy all conditions of our lemma.

To prove Theorem 127 let E be a set G_δ of real numbers. Then there exists an infinite sequence of open sets G_n $(n = 1, 2, \ldots)$ such that $E = G_1 . G_2 . G_3 \ldots$. Suppose that E is non-countable. Since $E \subset G_1$ and G_1 is open, we may apply our lemma on setting $U = G_1$, $N = E$, and $\eta = 1$. We thus obtain an infinite sequence $D_1, D_2, D_3, \ldots$ of non-overlapping open intervals; each D_n has length < 1, each $\bar{D}_n \subset G_1$, each $E . D_n$ is non-countable, and the set $E - (D_1 + D_2 + \ldots)$ is at most countable.

Let n_1 denote a natural number. Since the sets G_2 and D_{n_1} are open, the set $G_2 . D_{n_1}$ is open; since $E \subset G_2$ and $E . D_{n_1}$ is non-countable, the set $E . G_2 . D_{n_1}$ is non-countable. Applying the lemma to the sets $U = G_2 . D_{n_1}$, $N = E . D_{n_1}$, with $\eta = \frac{1}{2}$, we obtain an infinite sequence of non-overlapping open intervals $D_{n_1,1}, D_{n_1,2}, \ldots$; each $D_{n_1,n}$ has length less than $\frac{1}{2}$, each $\bar{D}_{n_1,n} \subset G_2 . D_{n_1}$, each $E . D_{n_1,n}$ is non-countable, and the set $E . D_{n_1} - (D_{n_1,1} + D_{n_1,2} + \ldots)$ is at most countable.

Further, let n_1, n_2, be two natural numbers. Since the sets G_3 and D_{n_1,n_2} are open and the set $E \,.\, G_3 \,.\, D_{n_1,n_2}$ is non-countable, we may apply our lemma to $U = G_3 \,.\, D_{n_1,n_2}$, $N = E \,.\, D_{n_1,n_2}$, with $\eta = 1/3$.

Continuing this argument, we obtain for every finite combination n_1, n_2, $\ldots$, n_k (abbreviated to $n(k)$) of natural numbers an open interval $D_{n(k)}$ such that

$$\text{(i)} \qquad \delta(D_{n(k)}) < 1/k,$$

$$\text{(ii)} \qquad D_{n(k-1),p} \,.\, D_{n(k-1),q} = 0, \qquad p \neq q,$$

$$\text{(iii)} \qquad \bar{D}_{n(k)} \subset G_k \,.\, D_{n(k-1)},$$

$$\text{(iv)} \qquad E \,.\, D_{n(k)} \text{ is non-countable},$$

$$\text{(v)} \qquad E \,.\, D_{n(k-1)} - (D_{n(k-1),1} + D_{n(k-1),2} + \ldots) \text{ is at most countable.}$$

Let N denote the set of all irrational numbers in the interval $(0, 1)$, x a given number of N, and let

$$x = \frac{1}{m_1 +} \, \frac{1}{m_2 +} \, \frac{1}{m_3 +} \cdots \tag{128}$$

be the development of x as a continued fraction. Put

$$F(x) = \bar{D}_{m_1} \,.\, \bar{D}_{m_1,m_2} \,.\, \bar{D}_{m_1,m_2,m_3} \,.\, \ldots. \tag{129}$$

It follows from (iii) and (iv) that the set (129) is the intersection of a descending sequence of closed non-empty intervals and is therefore non-empty; moreover, by (128), $F(x)$ is contained in an interval of length $< 1/k$, for $k = 1, 2, \ldots$; hence $F(x)$ consists of a single element which we denote by $f(x)$. From (129) and (iii) we have $f(x) \in G_k$ for $k = 1, 2, \ldots$; so $f(x) \in E$. The set T of all the numbers $f(x)$ for $x \in N$ is therefore a subset of the set E. We next show that the set $E - T$ is at most countable.

To prove this, let

$$R = (E - S) + \sum (E \,.\, D_{n(k)} - S_{n(k)}), \tag{130}$$

where the summation extends over all finite combinations $n(k)$ of natural numbers, and where $S = D_1 + D_2 + \ldots$, while

$$S_{n(k)} = D_{n(k),1} + D_{n(k),2} + \ldots. \tag{131}$$

It is evident from (131) and (v) that the terms of the sum (130) are at-most-countable sets; consequently, the set R is at most countable.

Let y denote a number of the set $E - R$. Then $y \in E$ and $y \notin R$; so, from (130), $y \notin E - S$. But $y \in E$; therefore $y \in S$ and, since $S = D_1 + D_2 + \ldots$, there exists an index m_1 such that $y \in D_{m_1}$. From $y \notin R$ and

(130), we find that $y \notin (E \,.\, D_{m_1} - S_{m_1})$; but since $y \in E \,.\, D_{m_1}$, we have $y \in S_{m_1}$; hence, from (131), there exists an index m_2 such that $y \in D_{m_1, m_2}$.

Continuing this argument, we obtain an infinite sequence $m_1, m_2, m_3, \ldots$ of indices such that

$$y \in D_{m(k)}, \qquad k = 1, 2, \ldots .$$

From (129) we have $y \in F(x)$, where x is the number defined by (128); in virtue of the definition of the set T, this proves that $y \in T$. Hence $E - R \subset T$; this gives $E - T \subset R$ and, since R is at most countable, the set $E - T$ is at most countable. It remains to show that $N \, h_f \, T$.

It follows from the definition of the set T that $f(N) = T$. Let x and x' be two different numbers of the set N. Suppose that the developments of the numbers x and x' as continued fractions first differ in the rth terms and that the denominator of this term for x' is $m'_r \neq m_r$. From (129) we have $F(x) \subset \bar{D}_{m(r+1)}$ and so, from (iii), $F(x) \subset D_{m(r)}$; similarly $F(x') \subset D_{m(r-1), m'}$ and so, from (ii), $F(x) \,.\, F(x') = 0$; consequently $f(x) \neq f(x')$. Hence f is (1, 1) in N.

We next show that the function f is continuous in N. Let x be a given number of the set N and η an arbitrary positive number. Choose k such that $1/k < \eta$. Corresponding to the numbers x and k there exists a positive number δ such that every number x' of N satisfying the inequality

$$|x - x'| < \delta \tag{132}$$

can be developed as a continued fraction with its first k terms the same as the first k terms of the corresponding development of x. From (128) and (129) we conclude that $F(x) \subset \bar{D}_{m(k)}$ and $F(x') \subset \bar{D}_{m(k)}$; this, because of $f(x) \in F(x), f(x') \in F(x')$, $1/k < \eta$, and (i), gives

$$|f(x) - f(x')| < \eta. \tag{133}$$

Thus, corresponding to every number x of N and every positive number η, there exists a number $\delta > 0$ such that the inequality (132) implies the inequality (133) for numbers $x' \in N$. This establishes the continuity of the function f in the whole set N.

Let $\phi(y)$ denote a function defined in T and inverse to the function f and let y be a given number of T. Then there exists a unique number x of N such that $f(x) = y$. If (128) is the development of x as a continued fraction then, from $f(x) \in F(x)$, (129), and (iii), we have

$$y \in D_{m(k)}, \qquad k = 2, 3, 4, \ldots . \tag{134}$$

Let δ denote a positive number. Corresponding to the numbers x and δ there exists an integer $k > 1$ such that every number x' which has the same first k terms as x in its development as a continued fraction satisfies the inequality

$$|x - x'| < \delta. \tag{135}$$

Let y' denote a number of the set T in the interval $D_{m(k)}$ and let

$$x' = \frac{1}{m'_1 +} \frac{1}{m'_2 +} \frac{1}{m'_3 +} \cdots$$

be the development of the number x' as a continued fraction. Suppose that the developments of the numbers x and x' first differ in their rth terms. Hence $m'_i = m_i$ for $i = 1, 2, \ldots, r - 1$, and $m'_r \neq m_r$. From (129) and (iii) we have

$$y \in D_{m(r-1), m_r} \text{ and } y' \in D_{m(r-1), m'_r};$$

so, by (ii), since $m_r \neq m'_r$, $y' \notin D_{m(r)}$. Consequently, from (iii), $y' \notin D_{m(i)}$ for $i \geqslant r$. But, by hypothesis, $y' \in D_{m(k)}$; hence $r > k$, i.e., the numbers x' and x have the same first k terms in their developments as continued fractions. The definition of the number k then gives the inequality (135). So, corresponding to every number y of T, there exists an open interval $D = D_{m(k)}$ containing y and such that every number y' of the set $T \,.\, D$ satisfies the inequality (135), where $x = \phi(y)$, $x' = \phi(y')$. This establishes the continuity of the function ϕ in the set T; hence the relation $N\, h_f\, T$. But the set N is homeomorphic with the set of all irrational numbers; hence Theorem 127 is proved.

It follows from Theorem 127 that *two non-countable linear sets G_δ are homeomorphic, except for a countable set of their elements.*

Thus we obtain the following corollary to Theorem 127.

COROLLARY. *A compact open set contained in a metric space is the sum of two sets; one of these, if not empty, is a continuous and* (1, 1) *image of the set of all irrational numbers and the other is at most countable.*

96. Biuniform and continuous images of Borel sets. Denote by **L** the family of all sets E contained in m-dimensional Euclidean space which satisfy the following condition:

The set E is the sum of two sets of which one, if not empty, is a continuous and (1, 1) image of the set of all irrational numbers and the other is at most countable.

We shall show that the sum of a countable aggregate of disjoint sets belonging to **L** belongs to **L**.

Let $E = E_1 + E_2 + \ldots$, where $E_n \in L$ for $n = 1, 2, \ldots$, and $E_i \,.\, E_j = 0$ for $i \neq j$. We may therefore write $E_n = P_n + Q_n$ where P_n is either the null set or a continuous image of the set of all irrational numbers and where Q_n is an at-most-countable set. But the set of all irrational numbers is homeomorphic with the set N_n of all irrational numbers in the interval $(n, n + 1)$. Hence P_n is either empty or a continuous and (1, 1) image of the set N_n. Denote by S the sum of all sets N_n extending over the indices n for which P_n is non-empty. The set S is obviously either empty or homeomorphic with the set of all irrational numbers and the set $P = P_1 + P_2 + \ldots$ is a continuous and (1,1) image of the set S. Moreover, since $E = P + Q$, where $Q = Q_1 + Q_2 + \ldots$ is an at-most-countable set, we have $E \in \mathbf{L}$.

We next show that the intersection of a countable aggregate of sets belonging to $\mathbf{L}$ is itself a set of $\mathbf{L}$.

Let $E = E_1 . E_2 . E_3 \ldots$, where $E_n \in \mathbf{L}$ for $n = 1, 2, \ldots$. We may therefore write $E_n = P_n + Q_n$ where the sets P_n and Q_n have the same meaning as above. If the set P_n is empty for some n, then the set E_n (and so the set E) is at most countable and therefore belongs to the family $\mathbf{L}$. Hence it may be assumed that $P_n \neq 0$ for $n = 1, 2, \ldots$.

We may now put $P_n = f_n(N)$ for every n, where f_n is a continuous and (1, 1) function in the set N of all irrational numbers in the interval (0, 1). Since $E = E_1 . E_2 . E_3 \ldots$, $E_n = P_n + Q_n$, and the sets Q_n $(n = 1, 2, \ldots)$ are at most countable, it follows that we may write $E = P + R$, where $P = P_1 . P_2 \ldots$, and R is at most countable. Hence, to prove that $E \in \mathbf{L}$, it will be sufficient to show that $P \in \mathbf{L}$.

Let n be a given natural number, t a given irrational number in the interval (0, 1), and

$$(136) \qquad t = \frac{1}{k_1 +} \frac{1}{k_2 +} \frac{1}{k_3 +} \cdots$$

the development of t as a continued fraction. Put

$$(137) \qquad \phi_n(t) = \frac{1}{k_{2^{n-1}} +} \frac{1}{k_{3.2^{n-1}} +} \frac{1}{k_{5.2^{n-1}} +} \cdots \frac{1}{k_{(2m-1)2^{n-1}} +} \cdots$$

It follows from the properties of continued fractions that the functions $\phi_n(t)$ are continuous in the set N and that $\phi_n(N) = N$. Put $F_n(t) = f_n(\phi_n(t))$ for $t \in N$; the functions $F_n(t)$ are continuous in the set N and $P_n = F_n(N)$ for $n = 1, 2, \ldots$.

It will be shown that $P = F_1(T)$, where T is the set of all numbers t of N for which

$$(138) \qquad F_n(t) = F_1(t), \qquad n = 1, 2, \ldots.$$

Assume that $x \in P$. Then for every natural n, $x \in P_n = f_n(N)$ and so there exists a number t_n such that $f_n(t_n) = x$. Let

$$(139) \qquad t_n = \frac{1}{k_{n,1} +} \frac{1}{k_{n,2} +} \frac{1}{k_{n,3} +} \cdots$$

be the development of the number t_n as a continued fraction.

Every natural number p may be expressed uniquely in the form

$$(140) \qquad p = (2m_p - 1)2^{n_p - 1}$$

where m_p and n_p are natural numbers defined by the number p. Put

$$(141) \qquad k_p = k_{n_p, 2m_p - 1}, \qquad p = 1, 2, \ldots$$

and let t be defined by (136). In order to show that $x \in F_1(T)$ it will be sufficient to show that $x \in F_n(t)$ for $n = 1, 2, \ldots$.

For all natural m and n we have, from (140) and (141),

$$k_{(2m-1)2^{n-1}} = k_{n,2m-1}$$

and so, from (137) and (139),

$$\phi_n(t) = t_n, \qquad n = 1, 2, \ldots;$$

hence

$$F_n(t) = f_n(\phi_n(t)) = f_n(t_n) = x, \qquad n = 1, 2, \ldots.$$

We have therefore proved that the relation $x \in P$ implies $x \in F_1(T)$; consequently $P \subset F_1(T)$.

Now let x denote an element of $F_1(T)$. It follows from the definition of the set T that there exists a number t of the set N such that

$$x = F_n(t), \qquad n = 1, 2, \ldots.$$

But $P_n = F_n(N)$; hence $x \in P_n$ for $n = 1, 2, \ldots$, and so $x \in P$. This gives $F_1(T) \subset P$ and since $P \subset F_1(T)$, we have $P = F_1(T)$. Since $T \subset N$ and $F_1(t)$ is continuous in N, $F_1(t)$ is also continuous in T. We next show that $F_1(t)$ is (1, 1) in T.

Let t and t' be two different numbers of T. Let

$$t' = \frac{1}{k'_1 +} \frac{1}{k'_2 +} \frac{1}{k'_3 +} \cdots \tag{142}$$

be the development of t' as a continued fraction. Since $t \neq t'$ the developments (136) and (142) must differ in some term, say the pth; hence $k_p \neq k'_p$. Thus, from (140) and (137),

$$\phi_{n_p}(t) \neq \phi_{n_p}(t')$$

and so, since f_n is (1, 1) in N, we obtain

$$f_{n_p}(\phi_{n_p}(t)) \neq f_{n_p}(\phi_{n_p}(t')), \text{ that is, } F_{n_p}(t) \neq F_{n_p}(t');$$

on account of (138), this gives $F_1(t) \neq F_1(t')$, as required. Consequently P is a continuous and (1, 1) image of the set T. We next show that T is a G_δ. Since N is a G_δ and $T \subset N$, it is sufficient to show that T is closed in the set N.

Let t_0 be an element of the set $T' \,.\, N$; we must show that $t_0 \in T$. Since $t_0 \in T'$ there exists an infinite sequence t_k of elements of T such that

$$\lim_{k \to \infty} t_k = t_0.$$

From $t_k \in T$ and the definition of the set T, we have $F_n(t_k) = F_1(t_k)$ for $n = 1, 2, \ldots$ and $k = 1, 2, \ldots$; so, since $F_n(t)$ is continuous in N, since $t_0 \in N$, and since $\lim_{k \to \infty} t_k = t_0$, we have

$$F_n(t_0) = F_1(t_0), \qquad n = 1, 2, \ldots;$$

hence $t_0 \in T$. Consequently the set T is a G_δ. Therefore, we may, by Theorem 127, write $T = X + Y$ where X is either empty or a homeomorphic image of the set of all irrational numbers and where Y is an at-most-countable set.

From $P = F_1(T)$, we get $P = F_1(X) + F_1(Y)$ where $F_1(Y)$ is at most countable. If X is the null set then P is at most countable and so $P \in \mathbf{L}$. If $X \neq 0$ we may write $X = \psi(N)$, where ψ is a continuous and (1, 1) function in the set N; since $P = F_1(X)$ and F_1 is continuous and (1, 1) in $X \subset T$, we have $P = F_1(\psi(N)) = \phi(N)$, where $\phi(t) = F_1(\psi(t))$ is a function continuous and (1, 1) in the set N. We conclude from this that $P \in \mathrm{L}$. This proves that the family $\mathbf{L}$ satisfies conditions 2″ and 3″ of § **81**.

It follows from the corollary at the end of § **95** that every compact open set belongs to the family $\mathbf{L}$. Let E be a compact set G_δ; we may then write $E = G_1 \,.\, G_2 \,.\, G_3 \ldots$, where the sets G_n are open. Since E is compact and therefore bounded, there exists a sphere S containing E. But in m-dimensional Euclidean space spheres are compact sets; the open sets $\Gamma_n = G_n \,.\, S$ are therefore compact. Since $E \subset S$ and $E = G_1 \,.\, G_2 \ldots$, we have $E = \Gamma_1 \,.\, \Gamma_2 \,.\, \Gamma_3 \ldots$. Each set Γ_n, $n = 1, 2, \ldots$, is open and compact and so belongs to $\mathbf{L}$; hence, by property 3″, the intersection of all these sets also belongs to $\mathbf{L}$. Thus every compact set G_δ belongs to $\mathbf{L}$.

Next let U be an open set (not necessarily compact). Let p denote an element of the space R_m under consideration; put $H_1 = S(p, 1)$ and $H_n = S(p, n) - S(p, n-1)$, for $n = 2, 3, \ldots$. Then $R_m = H_1 + H_2 + H_3 + \ldots$, where the sets H_n, $n = 1, 2, \ldots$, are compact and disjoint sets G_δ. Thus $U = U \,.\, H_1 + U \,.\, H_2 + \ldots$ is the sum of compact and disjoint sets G_δ. Since compact sets G_δ belong to $\mathbf{L}$, the set U belongs to $\mathbf{L}$ by property 2″.

Thus every open set belongs to $\mathbf{L}$, i.e., the family $\mathbf{L}$ satisfies condition 1″ of § **81**. Since the family $\mathbf{L}$ satisfies conditions 1″, 2″, and 3″ of § **81**, every Borel set belongs to $\mathbf{L}$.

On the other hand, it follows from Theorem 124 and the definition of the family $\mathbf{L}$ that every set belonging to $\mathbf{L}$ is a Borel set. The family $\mathbf{L}$ is therefore identical with the family of all Borel sets. This gives

THEOREM 128. *A set E contained in m-dimensional Euclidean space is a Borel set if and only if it is the sum of two sets of which one is either empty or a continuous and* (1, 1) *image of the set of all irrational numbers and the other is at most countable.*

Theorem 128 leads immediately to the result that every non-countable Borel set contained in R_m has the power of the continuum (a result obtained differently in Corollary 2, Theorem 118).

Let E be a given Borel set contained in m-dimensional Euclidean space and let f be a function continuous and (1, 1) in E. If E is at most countable, so is $T = f(E)$; T is therefore a Borel set. If E is non-countable then, by Theorem 128, we may write $E = \phi(N) + P$, where ϕ is continuous and (1, 1) in the set N of all irrational numbers and where P is at most countable. We then have $T = f(E) = f(\phi(N)) + f(P)$. Put $F(t) = f(\phi(t))$ for $t \in N$; the function F is continuous and (1, 1) in N. Since $T = F(N) + f(P)$ and $f(P)$ is at most countable it follows, from Theorem 128, that T is a Borel set. We thus get

THEOREM 129 (Lusin).[47] *A continuous and* (1, 1) *image of a Borel set contained in m-dimensional Euclidean space is a Borel set.*

97. The analytic set as a sum or an intersection of $\aleph_1$ Borel sets. Let E denote an analytic set which is the nucleus of the not necessarily regular system $[E_{n(k)}]$ of Borel sets. For every finite combination $n_1, n_2, \ldots, n_k$ (abbreviated to $n(k)$) of indices put

$$E^0_{n(k)} = E_{n(k)}; \tag{143}$$

$$E^{\alpha+1}_{n(k)} = E^{\alpha}_{n(k)} \cdot \sum_{n=1}^{\infty} E^{\alpha}_{n(k),n} \tag{144}$$

for every ordinal number $\alpha < \Omega$ of the first kind; and

$$E^{\alpha}_{n(k)} = \prod_{\xi<\alpha} E^{\xi}_{n(k)} \tag{145}$$

for every ordinal number $\alpha < \Omega$ of the second kind. The sets $E^{\alpha}_{n(k)}$, defined by transfinite induction, are Borel sets (§ **81**) for every finite combination $n(k)$ of indices and every ordinal number $\alpha < \Omega$. It follows from (144) and (145), by transfinite induction, that

$$E^{\alpha}_{n(k)} \subset E^{\beta}_{n(k)}, \qquad \alpha \geqslant \beta. \tag{146}$$

Put

$$S^{\alpha} = \sum_{n=1}^{\infty} E^{\alpha}_{n} \tag{147}$$

and

$$T^{\alpha} = \sum (E^{\alpha}_{n(k)} - E^{\alpha+1}_{n(k)}), \tag{148}$$

where the sum (148) extends over all finite combinations $n(k)$ of natural numbers. The sets (147) and (148) and their differences $S^{\alpha} - T^{\alpha}$ are all Borel sets for $\alpha < \Omega$ (§ **79**).

We shall show that

$$E = \sum_{\alpha<\Omega} (S^{\alpha} - T^{\alpha}) = \prod_{\alpha<\Omega} S^{\alpha}, \tag{149}$$

where the summation and intersection range over all ordinal numbers $\alpha < \Omega$.

Let $\alpha < \Omega$ be a given ordinal number and x an element of the set $S^\alpha - T^\alpha$. Then

$$x \in S^\alpha \tag{150}$$

and

$$x \notin T^\alpha. \tag{151}$$

From (150) and (147) we deduce the existence of a natural number m_1 such that $x \in E^\alpha_{m_1} = E^\alpha_{m(1)}$. Moreover, from (151) and (148), it follows that

$$x \notin (E^\alpha_{m(1)} - E^{\alpha+1}_{m(1)});$$

since $x \in E^\alpha_{m(1)}$, we have $x \in E^{\alpha+1}_{m(1)}$. But, from (144),

$$E^{\alpha+1}_{m(1)} = E^\alpha_{m(1)} \cdot \sum_{n=1}^{\infty} E^\alpha_{m(1),n}$$

and so, since $x \in E^{\alpha+1}_{m(1)}$ there exists an index m_2 such that $x \in E^\alpha_{m(2)}$. Furthermore, on account of (151) and (148), we have

$$x \notin (E^\alpha_{m(2)} - E^{\alpha+1}_{m(2)});$$

so, since $x \in E^\alpha_{m(2)}$, we get $x \in E^{\alpha+1}_{m(2)}$. Again, by (144),

$$E^{\alpha+1}_{m(2)} \subset \sum_{n=1}^{\infty} E^\alpha_{m(2),n}$$

and so, since $x \in E^{\alpha+1}_{m(2)}$, there exists an index m_3 such that $x \in E^\alpha_{m(3)}$. Continuing this argument we obtain an infinite sequence $m_1, m_2, m_3, \ldots$ of indices such that

$$x \in E^\alpha_{m(k)}, \qquad k = 1, 2, \ldots.$$

Consequently, from (143) and (146),

$$x \in E_{m(k)}, \qquad k = 1, 2, \ldots;$$

therefore (from the definition of the set E), $x \in E$.

We have thus proved that $(S^\alpha - T^\alpha) \subset E$ for every ordinal $\alpha < \Omega$; it follows that

$$\sum_{\alpha<\Omega} (S^\alpha - T^\alpha) \subset E. \tag{152}$$

Let x denote an element of the set E. Then there exists an infinite sequence $m_1, m_2, m_3, \ldots$ of indices such that

$$x \in E_{m(k)}, \qquad k = 1, 2, \ldots. \tag{153}$$

It will be shown that

$$x \in E^\alpha_{m(k)}, \qquad k = 1, 2, \ldots \tag{154}$$

for every ordinal $\alpha < \Omega$. For $\alpha = 0$, (154) is true because of (153) and (143). Let β denote an ordinal number $< \Omega$ and suppose that (154) is true for every ordinal number $\alpha < \beta$. If β is a number of the second kind, it follows from (145) that (154) is true for β. If β is a number of the first kind, we may put $\beta = \alpha + 1$, where $\alpha < \beta$; we then obtain from (144) (for every natural k)

$$E^{\beta}_{m(k)} = E^{\alpha}_{m(k)} \cdot \sum_{n=1}^{\infty} E^{\alpha}_{m(k),n} \supset E^{\alpha}_{m(k)} \cdot E^{\alpha}_{m(k+1)};$$

hence $x \in E^{\beta}_{m(k)}$ since, by (154), $x \in E^{\alpha}_{m(k)}$ and $x \in E^{\alpha}_{m(k+1)}$. Relation (154) is therefore proved by transfinite induction for every ordinal number $\alpha < \Omega$.

In particular, it follows from (154) that $x \in E^{\alpha}_{m_1}$ for $\alpha < \Omega$ and so, from (147), $x \in S^{\alpha}$ for $\alpha < \Omega$. Consequently $E \subset S^{\alpha}$ for all $\alpha < \Omega$; this gives

$$E \subset \prod_{\alpha<\Omega} S^{\alpha}. \tag{155}$$

Furthermore,

$$\prod_{\alpha<\Omega} T^{\alpha} = 0. \tag{156}$$

For suppose that (156) is not true. Then there exists an element x such that

$$x \in T^{\alpha} \qquad \text{for all } \alpha < \Omega. \tag{157}$$

It follows from (157) and (148) that, corresponding to every ordinal number $\alpha < \Omega$, there exists at least one set $n(k)$ of indices (dependent on α) such that

$$x \in (E^{\alpha}_{n(k)} - E^{\alpha+1}_{n(k)}).$$

But the set of all finite combinations $n_1, n_2, \ldots, n_k$ is countable whereas the set of all ordinal numbers $\alpha < \Omega$ is non-countable. Consequently there exists a set $p(r)$ of indices and two ordinal numbers $\xi < \Omega$ and $\eta < \Omega$, say $\eta < \xi$, such that

$$x \in (E^{\xi}_{p(r)} - E^{\xi+1}_{p(r)}) \tag{158}$$

and

$$x \in (E^{\eta}_{p(r)} - E^{\eta+1}_{p(r)}). \tag{159}$$

It follows from (158) and (159) that

$$x \in E^{\xi}_{p(r)} \text{ and } x \notin E^{\eta+1}_{p(r)};$$

this contradicts (146) since, from $\eta < \xi$, we have $\eta + 1 \leqslant \xi$. Hence (156) is proved.

Thus for x any element of the set E there exists an ordinal number $\alpha < \Omega$ such that $x \notin T^{\alpha}$. But, from (155), we have $x \in S^{\alpha}$; hence $x \in S^{\alpha} - T^{\alpha}$. Consequently

$$E \subset \sum_{\alpha<\Omega} (S^{\alpha} - T^{\alpha}). \tag{160}$$

Relations (152) and (160) give the first part of (149).

Let x denote an element of the set

$$(161) \qquad P = \prod_{\alpha<\Omega} S^\alpha.$$

From (156), there exists an ordinal $\alpha < \Omega$ such that $x \notin T^\alpha$. Since $x \in P$, we have $x \in S^\alpha$ from (161); consequently $x \in S^\alpha - T^\alpha$ and so, by (152), $x \in E$. Hence $P \subset E$; since, from (155) and (161), $E \subset P$, we have $E = P$. This completes the proof of (149).

It follows from (149) that *every analytic set contained in a metric space is both the sum and the intersection of $\aleph_1$ Borel sets.* On passing to complements we find that *the complement of an analytic set is both the intersection and the sum of $\aleph_1$ Borel sets.*

Let E denote an analytic set contained in a separable complete space. Relation (149) gives

$$(162) \qquad CE = \sum_{\alpha<\Omega} CS^\alpha.$$

If no term of the sum (162) is non-countable, then (since (162) contains $\aleph_1$ terms) the set CE has cardinal $\aleph_1$ at most; if, however, there exists among the terms of (162), a non-countable set then, being a Borel set, it must contain, by Theorem 120, a non-empty perfect subset. This gives

THEOREM 130. *In a separable complete space the complement of an analytic set which does not contain a perfect non-empty subset has cardinal at most $\aleph_1$.*

This theorem would be trivial if the power of the continuum were equal to $\aleph_1$. As it is, we do not know (without the continuum hypothesis) whether there exist analytic sets whose complements have cardinal $\aleph_1$. Furthermore, we are not able to establish (even with the continuum hypothesis) whether or not every non-countable complement of an analytic set contains a perfect non-empty subset.

It follows from Theorem 130 that in a separable complete space the complement of an analytic set cannot have a cardinal between $\aleph_1$ and $\mathbf{c}$.

*Let E denote an analytic set contained in a separable complete space and f a function continuous in the set $T = CE$; put $Q = f(T)$; then Q is a continuous image of the complement of an analytic set. From (162),

$$(163) \qquad Q = f(T) = f(CE) = \sum_{\alpha<\Omega} f(CS^\alpha).$$

The sets CS^α are Borel sets; their continuous images $f(CS^\alpha)$ are analytic sets (corollary, Theorem 114) and so, as shown above, are sums of $\aleph_1$ Borel sets. It therefore follows from (163) that the set Q is the sum of $\aleph_1$ Borel sets. Hence *in a separable complete space a continuous image of the complement of an analytic set is the sum of $\aleph_1$ Borel sets.*[48] From the above we deduce

THEOREM 131. *In a separable complete space a continuous image of the complement of an analytic set which does not contain a perfect non-empty subset has cardinal at most* $\aleph_1$.

We do not know, however, whether an analogous theorem is true for complements of continuous images of complements of analytic sets.*

98. Projection of closed sets. By a *projection* (parallel to an axis) of a point $(x_1, x_2, \ldots, x_m, x_{m+1})$ of a Euclidean $(m+1)$-dimensional space R_{m+1} (on an m-dimensional space R_m) we shall mean the point $(x_1, x_2, \ldots, x_m)$ of the space R_m. By a *projection of a set E* contained in the space R_{m+1} (on the space R_m) we shall mean the set of projections (on R_m) of all points of E. Hence *the projection of the sum of any aggregate of sets is the sum of the projections of these sets and the projection of a subset of a set is a subset of the projection of this set.* The projection of the intersection of two sets is contained in the intersection of the projections of these sets; the converse is not necessarily true (e.g., the case of the two unit sets $\{(0, 0)\}$ and $\{(0, 1)\}$). The projection of the difference of two sets contains the difference of their projections but not necessarily vice versa (see the above example).

It is obvious from the formula for distance in Euclidean space that the distance between the projections of two points contained in R_{m+1} is not greater than the distance between the points. Consequently, the projection of a set is a continuous image of the set.

If a linear set T is an image (continuous or not) of a linear set E, where $T = f(E)$, then T is the projection of the plane set Q consisting of all points (x, y), where $x = f(y)$.

We shall denote the projection of a set E either by $P(E)$ or by PE. There is a close connection between the *projection* and the *sum.* For every plane set E we have the relation

$$PE = \sum_y \mathrm{E}_x[(x, y) \in E],$$

where the summation extends over all real numbers y.

For, if $a \in PE$, there exists a point $(x, y) \in E$ whose projection is a, i.e., there exists a number b such that $(a, b) \in E$; hence $a \in \mathrm{E}_x [(x, y) \in E]$ which gives $a \in \sum_y \mathrm{E}_x [(x, y) \in E]$. Also, if $a \in \sum_y \mathrm{E}_x [(x, y) \in E]$, there exists a number b such that $a \in \mathrm{E}_x [(x, b) \in E]$ and so $(a, b) \in E$; consequently a is the projection of the point (a, b) and so $a \in PE$. This proves our formula.[49]

THEOREM 132. *The projection of a closed set contained in* R_{m+1} *is an* F_σ *in* R_m *and vice versa.*

Proof. We shall prove the theorem for $m = 1$; the proof for $m > 1$ is analogous to the above.

Let E be a plane closed set. If for any natural k we denote by E the set $E \,.\, \overline{S(0, k)}$, we obtain a decomposition of the set E into a sum of an infinite sequence of closed and compact sets, say

$$E = E_1 + E_2 + E_3 + \ldots ;$$

then

$$PE = PE_1 + PE_2 + PE_3 + \ldots . \tag{164}$$

But the projection of a set is a continuous image of the set; hence, by Theorem 57, the sets PE_k ($k = 1, 2, \ldots$) are closed and so, from (164), PE is an F_σ.

Next let T denote a linear set F_σ. Hence

$$T = T_1 + T_2 + \ldots , \tag{165}$$

where each T_k is closed. Put

$$E_k = \underset{(x,y)}{\mathrm{E}} [x \in T_k, y = k], \qquad k = 1, 2, \ldots ;$$

it is clear that these are plane closed sets and that E_k is congruent to T_k. The set

$$E = E_1 + E_2 + \ldots \tag{166}$$

is closed. For if p is a limit element of E then p is a point on the line $y = k$ (for some natural k) and so must be a limit element of the closed set E_k. It is therefore an element of E_k and so of E.

Finally, it is obvious that $PE_k = T_k$ for all k and so, since (166) implies (164), $PE = T_1 + T_2 + \ldots$; this gives $T = PE$ by (165); consequently T is the projection of the closed set E. This completes the proof of Theorem 132. It follows from this theorem that *the projection of an open set contained in R_{m+1} is an open set of R_m and vice versa.*

*A projection is said to be (1, 1) if every point of the projection of a set is the projection of only one point of the set. It can be shown that *every set F_σ contained in the space R_m is a (1, 1) projection of a closed set contained in R_{m+1}.**

99. Analytic sets as projections of sets G_δ. Every analytic linear set E is, by Theorem 113, a continuous image of the set N of all irrational numbers, i.e., $E = f(N)$; it is therefore the projection of the plane set

$$Q = \underset{(x,y)}{\mathrm{E}} [y \in N, x = f(y)].$$

It will be shown that the set Q is a G_δ.

Denote by T the set of all points in the plane with rational coordinates; this will be a set F_σ, since it is the sum of a countable aggregate of lines (parallel to the x-axis). It follows that

$$Q = \bar{Q} - T. \tag{167}$$

For, since $Q \,.\, T = 0$, we have $Q \subset \bar{Q} - T$. Next let $(x_0, y_0) \in \bar{Q} - T$. Then $y_0 \notin T$ and so $y_0 \in N$. Since $(x_0, y_0) \in \bar{Q}$ there exists a sequence (x_n, y_n), $n = 1, 2, \ldots$, such that $(x_n, y_n) \in Q$ for all n and

$$\lim_{n\to\infty} (x_n, y_n) = (x_0, y_0);$$

this gives

$$\lim_{n\to\infty} x_n = x_0 \text{ and } \lim_{n\to\infty} y_n = y_0.$$

Since $(x_n, y_n) \in Q$, we have $y_n \in N$ and $x_n = f(y_n)$; so, since $\lim\limits_{n\to\infty} y_n = y_0$ and the function f is continuous in N, we have

$$\lim_{n\to\infty} f(y_n) = f(y_0), \text{ i.e., } \lim_{n\to\infty} x_n = f(y_0).$$

This gives $x_0 = f(y_0)$. Now $y_0 \in N$ implies $(x_0, y_0) \in Q$. This proves (167). Since T is an F_σ it follows from (167) that the set Q is a G_δ. Thus every linear analytic set is the projection of a plane set G_δ. It can also be proved that every analytic set of R_m is the projection of a set G_δ contained in the space R_{m+1} and, since by the corollary to Theorem 114 the converse is also true, it follows that *analytic sets of R_m are identical with the projections (into R_m) of all sets G_δ contained in the space R_{m+1}.*

100. Projective sets. Let $\mathbf{R}$ denote a family of sets each of which is contained in some Euclidean space (not necessarily of the same dimension); denote by $P(\mathbf{R})$ the family of projections of all sets of $\mathbf{R}$ into a space of dimension one lower than the dimension of the space containing the given set; denote by $C(\mathbf{R})$ the family of the complements of all sets of the family $\mathbf{R}$ with respect to the space in which the given family is contained.

In particular, let $\mathbf{F}$ denote the family of all closed sets contained in Euclidean spaces (i.e., in R_1, in R_2, and so on). It follows from Theorem 132 that $P(\mathbf{F})$ is the family of all sets F_σ contained in Euclidean spaces and so $CP(\mathbf{F})$ is the family of all sets G_δ; consequently $PCP(\mathbf{F})$ is the family of all analytic sets contained in Euclidean spaces while, by Theorem 122, the intersection $PCP(\mathbf{F}) \,.\, CPCP(\mathbf{F})$ is the family of all Borel sets contained in Euclidean spaces.

The sets obtained from a closed set by applying alternately the operations P (of projection) and C (of taking the complement) are called *projective sets.*

Analytic sets (or sets $PCP(\mathbf{F})$) are also called sets P_1; their complements are called sets C_1. Sets P_1 and C_1 form the first class of projective sets. Sets P_n and C_n, for $n = 2, 3, \ldots$, are defined by induction as follows: $P_n = PC_{n-1}$ and $C_n = CP_n$. The sets P_n and C_n constitute the nth class of projective sets. Sets F_σ may be denoted by P_0, sets G_δ by C_0, open sets by P_{-1}, and closed sets by C_{-1}. Sets which are both a P_n and a C_n are called sets B_n. It will be shown that each P_n and each C_n is a set B_{n+1} for $n = 1, 2, \ldots$.

Let E denote a set P_1; then $E = PH$, where H is a G_δ. The set CH, being an F_σ, is a set P_1 and so H is a set C_1; the set $E = PH$ is therefore a P_2. Hence every set P_1 is a set P_2. But every C_1, as its own projection, is a set P_2. Thus all sets P_1 and all sets C_1 are sets P_2 and so, passing to complements, we conclude that all sets C_1 and all sets P_1 are sets C_2. Hence each P_1 and each C_1 is both a P_2 and a C_2 and therefore a B_2.

Next, suppose that for a given n we have proved that each set P_n and each set C_n is a set B_{n+1}; let E denote a set P_{n+1}. Then $E = PH$, where H is a C_n and so a B_{n+1} and, therefore, a C_{n+1}. Hence the set $E = PH$ is a P_{n+2}. Thus every P_{n+1} is a P_{n+2}. But the sets C_{n+1}, as their own projections, are also sets P_{n+2}. Hence each P_{n+1} and each C_{n+1} is a P_{n+2}; so each C_{n+1} and each P_{n+1} is a C_{n+2}; it follows that sets P_{n+1} and sets C_{n+1} are sets B_{n+2}. This establishes by induction the property of the sets P_n and C_n, for $n = 1, 2, \ldots$. It can be shown that the sets PB_n are identical with the sets P_n, for $n = 1, 2, \ldots$.

It has been proved[50] that the sum and the intersection of a countable aggregate of sets $P_n(C_n)$ is a set $P_n(C_n)$ for $n = 1, 2, \ldots$. It can also be shown that sets P_n and sets C_n are topological invariants.

101. Universal sets.[51] A plane set U is called *universal* with respect to a given family $\mathbf{R}$ of linear sets if its intersection with lines parallel to the y-axis gives all the sets of $\mathbf{R}$ and no others. A set contained in 3-dimensional space is called universal with respect to a given family $\mathbf{R}$ of plane sets if, on intersecting it with planes parallel to the yz-plane, we obtain all the sets of the family $\mathbf{R}$ and only such sets. We similarly define universal sets in m-dimensional space with respect to a given family R of sets contained in R_{m-1}.

The following theorem illustrates the use of universal sets to prove the existence of different classes of sets.

THEOREM 133. *Let* $\mathbf{R}$ *denote the family of all linear and plane sets possessing the following two properties:*

(a) *The intersection of a plane set of* $\mathbf{R}$ *with a line is a set of* $\mathbf{R}$.

(b) *A set similar* (in the sense of elementary geometry) *to a linear set of* $\mathbf{R}$ *belongs to* $\mathbf{R}$.

Then, if D is the set of all points of the line $y = x$ and U is a plane set of $\mathbf{R}$ *which is universal with respect to all linear sets of* $\mathbf{R}$, *we have*

$$D \,.\, U \in \mathbf{R} \text{ and } (D - U) \notin \mathbf{R}.$$

Proof. That $D \,.\, U \in \mathbf{R}$ follows from the hypothesis and property (a). It remains to show that $(D - U) \notin \mathbf{R}$.

Suppose that $(D - U) \in \mathbf{R}$. In view of property (b), the projection H of the set $D - U$ on the y-axis belongs to $\mathbf{R}$ (since it is a set similar to $D - U$); so, since U is a universal set, there exists a real number a such that the line $x = a$ meets U in a set E whose projection on the y-axis is the set H. Denote by Q the projection of the set $D \,.\, U$ on the y-axis. The set H is obviously the complement of Q with respect to the y-axis.

Denote by p the point (a, a) and consider two cases:

1. $p \in D \, . \, U$. This implies that $a \in Q$ and so $a \notin H$; hence $p \notin E$ (since H is the projection of E on the y-axis). But E is the set of points in which the line $x = a$ meets U; hence from $p \notin E$ it follows that $p \notin U$ contrary to assumption.

2. $p \notin D \, . \, U$. Since $p \in D$ it follows that $p \in (D - U)$ and so, since H is the projection of $(D - U)$ on the y-axis, $a \in H$; hence the point $(a, a) \in E$. Since $E \subset U$ we have $p \in U$ and, since $p \in D$, we find that $p \in D \, . \, U$ contrary to assumption. The assumption that $(D - U) \in \mathbf{R}$ thus leads to a contradiction. This completes the proof of the theorem.

Furthermore, it is clear that $CU \notin \mathbf{R}$; for if $CU \in \mathbf{R}$ then, by property (a), $D \, . \, CU \in \mathbf{R}$, i.e., $(D - U) \in \mathbf{R}$; this is impossible.

It follows from Theorem 133 that, if $\mathbf{R}$ is a given family of linear and plane sets, in order to prove the existence of a linear and a plane set of $\mathbf{R}$ whose complements do not belong to $\mathbf{R}$ it is sufficient to show that the family $\mathbf{R}$ possesses properties (a) and (b) and that there exists a plane set of $\mathbf{R}$ which is universal with respect to all linear sets of the family.

102. Universal sets P_n and C_n. Let

$$\delta_1, \delta_2, \delta_3, \ldots \tag{168}$$

be an infinite sequence of open intervals with rational endpoints (such a sequence can actually be constructed). Then every open linear set is the sum of all those intervals of the sequence (168) which are contained in the set. For if x is any irrational number of the interval $(0, 1)$ let

$$x = \frac{1}{\nu(1, x) +} \frac{1}{\nu(2, x) +} \cdots$$

represent the development of x as a continued fraction. Put

$$G(x) = \delta_{\nu(1,x)} + \delta_{\nu(2,x)} + \ldots;$$

this is an open linear set defined for every irrational x of the interval $(0, 1)$.

Let E be the set of all points (x, y) in the plane such that either x is a rational number of the interval $(0, 1)$ and y is any real number or x is an irrational number of the interval $(0, 1)$ and $y \in G(x)$. Let U be the interior of E. It will be shown that the open plane set U is universal with respect to all open linear sets.[52]

Let V denote a given open linear set. If $V = 0$, then the line $x = 2$ meets the set U in the set V. If $V \neq 0$, let $n_1, n_2, \ldots$ denote natural numbers n such that $\delta_n \subset V$ (there will be infinitely many of them); then

$$V = \delta_{n_1} + \delta_{n_2} + \ldots. \tag{169}$$

Put

$$a = \frac{1}{n_1 +} \frac{1}{n_2 +} \cdots; \tag{170}$$

thus a is an irrational number in the interval (0, 1). It follows from the definition of the set $G(x)$ and from (165) that $G(a) = V$. Hence to prove the required property of the set U it will be sufficient to show that the set H, in which the line $x = a$ meets U, is identical with the set $G(a)$. It follows from the definition of E that the line $x = a$ meets E in the set $G(a)$ and, since $E \supset U$, we have $H \subset G(a)$. It remains to show that $G(a) \subset H$.

Let $y_0 \in G(a)$; by (169), there exists a natural number k such that $y_0 \in \delta_{n_k}$. From (170) and the properties of continued fractions it follows that there exists a number $\eta > 0$ such that $0 < a - \eta < a + \eta < 1$ and, if x is an irrational number such that $|x - a| < \eta$, then $\nu(k, x) = n_k$; this gives

$$\delta_{n_k} = \delta_{\nu(k,x)} \subset G(x).$$

Now let (x, y) be a point in the plane such that

$$|x - a| < \eta \text{ and } y \in \delta_{n_k}. \tag{171}$$

If x is irrational, then

$$y \in \delta_{n_k} \subset G(x)$$

and so, from the definition of E, $(x, y) \in E$; if x is rational, $(x, y) \in E$. Thus every point (x, y) which satisfies (171) belongs to E, i.e., the open rectangle (171) is contained in E. But the point (a, y_0) is an interior point of this rectangle; hence (a, y_0) is contained in the interior of E, i.e., in the set U. Since $(a, y_0) \in U$ it follows that $y_0 \in H$; hence $G(a) \subset H$. Consequently the open plane set U constructed above is universal with respect to all open linear sets.

The above proof can easily be modified to show that *it is possible to construct an open set U in R_{m+1} which is universal with respect to all open sets of the space R_m.* It is clear that the complement CU (with respect to R_{m+1}) of such a set is a closed set in R_{m+1} and is universal with respect to all closed sets of R_m.[53]

Since sets F_σ contained in R_{m-1} are, by Theorem 132, identical with the projections of closed sets contained in R_m it follows that in R_m the set PCU is a set F_σ which is universal with respect to all sets F_σ contained in R_{m-1}; furthermore the set $CPCU$ is a set G_δ in R_m which is universal with respect to all sets G_δ contained in R_{m-1}. Since an analytic set is the projection of a G_δ and vice versa (§ 99), it follows that $PCPCU$ is an analytic set (i.e., a P_1) contained in R_{m-1} which is universal with respect to all analytic sets contained in R_{m-2}. This implies further that $CPCPCU$ is a set C_1 contained in R_{m-1} which is universal with respect to all sets C_1 contained in R_{m-2}; consequently $PCPCPCU$ is a set P_2 in R_{m-2} which is universal with respect to all sets P_2 contained in R_{m-3}, and so on. Thus

For every pair m, n of natural numbers there exists in space R_{m+1} a set P_n (C_n) which is universal with respect to all sets P_n (C_n) contained in space R_m.

103. The existence of projective sets of any given class. It is easy to prove by induction that the family of all linear and plane sets P_n (C_n) possesses properties (a) and (b) of Theorem 133. Consequently if we intersect the plane universal set P_n with the line $y = x$ we obtain a (linear) set E_n which is a P_n but its complement (with respect to the line $y = x$) is not a P_n. Thus *for every natural number n there exists a linear set C_n which is not a P_n.*

It is clear that the set CE_n is a C_n which is not a P_n. As a consequence the sets E_n and CE_n do not belong to a projective class lower than n. Therefore *for every natural number n there exists a linear set P_n (C_n) which is not a projective set of class lower than n.* All these sets can actually be constructed.

Furthermore the plane set P_n, which is universal with respect to all linear sets P_n, is not a C_n (for then its intersection with the line $y = x$ would be a C_n). Thus for $n = 1$ the set E_1 is an analytic set whose complement is not analytic. Hence *there exist linear analytic sets which are not Borel sets.*

A simple example of such a set (although the proof is far from simple) was given by Hurewicz.[54] It is the set of all irrational numbers x in the interval $(0, 1)$ for which the set $CG(x)$ (where $G(x)$ is the open set defined in § **100**) is non-countable.

An arithmetical example of a linear analytic set which is not a Borel set was given by Lusin.[55] It is the set of all those irrational x of the interval $(0, 1)$ for which the sequence $\nu(1, x), \nu(2, x), \ldots$ (defined in § **102**) contains an infinite subsequence with the property that each of its terms divides the following one.

Denote by $E_n^{(k)}$ the subset of E_n for which $k \leqslant x < k + 1$. Then

$$E_n = \sum_{k=-\infty}^{\infty} E_n^{(k)}.$$

If each of the sets $E_n^{(k)}$ were a set C_n then the set E_n would be a C_n which is impossible. Hence there exists an integer k such that the set $E_n^{(k)}$ is a P_n but not a C_n. Hence there exists in the interval $k \leqslant x < k + 1$ (and so also in the interval $n \leqslant x < n + 1$) a set T_n which is a P_n but not a C_n.

Next put $T = T_1 + T_2 + \ldots$. The set T is not projective. For if T were a projective set then it would be a P_n for some natural n; so the subset of T corresponding to the interval $n + 1 \leqslant x < n + 2$ would be a set P_n. But this subset is the set T_{n+1} which is not a C_{n+1} while every P_n is a C_{n+1}. We therefore have the result that *the sum of a countable aggregate of projective sets may not be a projective set.* Passing to complements we obtain the result that *the intersection of a countable aggregate of projective sets may not be a projective set* (this is not so in the case of projective sets of the same class (§ **100**)).

It is clear from the above that *it is possible actually to construct a linear set which is not a projective set.*

104. A universal set $F_{\sigma\delta}$. In § **102** we have constructed a plane set U (hence an F_σ) which is universal with respect to all linear sets F_σ. Denote by

U_k the subset of U contained in the strip $k \leqslant y < k + 1$; denote by V_k the set obtained by translation of U_k through $-k$ in the direction of the y-axis (i.e., in the strip $0 \leqslant y < 1$).

It will be shown that for every infinite sequence $E_1, E_2, \ldots$ of sets F_σ contained in the interval $0 \leqslant y < 1$ there exists a real number a such that the line $x = a$ meets the set V_k in the set E_k, for $k = 1, 2, \ldots$.

Denote by T_k the set obtained from E_k by a translation through the length k and put $T = T_1 + T_2 + \ldots$; T will be a set F_σ. Since U is universal, there exists a real number a such that the line $x = a$ meets U in the set T and the set U_k in the set T_k; it therefore meets the set V_k in the set E_k for $k = 1, 2, \ldots$.

Put $V = V_1 . V_2 . V_3 \ldots$; hence V is an $F_{\sigma\delta}$, since V_k is an F_σ for $k = 1, 2, \ldots$. Let E denote a set $F_{\sigma\delta}$ contained in the interval $0 \leqslant y < 1$. We may write $E = E_1 . E_2 \ldots$, where E_k $(k = 1, 2, \ldots)$ is an F_σ contained in the interval $0 \leqslant y < 1$. Since the line $x = a$ meets the set V_k in the set E_k for $k = 1, 2, \ldots$, it meets the set $V = V_1 . V_2 . V_3 \ldots$ in the set $E = E_1 . E_2 . E_3 \ldots$ (for if D denotes the line $x = a$ then $D . V_k = E_k$ for $k = 1, 2, \ldots$; so $D . V = D . V_1 . D . V_2 \ldots = E_1 . E_2 \ldots = E$). Consequently the set V is universal with respect to all sets $F_{\sigma\delta}$ contained in the interval $0 \leqslant y < 1$.

Under a homeomorphic mapping of the interval $0 < y < 1$ on the set of all real numbers we obtain from the set V a plane set $F_{\sigma\delta}$ which is universal with respect to all linear sets $F_{\sigma\delta}$ and such that the line $x = a$ meets it in a set $F_{\sigma\delta}$ which is not a $G_{\delta\sigma}$.

Starting with the above universal set $F_{\sigma\delta}$ we can similarly prove the existence of a universal set $F_{\sigma\delta\sigma}$, then of a universal set $F_{\sigma\delta\sigma\delta}$, and so on.

In the above construction we have employed so-called *countably-universal* sets. An infinite sequence $V_1, V_2, \ldots$ of plane sets is said to be countably universal[56] with respect to a family $\mathbf{R}$ of linear sets if the line $x = c$, c a constant, meets the sets V_k $(k = 1, 2, \ldots)$ in the sets of $\mathbf{R}$ and if, for every infinite sequence $E_1, E_2, \ldots$ of linear sets of $\mathbf{R}$, there exists a real number a such that the line $x = a$ meets the set V_k in the set E_k for $k = 1, 2, \ldots$.

By modifying the above proof it can be shown that there exists an infinite sequence of analytic plane sets which is countably universal with respect to all analytic linear sets.

If $V_1, V_2, \ldots$ is a sequence of plane sets which is countably universal with respect to the family $\mathbf{R}$ of linear sets then $V_1 + V_2 + \ldots$ $(V_1 . V_2 \ldots)$ is a universal set with respect to the family $\mathbf{R}_\sigma$ $(\mathbf{R}_\delta)$; if $0 \in \mathbf{R}$, then $V_1 - V_2$ is a universal set with respect to the family $\mathbf{R}_\rho$ ($\mathbf{R}_\rho$ is the family of differences); $(V_1 - V_2) - (V_3 - V_4)$ is universal with respect to the family $\mathbf{R}_{\rho\rho}$, and so forth.

APPENDIX

1. It is assumed that we know what is meant by a set of objects, e.g., the set of books in a certain library, a set of chairs in a hall, a set of ideas, or even a set of sets. The objects constituting a set are said to be its elements, and the notation $p \in E$, $p \notin E$ is used to denote that p is or is not an element of E.

A set E is defined when of every element p it can be said whether $p \in E$ or $p \notin E$.

A set A is a subset of a set B, i.e., $A \subset B$ or $B \supset A$, if, whenever $p \in A$, then $p \in B$. If $A \subset B$ and $B \subset A$, A and B are identical, i.e., $A = B$. If $A \subset B$ and $A \neq B$, A is said to be a *proper* subset of B.

The set containing no elements is called the *null* set. We postulate that the null set is contained in every set. Sets which have no elements in common are said to be *disjoint*.

If two sets A and B are such that a (1, 1) correspondence can be established between their elements, then A and B are said to have the same *power* or to be *equivalent*. For example, the set of all positive odd integers less than 100 and the set of all positive even integers not greater than 100 have the same power, for to every odd number may be correlated the even number greater by unity. The idea of power may be extended to sets which are not finite; e.g., the set of all natural numbers

$$1, 2, 3, \ldots, n, \ldots$$

and the set of all positive even integers

$$2, 4, 6, \ldots, 2n, \ldots$$

have the same power.

Two finite sets have the same power if and only if the number of elements in each set is the same. In the last example the given set and its subset have the same power. A finite set cannot have the same power as any of its subsets. *A set which has the same power as one of its proper subsets is said to be infinite in the Dedekind sense.*

2. A set which has the same power as the set of all natural numbers is said to be *countable*. The elements of a countable set can, therefore, be enumerated as a sequence

$$u_1, u_2, u_3, \ldots$$

with increasing indices. Conversely, the set of all terms of an infinite sequence is countable.

A subset of a countable set, if not finite, is obviously countable (since any subset of a sequence may be arranged as a sequence with increasing indices). Thus the sets of all odd numbers, all prime numbers, all squares are each countable.

The *sum* $A + B$ of two sets A and B consists of elements p such that either $p \in A$, or $p \in B$; the *intersection* $A.B$ consists of elements p such that $p \in A$ and $p \in B$. The difference $A - B$ is the set of all elements p such that $p \in A$ and $p \notin B$. The set $A - B$ where $B \subset A$ is called the *complement* of B with respect to A.

The sum of a finite set and a countable set is a countable set. For the sum of the sets

$$u_1, u_2, \ldots, u_m \text{ and } v_1, v_2, \ldots$$

may be written as the infinite sequence

$$u_1, u_2, \ldots, u_m, v_1, v_2, v_3, \ldots .$$

The sum of two countable sets is a countable set. In fact, the sum of the countable set

$$u_1, u_2, u_3, \ldots$$

and the countable set

$$v_1, v_2, v_3, \ldots$$

may be written as the infinite sequence

$$u_1, v_1, u_2, v_2, u_3, v_3, \ldots .$$

The definition of a sum of two sets may be easily extended to a sum of a finite or infinite sequence of sets. Given an infinite sequence of sets E_1, E_2, $E_3, \ldots$, the sum

$$S = E_1 + E_2 + E_3 + \ldots$$

is a set consisting of all elements p such that $p \in E_i$ for at least one value of i. *The sum of a countable aggregate of countable sets is countable.* For, if E_1, $E_2, \ldots$ be an infinite sequence of countable sets, the elements of

$$S = E_1 + E_2 + E_3 + \ldots$$

may be written down as a double sequence

$$\begin{array}{l} a_{11}, a_{12}, a_{13}, \ldots \\ a_{21}, a_{22}, a_{23}, \ldots \\ a_{31}, a_{32}, a_{33}, \ldots \\ \ldots \qquad \ldots, \end{array}$$

where $a_{k_1}, a_{k_2}, \ldots$ are the elements of E_k. Arranging the elements of the double sequence into groups such that the nth group consists of all a_{kl},

$k + l = n + 1$, we obtain the infinite sequence

$$a_{11}, a_{21}, a_{12}, a_{31}, a_{22}, a_{13}, a_{41}, a_{32}, \ldots$$

containing all terms of S.

The set of all rational numbers is countable. For let S be the set of all positive rational numbers. Denote by E_n the set of all positive rational numbers in lowest terms with n in the denominator; then

$$S = E_1 + E_2 + E_3 + \ldots,$$

where E_n is countable. Hence S is countable. Similarly, the set T of all negative rational numbers is countable and, therefore, also the set of all rational numbers.

The set E of all finite sequences of natural numbers is countable. For a finite sequence $n_1, n_2, \ldots, n_k$ may be correlated in a unique way with the number

$$N = 2^{n_1-1} + 2^{n_1+n_2-1} + \ldots + 2^{n_1+n_2+\ldots+n_k-1}.$$

The set of all polynominals with rational coefficients is countable, since a (1, 1) correspondence may be established between a polynominal and the finite sequence consisting of the coefficients. All such polynominals may, therefore, be represented as an infinite sequence

$$P_1, P_2, P_3, \ldots.$$

A polynominal has at most a finite number of roots; writing down all the roots of P_1, then those of P_2 and so on, we obtain an infinite sequence

$$x_1, x_2, x_3, \ldots$$

consisting of the roots of all polynomials with rational coefficients, i.e., of all algebraic numbers. Hence, *the set of all algebraic numbers is countable.*

3. A set which is neither finite nor countable is said to be *non-countable.*

The set E of all infinite sequences of natural numbers is non-countable. For, if it were countable, it could be written as a double sequence

$$\begin{array}{lll} n_{11}, & n_{12}, & n_{13}, \ldots \\ n_{21}, & n_{22}, & n_{23}, \ldots \\ n_{31}, & n_{32}, & n_{33}, \ldots \\ \ldots & & \ldots \end{array} \tag{1}$$

But the infinite sequence

$$n_{11} + 1, n_{22} + 1, n_{33} + 1, \ldots, n_{kk} + 1, \ldots \tag{2}$$

differs from each of the sequences (1) and so does not belong to E, which is contrary to the hypothesis that E consists of all infinite sequences of natreal numbers. Similarly, it can be shown that for every infinite sequence of ural numbers there exists a real number which is not a number of the sequence. Let

$$x_1, x_2, \ldots$$

be a sequence of real numbers. Every real number can be expressed in one sometimes two, ways as an infinite decimal. Writing down these developments (one or both if there are two) of the successive terms of the sequence, we obtain an infinite sequence of infinite decimals

$$\begin{array}{ll} c_1, & c_{11}c_{12}c_{13}\ldots \\ c_2, & c_{21}c_{22}c_{23}\ldots \\ c_3, & c_{31}c_{32}c_{33}\ldots \\ \ldots & \ldots \end{array}$$

We now construct the decimal

$$0.a_1a_2a_3\ldots,$$

where $a_1 \neq c_{11}$, $a_2 \neq c_{22}$, ... generally $a_k \neq c_{kk}$ (we may choose, for example, $a_k = c_{kk} + 1$ if $c_{kk} < 9$ and $a_k = 0$ if $c_{kk} = 9$).

The real number thus constructed is obviously different from every term of the given sequence. It follows, therefore, that *the set of all real numbers is non-countable.*

There exists, as is well known, a (1, 1) correspondence between the set of real numbers and the points on a straight line; hence, the set of points on a straight line is non-countable.

If a finite or countable set be removed from a non-countable set, the remaining set is non-countable. For, let P be a non-countable set, Q finite or countable, and R the remainder. Hence, $P = Q + R$; if R were finite or countable then P would be finite or countable, contrary to the assumption that it is non-countable. The set R is, therefore, neither finite nor countable and so must be non-countable. After removing from the non-countable set of real numbers the countable set of algebraic numbers, there remains a non-countable set of real numbers which are known as the transcendental numbers.

4. Let all sets be divided into classes, two sets belonging to the same class if and only if they have the same power; then all sets of a given class have a common characteristic. The symbols used to designate classes of sets of equal power are called *cardinal numbers.* The cardinal number corresponding to the class of all countable sets is denoted by $\aleph_0$ and the one corresponding to the class of all sets of the same power as the set of all real numbers by $\mathbf{c}$. It follows from the definition that to every set corresponds a cardinal number (namely that number which serves to designate the class containing the given set). The cardinal number corresponding to a set E is frequently denoted by $\bar{\bar{E}}$ and is called *the power of the set E.* Sets with cardinal $\mathbf{c}$ are said to have *the power of the continuum.*

Cardinal numbers different from the natural numbers are called *transfinite numbers*. There exist different transfinite numbers, e.g., $\aleph_0$ and $\mathbf{c}$.

The sum $\mathbf{m} + \mathbf{n}$ of two cardinal numbers is the cardinal number of the set $M + N$, where M and N are disjoint sets and $\bar{\bar{M}} = \mathbf{m}$ and $\bar{\bar{N}} = \mathbf{n}$. It is easily seen that addition of cardinal numbers is commutative and associative.

It follows from § **2** that

$$\mathbf{n} + \aleph_0 = \aleph_0, \tag{3}$$

$$\aleph_0 + \aleph_0 = \aleph_0, \tag{4}$$

$$\aleph_0 + \aleph_0 + \ldots = \aleph_0 \qquad (\aleph_0 \text{ terms}).$$

Consider the set E of all real numbers and let N be the set of all rational numbers and M the remainder. Then

$$\bar{\bar{E}} = \bar{\bar{M}} + \bar{\bar{N}};$$

but $\bar{\bar{E}} = \mathbf{c}$, $\bar{\bar{N}} = \aleph_0$, and let $\bar{\bar{M}} = \mathbf{m}$; then

$$\mathbf{c} = \mathbf{m} + \aleph_0 \tag{5}$$

$$\mathbf{c} + \aleph_0 = (\mathbf{m} + \aleph_0) + \aleph_0 = \mathbf{m} + (\aleph_0 + \aleph_0) = \mathbf{m} + \aleph_0 = \mathbf{c};$$

hence

$$\mathbf{c} + \aleph_0 = \mathbf{c}. \tag{6}$$

For n a natural number

$$\mathbf{c} + \mathbf{n} = (\mathbf{c} + \aleph_0) + \mathbf{n} = \mathbf{c} + (\aleph_0 + \mathbf{n}) = \mathbf{c} + \aleph_0 = \mathbf{c}. \tag{7}$$

The relation $y = x/(1 + |x|)$ established a (1, 1) correspondence between the set of all real numbers x and the set of all real numbers y in the interval $(-1, 1)$. The cardinal of the latter set is, therefore, $\mathbf{c}$. Let a and $b > a$ be two given real numbers. The relation $z = \frac{1}{2}(b - a)y + \frac{1}{2}(a + b)$ establishes a (1, 1) correspondence between the set of all real numbers y which satisfy the inequality $-1 < y < 1$ and the set of all real numbers z satisfying the inequality $a < z < b$; hence, the two sets have the same cardinal i.e., the cardinal $\mathbf{c}$. This cardinal will not change if we add one element to the set. Hence, for every a and $b > a$, the set of all real numbers in the interval (a, b) has cardinal $\mathbf{c}$ (the end-points being included or excluded).

In particular, the set M of all real numbers x satisfying the inequality $0 \leqslant x < 1$, the set N of all real numbers x such that $1 \leqslant x < 2$, and the set S of all real numbers x such that $0 \leqslant x < 2$ have all cardinal $\mathbf{c}$. But M and N are disjoint sets and $M + N = S$; therefore $\bar{\bar{M}} + \bar{\bar{N}} = \bar{\bar{S}}$, and so

$$\mathbf{c} + \mathbf{c} = \mathbf{c}. \tag{8}$$

The definition of a sum of cardinal numbers may be extended to an infinite sequence of cardinal numbers. If

$$\mathbf{m}_1, \mathbf{m}_2, \mathbf{m}_3, \ldots$$

is an infinite sequence of cardinal numbers and $M_k (k = 1, 2, \ldots)$ are disjoint sets such that $\bar{\bar{M}}_k = \mathbf{m}_k$, then the sum of the infinite series

$$\mathbf{m}_1 + \mathbf{m}_2 + \mathbf{m}_3 + \ldots$$

is the cardinal number of the set

$$M_1 + M_2 + M_3 + \ldots.$$

Thus it may be easily seen that

$$\aleph_0 = 1 + 1 + 1 + 1 + \ldots$$

(it is sufficient here to take M_k to consist of one number k). Also

$$\aleph_0 = 1 + 2 + 3 + 4 + \ldots$$

(here, M_k may be taken to consist of k natural numbers n satisfying the inequality $\frac{1}{2}(k - 1)k < n \leqslant \frac{1}{2}k(k + 1)$, for $k = 1, 2, \ldots$). Similarly,

$$\aleph_0 = \aleph_0 + \aleph_0 + \aleph_0 + \ldots, \tag{9}$$

and

$$\mathbf{c} = \mathbf{c} + \mathbf{c} + \mathbf{c} + \ldots, \tag{10}$$

for it would be sufficient to consider in the first case the (countable) set M_k consisting of the natural numbers

$$2k - 1,\ 2(2k - 1),\ 2^2(2k - 1),\ 2^3(2k - 1), \ldots$$

and in the second case the set M_k of all real numbers x satisfying the inequality $k - 1 < x < k$ $(k = 1, 2, \ldots)$, noting that the set of all real positive numbers has cardinal $\mathbf{c}$.

The product $\mathbf{m}.\mathbf{n}$ of the cardinal numbers $\mathbf{m}$ and $\mathbf{n}$ is the cardinal number of the set P consisting of all pairs (x, y) where $x \in M$, $y \in N$, and where $\bar{\bar{M}} = \mathbf{m}$, $\bar{\bar{N}} = \mathbf{n}$. It is easily seen that multiplication of cardinal numbers is commutative and associative. It follows that

$$\aleph_0.\aleph_0 = \aleph_0.$$

For let M and N be each the set of natural members. The product P will then be the set of all pairs (m, n) of natural numbers, i.e., the set of all elements of the double sequence

$$\begin{array}{l} (1, 1), (1, 2), (1, 3), \ldots \\ (2, 1), (2, 2), (2, 3), \ldots \\ (3, 1), (3, 2), (3, 3), \ldots \\ \ldots \qquad\qquad\quad \ldots \end{array}$$

which may be ordered as a single infinite sequence by the diagonal method. Hence $\bar{\bar{P}} = \aleph_0$, and since $\bar{\bar{M}}.\bar{\bar{N}} = \bar{\bar{P}}$, we have $\aleph_0.\aleph_0 = \aleph_0$.

If $\mathbf{m}$ be a finite cardinal and $\mathbf{n}$ any cardinal number, we have

(11) $$\mathbf{m}.\mathbf{n} = \mathbf{n} + \mathbf{n} + \ldots + \mathbf{n} \qquad (m \text{ terms});$$

for, let M be the set of natural numbers $1, 2, \ldots, m$, N a set such that $\bar{\bar{N}} = \mathbf{n}$, and P the set of all pairs (m, n), where $m \in M$ and $n \in N$. Hence $\bar{\bar{P}} = \mathbf{m}.\mathbf{n}$. Denoting for a given k the set of all pairs (k, n) by P_k we get $\bar{\bar{P}}_k = \mathbf{n}$, for $k = 1, 2, \ldots, m$, and $P = P_1 + P_2 + \ldots + P_m$, where the sets P_k are disjoint. Since $\bar{\bar{P}} = \mathbf{m}.\mathbf{n}$, (11) follows at once.

In particular, for $\mathbf{n} = \aleph_0$, (11) and (4) give

$$\mathbf{m}.\aleph_0 = \aleph_0 \qquad (\mathbf{m} = 1, 2, \ldots).$$

Similarly, for $\mathbf{n} = \mathbf{c}$, (11) and (8) give

$$\mathbf{m}.\mathbf{c} = \mathbf{c} \qquad (\mathbf{m} = 1, 2, \ldots).$$

Changing slightly the proof of (11), it can be easily shown that, for every cardinal number $\mathbf{n}$,

(12) $$\aleph_0.\mathbf{n} = \mathbf{n} + \mathbf{n} + \mathbf{n} + \ldots$$

and so for $\mathbf{n} = \mathbf{c}$,

$$\aleph_0.\mathbf{c} = \mathbf{c} + \mathbf{c} + \mathbf{c} + \ldots = \mathbf{c},$$

by (10).

To obtain the product $\mathbf{c}.\mathbf{c}$, let M and N be each the set of all positive real numbers not exceeding 1. The set P will, therefore, be the set of all pairs (x, y), where x and y are real numbers satisfying the inequalities $0 < x \leqslant 1$ and $0 < y \leqslant 1$. It can be shown that the set P has the same cardinal as the set E of all real numbers z satisfying the inequality $0 < z \leqslant 1$.

In fact, let (x, y) be an element of P. Develop x and y as infinite decimal fractions, e.g.

$$x = 0.\,4\,3\,0\,0\,7\,0\,9\,5\,0\,0\,0\,8\,3\ldots,$$
$$y = 0.\,0\,5\,6\,0\,0\,3\,0\,0\,0\,1\,4\,0\,2\ldots.$$

Divide the digits to the right of the decimal point into groups by means of a stroke after each significant figure; we thus get an infinite sequence of groups:

$$4\,|\,3\,|\,0\,0\,7\,|\,0\,9\,|\,5\,|\,0\,0\,0\,8\,|\,3\,|\ldots$$
$$0\,5\,|\,6\,|\,0\,0\,3\,|\,0\,0\,0\,0\,1\,|\,4\,|\,0\,2\,|\ldots.$$

Place the groups of the second sequence between the successive groups of the first, and so get a new sequence of groups:

$$4\,|0\,5\,|3\,|6\,|0\,0\,0\,7\,|0\,0\,3\,|0\,9\,|0\,0\,0\,1\,|5\,|4\,|0\,0\,0\,8\,|0\,2\,|3\,|\ldots;$$

omitting the strokes we get an infinite sequence of digits, which is the decimal part of a certain number

$$z = 0.4053600700309000154000802 3\ldots,$$

which we correlate with the pair (x, y).

It is easily seen that such a correlation establishes a (1, 1) correspondence between the elements of the set P and those of E. But $\bar{\bar{E}} = \mathbf{c}$; hence, $\bar{\bar{P}} = \bar{\bar{E}} = \mathbf{c}$, and since $\bar{\bar{P}} = \bar{\bar{M}}.\bar{\bar{N}} = \mathbf{c}.\mathbf{c}$ we have

$$\mathbf{c}.\mathbf{c} = \mathbf{c}. \tag{13}$$

It follows from the above that the set of all pairs (x, y) of real numbers x, y has the same cardinal as the set of all real numbers. Geometrically, this means that the set of all points in the plane has the same cardinal as the set of all points in a straight line and, therefore, also as the set of all points in a finite segment.

The definition of a product of two cardinal numbers may be extended to an infinite sequence. It follows readily that if $\mathbf{m}_1, \mathbf{m}_2, \ldots, \mathbf{m}_n$ are given cardinal numbers and $M_1, M_2, \ldots, M_n$ sets such that $\bar{\bar{M}}_k = \mathbf{m}_k$, for $k = 1, 2, \ldots, n$, then the cardinal number $\mathbf{m}_1.\mathbf{m}_2 \ldots \mathbf{m}_n$ is the cardinal of the set of all combinations $(m_1, m_2, \ldots, m_n)$, where $m_k \in M_k$, for $k = 1, 2, \ldots, n$. Similarly, the infinite product

$$\mathbf{m}_1.\mathbf{m}_2.\mathbf{m}_3 \ldots$$

of cardinal numbers is the cardinal of the set P of all the infinite sequences

$$m_1, m_2, m_3, \ldots,$$

where $m_k \in M_k$ and $\bar{\bar{M}}_k = \mathbf{m}_k$ $(k = 1, 2, \ldots)$.

In particular, let M_k be the set consisting of the numbers 0 and 1. The set P will, therefore, be the set of all infinite sequences

$$a_1, a_2, a_3, \ldots \tag{14}$$

consisting of the numbers 0 and 1. Denote by Q the set of the sequences belonging to P in which there is an infinite number of ones, and by R the remainder of P. Therefore R consists of all those sequences in which, from a certain stage onwards, there are only zeros and so has the same cardinal as the set of all finite sequences consisting of 0 and 1, which is a countable set. The set Q, however, has the same cardinal as the set X of all positive real numbers $\leqslant 1$ and so cardinal $\mathbf{c}$. The (1, 1) correspondence between the elements of Q and those of X may be easily established if we correlate the sequence (12) with the number

$$\frac{a_1}{2} + \frac{a_2}{2^2} + \frac{a_3}{2^3} + \ldots$$

(which obviously belongs to X).

Hence $\bar{\bar{P}} = \mathbf{c}$, and so

$$\mathbf{c} = 2.2.2\ldots. \tag{15}$$

Similarly, it can be shown that

$$\mathbf{c} = \aleph_0.\aleph_0.\aleph_0\ldots \tag{16}$$

for, it is sufficient to take as M_k ($k = 1, 2, \ldots$) the set of all natural numbers, and so P will be the set of all infinite sequences

$$n_1, n_2, n_3, \ldots \tag{17}$$

of natural numbers, which has the same cardinal as the set X of all positive real numbers $\leqslant 1$. To establish a (1, 1) correspondence between the elements of P and X it is sufficient to correlate the number

$$\frac{1}{2^{n_1}} + \frac{1}{2^{n_1+n_2}} + \frac{1}{2^{n_1+n_2+n_3}} + \ldots$$

of the set X with the sequence (17).
Furthermore,

$$\mathbf{c} = \mathbf{c}.\mathbf{c}.\mathbf{c}.\ldots \tag{18}$$

To prove this let $M_k = X$, for $k = 1, 2, \ldots$; we show that the set P of all infinite sequences consisting of elements of X has the same cardinal as the set X. To establish a (1, 1) correspondence between the elements of P and X, correlate the sequence

$$x_1, x_2, x_3, \ldots \tag{19}$$

which belongs to P with a number x in the form

$$g'_1\, g'_2\, g''_1\, g'_3\, g''_2\, g'''_1\, g'_4\, g''_3\ldots, \tag{20}$$

where $g'_n\, g''_n\, g'''_n\ldots$ designates x_n, and the sequence (19) is designated by the double sequence

$$\begin{array}{l} g'_1\, g''_1\, g'''_1\ldots \\ g'_2\, g''_2\, g'''_2\ldots \\ g'_3\, g''_3\, g'''_3\ldots \\ \ldots \qquad \ldots, \end{array}$$

from which (20) is obtained by the diagonal method.

5. Let P and Q be two given sets. If with every element of P there is correlated an element of Q, where the same element of Q may be correlated with several elements of P, we obtain a mapping of the set P on the set Q. Let now $\mathbf{m}$ and $\mathbf{n}$ be two cardinal numbers and M, N two sets such that $\bar{\bar{M}} = \mathbf{m}$, $\bar{\bar{N}} = \mathbf{n}$; then the power $\mathbf{m}^{\mathbf{n}}$ is defined to be the cardinal of the set of all mappings of the set N on the set M. It can be easily shown that for any three cardinal numbers $\mathbf{m}$, $\mathbf{n}$, $\mathbf{p}$ we have

$$\mathbf{m}^{\mathbf{n}+\mathbf{p}} = \mathbf{m}^{\mathbf{n}}.\mathbf{m}^{\mathbf{p}}$$

$$(\mathbf{mn})^{\mathbf{p}} = \mathbf{m}^{\mathbf{p}}.\mathbf{n}^{\mathbf{p}}$$

$$(\mathbf{m}^{\mathbf{n}})^{\mathbf{p}} = \mathbf{m}^{\mathbf{np}}.$$

If n be a natural number, we have obviously

$$\mathbf{m}^{n} = \mathbf{m}.\mathbf{m}\ldots\mathbf{m} \qquad (n \text{ factors}).$$

It follows also readily from the definitions of a power and of an infinite product of cardinal numbers (§ 4), that

$$\mathbf{m}^{\aleph_0} = \mathbf{m}.\mathbf{m}.\mathbf{m}.\ldots \tag{21}$$

In particular, for $\mathbf{m} = 2$ we obtain from (15)

$$2^{\aleph_0} = \mathbf{c}. \tag{22}$$

From (21), (16), and (18) we obtain

$$\aleph_0^{\aleph_0} = \mathbf{c}^{\aleph_0} = \mathbf{c}.$$

Let N be a set of cardinal $\mathbf{n}$; then $2^{\mathbf{n}}$ will be the cardinal number of the set E of all subsets of N, the null set and the set N being included. Thus $2^{\aleph_0}$ or $\mathbf{c}$ is the cardinal number of all subsets of the set of natural numbers, and $2^{\mathbf{c}}$ is the cardinal of the set of all subsets of the set of all real numbers and so the cardinal of the set of all functions of a real variable.

6. Given two cardinal numbers $\mathbf{m}$ and $\mathbf{n}$, we say that $\mathbf{m} < \mathbf{n}$ if the set M of cardinal $\mathbf{m}$ has equal power with a subset of the set N whose cardinal is $\mathbf{n}$, and if there is no subset of M of equal power with N. We cannot, however, as yet state that every two cardinal numbers $\mathbf{m}$ and $\mathbf{n}$ are related to each other by one of the three signs $>$, $=$, $<$.

It follows at once from the definition of inequality of cardinal numbers, that

$$\mathbf{n} < \aleph_0 \qquad \text{for } \mathbf{n} = 1, 2, 3, \ldots,$$

and

$$\aleph_0 < \mathbf{c}.$$

For, if M be the set of all natural numbers and N the set of all real numbers, then $\bar{\bar{M}} = \aleph_0$, $\bar{\bar{N}} = \mathbf{c}$, where M has the same power as a certain subset of N but not conversely (since N is non-countable).

It is, however, still unknown whether there are cardinal numbers $\mathbf{m}$ satisfying the inequality $\aleph_0 < \mathbf{m} < \mathbf{c}$. The assumption that there are no such cardinal numbers is known as the *continuum hypothesis*. The assumption that there is no cardinal number between $\mathbf{m}$ and $2^{\mathbf{m}}$, whatever the transfinite number $\mathbf{m}$, is known as the *Cantor aleph-hypothesis.* It can be shown that every cardinal number $\mathbf{m}$ satisfies the inequality

$$2^{\mathbf{m}} > \mathbf{m}; \tag{23}$$

in other words, *the set of all subsets of a given aggregate has power greater than that of the aggregate.* From (23) we get at once the infinite sequence of inequalities

$$\aleph_0 < \mathbf{c} < 2^{\mathbf{c}} < 2^{2^{\mathbf{c}}} < \ldots,$$

which shows that there is an infinite number of transfinite cardinal numbers.

7. The following is the so-called axiom of choice stated by Zermelo in 1904 and tacitly implied in several of the preceding results:

For every aggregate M consisting of non-empty disjoint sets E, there exists (at least one) aggregate N containing one, and one only, element of each set E.

The meaning of this axiom may be explained by the following examples:

Divide all real numbers into sets assigning two numbers to the same set if and only if their difference is rational. We thus get an aggregate M of disjoint non-empty sets. By the axiom of choice, there exists a set N containing one and only one number of each set E. No one, however, has been able so far to construct the set N, for it is impossible in this case to put down a law of selection which would pick out a certain element of the set E. This has lead some mathematicians to doubt even the probability of the truth of the axiom. Consider another example. Divide all countable sets of points on a straight line which are not symmetrical with respect to the point 0 into classes, assigning to the same class those sets which are symmetrical images of each other with respect to the point 0. There will obviously be two sets in each class. By the axiom of choice there exists a set N containing one set only of each pair, but we cannot devise any rule which would enable us to select this set. The existence of the set N is, therefore, deduced only on the basis of the axiom of choice.

If, however, all points of a straight line be divided into classes, assigning to the same class two sets if and only if they are disjoint and their sum gives the whole line, then the set N may be actually constructed; for it is sufficient to assign to N that set of each class which contains the point 0.

We shall next consider some of the applications of this axiom. Let M be an aggregate of cardinal $\mathbf{m}$ consisting of disjoint, non-empty sets. By the axiom of choice there exists a set N containing one and only one element of each set belonging to M. Evidently $\bar{\bar{N}} = \bar{\bar{M}}$; hence $\bar{\bar{N}} = \mathbf{m}$. On the other hand, N being a subset of the sum S of all sets constituting M, we have

$$\bar{\bar{S}} \geqslant \bar{\bar{N}}, \text{ and so } \bar{\bar{S}} \geqslant \mathbf{m}.$$

We thus arrive at the following result: *If any aggregate be divided into disjoint sets, the set of these sets has cardinal* $\leqslant$ *the cardinal of the original aggregate.*

It follows readily from the above that the cardinal of any set of points in the plane is not less than the cardinal of the set of its projections. For the given set may be divided into subsets consisting of all points which project into the same point.

Furthermore, if a set of cardinal $\mathbf{c}$ be divided into two parts, one of them at least has cardinal $\mathbf{c}$. For let P be the set of all points in the plane and so of cardinal $\mathbf{c}$. It will be sufficient to show that there is no division $P = A + B$, where both A and B have cardinal $< \mathbf{c}$. Suppose, on the contrary, that such a division exists. The projection of the set A on the x-axis has by the above cardinal $\leqslant$ the cardinal of A, hence $< \mathbf{c}$. There exists, therefore, an abscissa x_0 such that the straight line $x = x_0$ does not contain points of A. We conclude, similarly, that there exists an ordinate y_0 such that the straight line $y = y_0$ does not contain points of B. Hence the point (x_0, y_0) belongs neither to A nor to B, contrary to the fact that $P = A + B$. The above statement is therefore proved.

There are other more general forms of the axiom of choice, e.g., the following (Hilbert):

There exists a correspondence which correlates to each property W possessed by at least one object a certain element $\tau(W)$ possessing the property W.

This axiom leads to the so-called *general principle of selection* (Zermelo). E being any set, denote by W_E the property of belonging to the set E. If E is not the null set there exists at least one object which has the property W_E, whereas $\tau(W_E)$ will be an element of E. Hence, *there exists a correlation which assigns to every non-empty set an element of that set.*

There exists, therefore, for every given set a correlation which assigns to every non-empty subset of the given set a certain element belonging to that subset.

It can be shown, proceeding from the above, that *every non-empty set which is not finite contains a countable subset.* In fact, let E be a given non-empty set, which is not finite. Then to each non-empty subset C of E corresponds a certain element $\alpha(C)$ of C. Put $p_1 = \alpha(E)$, and let E_1 be the set obtained by removing p_1 from E. If E_1 were the null set or finite, then the set E would be finite. Hence E_1 is neither empty nor finite. Let further $p_2 = \alpha(E_1)$, and let E_2 be the set obtained from E_1 after removing p_2. As above, E_2 is neither empty nor finite. Let now $p_3 = \alpha(E_2)$, and so on.

We thus obtain an infinite sequence of different elements of the set E,

$$p_1, p_2, p_3, \ldots, \tag{24}$$

which forms a countable subset of the set E.

A set which contains a countable subset has the same cardinal as certain of its subsets (i.e., it is infinite in the Dedekind sense). For, retaining the above notation, we can establish a (1, 1) correspondence between the sets E and E_1 as follows: correlate every element of E which is not in (24) with itself and every element which belongs to (24) with its successor in (24).

Let $\mathbf{u}$ be a cardinal number which is not finite and U a set of cardinal $\mathbf{u}$. The set U is, therefore, neither empty nor finite and so must contain a subset of cardinal $\aleph_0$. Hence,

$$\mathbf{u} \geqslant \aleph_0$$

for every cardinal number $\mathbf{u}$ which is not finite. For a finite cardinal number $\mathbf{n}$ we have obviously the inequality $\mathbf{n} < \aleph_0$; hence, every cardinal number is $\geqslant$ or $< \aleph_0$.

If a set E is such that $\bar{\bar{E}} < \aleph_0$, then E is a finite. If $\bar{\bar{E}} = \aleph_0$, then E is countable; and if finally $\bar{\bar{E}} > \aleph_0$, then E is non-countable.

Let E be a set which is neither empty nor finite. Hence E contains a countable subset P. Remove from E the elements belonging to P, and denote the remainder by R; then $E = P + R$, and (since P and R are disjoint)

$$\bar{\bar{E}} = \bar{\bar{P}} + \bar{\bar{R}}. \tag{25}$$

Add to the set E any countable set Q distinct from E; therefore,

$$\bar{\bar{P}} + \bar{\bar{Q}} = \aleph_0 = \bar{\bar{P}}, \tag{26}$$

and putting $E + Q = S$, we shall get from (25) and (26)

$$\bar{\bar{S}} = \bar{\bar{E}} + \bar{\bar{Q}} = \bar{\bar{P}} + \bar{\bar{R}} + \bar{\bar{Q}} = \bar{\bar{P}} + \bar{\bar{R}} = \bar{\bar{E}}.$$

Hence, *the cardinal of a set which is neither empty nor finite does not change if we add to it a countable set of elements.*

Let E denote a non-countable set, P its countable subset, R a set defined as above. Hence R is neither the null set nor finite (since then $E = P + R$ would be countable), and so it will not change its cardinal if we add to it the countable set P, which gives $\bar{\bar{E}} = \bar{\bar{R}}$. We have thus proved that *the cardinal of a non-countable set does not change if we remove from it a countable set of elements.*

8. A set E is said to be *ordered* if there exists a convention according to which it can be said of any two different elements of the set that one element precedes the other in the set. This is expressed in writing by $a \prec b$, i.e., a precedes b or $a \succ b$, i.e., a follows b. Whatever this convention may be the following two conditions must be satisfied:

1. Relation $a \prec b$ excludes the relation $b \prec a$ (asymmetry).
2. If $a \prec b$ and $b \prec c$ then $a \prec c$ (transitivity).

An element of E, which is not preceded by any other, is said to be the *first* element; and one which is not followed by any other is called the *last* element of the set E.

The set of natural numbers apart from its usual order may be also ordered according to the following convention. Of two numbers the one with the least number of different prime factors will come first, and in case of an equal number of different prime factors the one of smaller value. It is easily seen that this agreement orders the set of natural numbers (i.e., conditions 1 and 2 are satisfied). Hence we get

$$1 \prec 2 \prec 3 \prec 4 \prec 5 \prec 7 \prec 6 \prec 34 \prec 35 \prec 30 \prec \ldots.$$

Two ordered sets G and Γ are said to be *similar*, i.e., $G \simeq \Gamma$, if there exists a (1, 1) correspondence between their elements which leaves the order relations between corresponding pairs of elements unchanged. Thus if a, b are any two elements of G and α, β their corresponding elements in Γ, then the relation

$$a \prec b$$

implies the relation

$$\alpha \prec \beta$$

and conversely.

It is easily seen that an ordered set is similar to itself and two sets similar to a third are similar to each other. (The relation of similarity is, therefore, reflexive, symmetrical, and transitive.)

Divide all ordered sets into classes assigning two sets to the same class if and only if they are similar. Then sets belonging to the same class are said to be of the same *ordinal type*. Ordinal types thus serve as symbols to designate the various classes.

Two ordered sets of the same type have obviously the same cardinal, but the converse is not necessarily true. The set of all natural numbers and the set of all rational numbers have the same cardinal (both countable), but when ordered according to their magnitude, are evidently of different types.

The ordinal type of a set E is denoted after Cantor by $\bar{E}$. If n be a natural number, then all ordered sets consisting of n elements are easily seen to be similar to the set of the first n natural numbers. We are, therefore, led to assume n for the symbol of the corresponding ordinal type.

The ordinal type of the class which contains the set of all natural numbers in their successive order is denoted by ω. The set of all negative integers $\ldots -4 \prec -3 \prec -2 \prec -1$ ordered according to their algebraic magnitude belongs to a different type ordered in the opposite direction to that of ω and is denoted by ω^*.

Generally, if α be a given type, then the type reversed in order to that one is denoted by α^*. It may happen that $\alpha^* = \alpha$; this is the case for every finite type, also for the type η of the set of all rational numbers ordered according to their magnitude, as well as for the type λ of the set of all real numbers ordered according to their magnitude.

9. A set E is said to be *dense* if between every two of its elements there is at least one element of E and, therefore, an infinite number of them. Thus the set of all rational numbers, and the set of all real numbers, each ordered according to magnitude, are both dense.

It can be proved that *two countable, ordered, dense sets, which have neither a first nor a last element, are similar, and are, therefore, of type* η. Similarly, it can be proved that *every countable ordered set is similar to a certain set of rational numbers which are ordered according to their magnitude.*

A *cut* of an ordered set E is a division of all the elements of the set into two non-empty classes A and B such that every element of the class A precedes every element of the class B. Such a division is denoted by $[A, B]$.

If in a given cut $[A, B]$ the class A has a last element and the class B a first element, then this cut is said to give rise to a *jump*. Thus in the set of natural numbers each cut supplies a jump. Obviously, in order that an ordered set be dense it is necessary and sufficient that none of its cuts gives rise to a jump.

If in a cut $[A, B]$ the class A has no last term and the class B no first term the cut is said to produce a *gap*. Thus in the set of all rational numbers. different from zero the cut into the class of negative rational numbers and the class of positive rational numbers produces a gap.

A set which has neither jumps nor gaps is said to be *continuous*.

If a given ordered set E has gaps, these may be removed by the addition of new elements in the following way. To each cut $[A, B]$ which produces a gap, we assign a new element not contained in E which is considered as following all the elements of A and preceding all those of B. Of two elements assigned to different cuts $[A, B]$, $[A_1, B_1]$ we consider the first as preceding the second when A is a proper subset of A_1 and as following the second when B is a proper subset of B_1.

It can be easily shown that adding such new elements to E we obtain a new ordered set F which has no gaps.

10. Let ϕ_1 and ϕ_2 be two ordinal types, O_1 and O_2 two disjoint ordered sets such that $\bar{O}_1 = \phi_1$ and $\bar{O}_2 = \phi_2$. Put $O = O_1 + O_2$ and order O as follows: two elements of O which belong both to O_1 or both to O_2 are to retain the ordinal relation which they had in their respective sets. Of two elements of O, one belonging to O_1 and one to O_2, the one belonging to O_1 will precede the other.

The set O is thus easily seen to be ordered and its type $\phi = \bar{O}$ will depend solely on the types ϕ_1, ϕ_2 and not on the sets O_1, O_2 which correspond to these types. We call ϕ the sum of ϕ_1 and ϕ_2 and write

$$\phi = \phi_1 + \phi_2.$$

It follows from the definition of the types ω and ω^* that

$$\omega^* + \omega$$

is the ordinal type of the set of all integers ordered according to their algebraic magnitude, i.e.

$$\ldots \prec -2 \prec -1 \prec 0 \prec 1 \prec 2 \prec \ldots,$$

while the sum

$$\omega + \omega^*$$

is the ordinal type of the class containing the set of the reciprocals of all the integers (zero excluded) ordered according to their algebraic magnitudes, i.e.

$$-\frac{1}{1} \prec -\frac{1}{2} \prec -\frac{1}{3} \prec \ldots \prec \frac{1}{3} \prec \frac{1}{2} \prec \frac{1}{1}.$$

The ordinal types $\omega^* + \omega$ and $\omega + \omega^*$ are different, for the first one does not contain a first nor last element whereas the second has both. The first type has no gaps, the second has a gap. Hence

$$\omega^* + \omega \neq \omega + \omega^*,$$

and so addition of ordinal types is not necessarily commutative.

Similarly, it may be shown that

$$1 + \omega \neq \omega + 1,$$

but if we put $\xi = \omega + \omega^*$ we find that

$$1 + \xi = \xi + 1$$

(since each sum is equal to ξ).

Furthermore, it is easily seen that

$$\eta + \eta = \eta, \quad \lambda + \lambda \neq \lambda,$$

and the relation

$$(\alpha + \beta)^* = \beta^* + \alpha^*$$

is true for every type α and β.

The definition of a sum of ordinal types may be extended immediately to any finite number of types, and such a sum is easily seen to justify the associative law. Thus

$$(\omega + 1) + \omega = \omega + (1 + \omega) = \omega + \omega.$$

Similarly,

$$\eta + 1 + \eta = \eta, \quad \lambda + 1 + \lambda = \lambda.$$

The sum of ordinal numbers may be further extended to infinite sequences. Let

(27) $$\alpha_1, \alpha_2, \alpha_3, \ldots$$

be an infinite sequence of ordinal types and

(28) $$O_1, O_2, O_3, \ldots$$

be mutually exclusive ordered sets such that $\bar{O}_n = \alpha_n$, for $n = 1, 2, 3, \ldots$. Put

(29) $$O = O_1 + O_2 + O_3 + \ldots$$

and let O be ordered as follows: if two elements of O belong to the same set O_n, then they retain in O the ordinal relation which they had in O_n, but if two elements of O belong to different sets, that element will come first which belongs to the earlier set in the sequence. It is easily seen that the set O will be ordered by the above procedure and its type will depend solely on the sequence (27) of types and not on the sets of sequence (28) corresponding to those types.

We may, therefore, say that every infinite series of ordinal types has a definite (well-defined) sum. E.g.,

$$\omega = 1 + 1 + 1 + \ldots$$

but also

$$\omega = 2 + 2 + 2 + \ldots = 1 + 2 + 3 + 4 + \ldots = 2 + 2^2 + 2^3 + \ldots.$$

We also note that

$$\eta = \eta + \eta + \eta + \ldots, \lambda = \lambda + 1 + \lambda + 1 + \lambda + 1 + \lambda + \ldots.$$

Let now ϕ and ψ be two ordinal types, U and V two ordered sets such that $\bar{U} = \phi$, $\bar{V} = \psi$. Denote by P the set of all pairs (u, v), where $u \in U$ and $v \in V$, and order P, assuming that

$$(u, v) \prec (u_1, v_1)$$

if $v \prec v_1$ (in V) or if $v = v_1$ then $u \prec u_1$ (in U).

It is easily seen that such an agreement will order P (i.e., conditions 1 and 2 of § 8 will be satisfied) and that the type of P will depend only on the types ϕ and ψ. The ordinal type of P is defined to be the product of the types ϕ and ψ and is written $\phi\psi$.

In order to obtain the product 2ω, consider the set U of type 2 consisting of the numbers 1 and 2 and the set of natural numbers ordered according to increasing magnitude. The set P will consist of all pairs (u, v) where u is 1 or 2 and v a natural number; ordering P as above we obtain the sequence

$$(1, 1) \prec (2, 1) \prec (1, 2) \prec (2, 2) \prec (1, 3) \prec (2, 3) \prec (1, 4) \prec (2, 4) \prec \ldots$$

which is of type ω. Hence

$$2\omega = \omega.$$

Similarly,

$$n\omega = \omega \qquad \text{for every natural } n.$$

The product $\omega.2$ is the type of the set

$$(1, 1) \prec (2, 1) \prec (3, 1) \prec (4, 1) \prec \ldots \prec (1, 2) \prec (2, 2) \prec (3, 2) \prec \ldots$$

and is, therefore, of type $\omega + \omega$. Hence

$$\omega 2 = \omega + \omega,$$

and, therefore,

$$\omega 2 \neq 2\omega.$$

Multiplication of ordinal types is thus seen to be non-commutative. We also note that $\eta 2 = \eta + \eta = \eta$, but $2\eta \neq \eta$ (since the type 2η contains jumps). Similarly $2\lambda \neq \lambda 2$, $\eta\lambda \neq \lambda\eta$.

Multiplication of ordinal types is, however, associative and distributive if the second factor is a sum. Thus

$$(\phi\psi)\theta = \phi(\psi\theta),$$

$$\phi(\psi + \theta) = \phi\psi + \phi\theta,$$

but

$$(1 + 1)\omega \neq 1\omega + 1\omega,$$

since the left-hand side is equal to ω, whereas on the right we have $\omega + \omega$.

We have obviously for every ordinal type ϕ and every natural number n the product ϕn equal to the sum of n terms each equal to ϕ. Similarly,

$$\phi\omega = \phi + \phi + \phi + \ldots.$$

11. An ordered set is said to be *well-ordered* if each of its non-empty subsets has a first element.

Every finite ordered set is well-ordered. Sets whose types are ω, $\omega + 1$, $\omega + \omega$, $\omega.\omega$ are evidently well-ordered; but the sets whose types are ω^*, η, λ are not well-ordered.

A well-ordered set cannot contain an infinite subset

$$a_1 \succ a_2 \succ a_3 \succ \ldots,$$

i.e., one of type ω^*; for it would then contain a subset without a first element, contrary to definition.

A non-empty subset of a well-ordered set is obviously well-ordered.

Well-ordered sets have the following important property which is known as the principle of transfinite induction:

If a certain theorem T

1, *is true for the first element of a well-ordered set W,*

2, *is true for an element a of W, if it is true for every element preceding a, then T is true for every element of W.*

Indeed, suppose that a certain theorem T satisfies conditions 1 and 2 but that there exist elements of W for which it is not true; let N be the set of such elements. N will, therefore, be a non-empty subset of a well-ordered set and so will have a first element say, a. It follows from the definition of N that T must be true for every element x of W which is such that $x \prec a$; but by condition 2, T must be true for a, which is contrary to the fact that $a \in N$. The principle of transfinite induction for well-ordered sets is, therefore, proved.

A well-ordered set may be similar to a proper subset of itself, e.g., the set of all natural numbers ordered according to their increasing magnitudes is similar to its subset consisting of the even numbers. We shall now prove that *if a well-ordered set W is similar to a proper subset S of itself, then an element of W cannot be correlated with an element of S which precedes it.*

For suppose that in the correlation of W and its subset S to the element a_1 of W corresponds a_2 of S such that $a_2 \prec a_1$; let a_3 be the element of S which corresponds to a_2 of W, hence $a_3 \prec a_2$, since $a_2 \prec a_1$ in W. Let now a_4 be the element of S corresponding to a_3 in W, and, since $a_3 \prec a_2$, we have $a_4 \prec a_3$. Arguing thus repeatedly we are led to an infinite sequence

$$a_1 \succ a_2 \succ a_3 \succ \dots$$

of W, which is impossible, and so the above statement is true.

Let W be a well-ordered set and a one of its elements. The set of all elements of W preceding a is called a *section* of W determined by the element a and denoted by $A(a)$. It follows from the above theorem that *a well-ordered set cannot be similar to a section of itself nor to any subsets of such a section*; for, in the similar correlation between the set W and a set $S \subset A\ (a)$ to an element a of W, there would have to correspond an element a' of the section $A\ (a)$, and so an element preceding a, contrary to the above theorem.

Given two well-ordered sets A and B it can be easily shown that either $A \simeq B$ or $A\ (a_0) \simeq B$ or else $A \simeq B(b_0)$, i.e., two well-ordered sets are either similar, or one of them is similar to a section of the other.

12. Ordinal types of well-ordered sets are called *ordinal numbers.*

If ϕ and ψ be two different ordinal numbers, then, as seen above, of two sets of these types just one is similar to a section of the other. In one case we write $\phi < \psi$, in the other $\psi < \phi$ (or $\phi > \psi$). It is convenient to include the number zero in the set of ordinal numbers, it being defined as the smallest of all ordinal numbers.

Let W be a well-ordered set of type ϕ. Let a be an element of W, and $\psi(a)$ the ordinal type of the section $A(a)$, where $\psi(a) = 0$ if a is the first element of W; we shall have obviously $\psi(a) < \phi$ and $\psi(a_1) < \psi(a_2)$, for $a_1 \prec a_2$. Hence, to every element of W there corresponds an ordinal number $\psi < \phi$, and to a later element corresponds a larger number. Conversely, every ordinal number $\psi < \phi$ corresponds to some element of W; in fact, if $\psi < \phi$, then the set W_1 of type ψ is similar to a certain section $A(a)$ of W, and so $\psi = \psi(a)$. Hence, *a well-ordered set of type* ϕ *is similar to the set of all ordinal numbers* $< \phi$ (0 *included*), *which are ordered according to increasing magnitudes.*

The elements of a well-ordered set may, therefore, be denoted by the symbol a_ψ, where the subscripts $\psi = \psi(a)$ are ordinal numbers (including 0 which is the subscript of the first element a_0). Thus, the n elements of a finite set may be denoted by

$$a_0, a_1, \ldots, a_{n-1},$$

the elements of a set of type ω by

$$a_0, a_1, a_2, \ldots,$$

the elements of a set of type $\omega + n$ (n a natural number) by

$$a_0, a_1, \ldots, a_\omega, a_{\omega+1}, \ldots, a_{\omega+n-1},$$

and so on. Generally, the elements of a well-ordered set of type ϕ may be written down as a transfinite sequence of type ϕ, i.e.,

$$a_0, a_1, \ldots, a_\omega, \ldots, a_\xi, \ldots \qquad (\xi < \phi).$$

Every set E of ordinal numbers is well-ordered. For let ϕ be any number of E; the set E_1 of all ordinal numbers $< \phi$ is, as previously shown, well-ordered of type ϕ. If ϕ is not the smallest number of E then the set of all numbers of E which belong to E_1 is non-empty, and so, as a subset of a well-ordered set, will have a first element, α say. It is easily seen that α is the smallest element of E. Hence, every set E of ordinal numbers has a smallest number, and this proves the above statement.

It can be shown that *the sum of two ordinal numbers, the second* > 0, *is always greater than the first number and* $\geqslant$ *either of the numbers.* From this it follows at once that for every ordinal number α,

$$\alpha + 1 > \alpha.$$

The number $\alpha + 1$ is said to be the *successor* of α. It can be shown that there is no ordinal number ξ satisfying the inequalities of $\alpha < \xi < \alpha + 1$. Hence, every ordinal number has a successor. But not every ordinal number has a *predecessor*, i.e., a number for which the given one is a successor. Thus, the numbers ω, $\omega + \omega$ have no predecessors. Ordinal numbers which have predecessors, i.e., those of the form $\alpha + 1$ are said to be of the first *kind*, and those without a predecessor are of the *second kind.*

Let E denote a well-ordered set of type α whose elements are ordinal numbers; we may, therefore, represent these elements by the symbols ϕ_ξ, where ξ is any ordinal number $< \alpha$ (0 included). In other words, the elements of the set E may be represented by the transfinite sequence

$$\phi_0, \phi_1, \phi_2, \ldots, \phi_\omega, \phi_{\omega+1}, \ldots, \phi_\xi, \ldots \qquad (\xi < \alpha), \tag{30}$$

of type α.

If for $\xi < \eta < \alpha$ we have $\phi_\xi < \phi_\eta$, the sequence is said to be *increasing*. In such a case the smallest ordinal number λ which exceeds every term of (30) is called the limit of the sequence, and we write

$$\lambda = \lim_{\xi<\alpha} \phi_\xi.$$

Thus,

$$\omega = \lim_{n<\omega} n = \lim_{n<\omega} n^2 = \lim_{n<\omega} 2^n;$$

$$\omega + \omega = \lim_{n<\omega} (\omega + n);$$

every number α of the second kind may, therefore, be written as

$$\alpha = \lim_{\xi<\alpha} \xi,$$

i.e., every number of the second kind is the limit of all ordinal numbers less than it.

All finite ordinal numbers (0 included) are said to be numbers of the *first class*. All ordinal numbers which are ordinal types of countable sets constitute the *second class* of ordinal numbers.

The set E of all numbers of the first and second classes is non-countable. Indeed, suppose it is countable; E being a set of ordinal numbers is, as shown previously, a set well-ordered according to the magnitude of the numbers. Let Ω be its type, and so, as the type of a well-ordered countable set, it would be a number of the second class, i.e., an element of E, ϕ say.

But every ordinal number is the type of the set of all ordinal numbers less than it; hence $\Omega = \phi$ would be the type of a section of the set E determined by the element ϕ of this set. Thus, the set E (which is of type Ω) would be similar to a section of itself, which is impossible.

Hence, the set of all numbers of the first and second classes is non-countable. The cardinal of this set is denoted by $\aleph_1$. Obviously $\aleph_0 < \aleph_1$, and it may be easily seen that there is no cardinal number between $\aleph_0$ and $\aleph_1$. For, suppose $\mathbf{m}$ is a cardinal number such that

$$\aleph_0 < \mathbf{m} < \aleph_1;$$

then there exists a subset E_1 of the set E such that $\bar{E}_1 = \mathbf{m}$. But since $\mathbf{m} < \aleph_1$,

E_1 must be similar to a section of E determined by some element, say ϕ; since $\phi \in E$ and so is a number of the first or second class, the section of E determined by ϕ is at most countable. We have, therefore, $\mathbf{m} \leqslant \aleph_0$, contrary to hypothesis.

The cardinal number $\aleph_1$ follows therefore immediately after $\aleph_0$. It is, however, still unknown whether $\aleph_1 = \mathbf{c}$, or $\aleph_1 \neq \mathbf{c}$. The assumption that $\aleph_1 = \mathbf{c}$ is known as *the continuum hypothesis.*

All ordinal numbers which are types of well-ordered sets of cardinal $\aleph_1$ constitute the numbers of the *third class*. The smallest of them is easily seen to be Ω.

It can be shown that the set of all the numbers of the third class has cardinal $> \aleph_1$; its cardinal is denoted by $\aleph_2$. The cardinal of a well-ordered set is generally called *aleph* ($\aleph$), and it can be shown that if a cardinal number is an aleph, then

$$\aleph + \aleph = \aleph\aleph = \aleph.$$

NOTES

THE following abbreviations are used for references occurring most frequently throughout the book.

[AH] P. Alexandroff and H. Hopf, *Topologie* I (Berlin, 1935).
[B] S. Banach, *Théorie des opérations linéaires*, Monografje Matematyczne (Warszawa, 1932).
[FM] Fundamenta Mathematicae.
[H] F. Hausdorff, *Grundzüge der Mengenlehre* 1st ed. (Leipzig, 1914).
[H_1] F. Hausdorff, *Mengenlehre*, 3rd ed. (Berlin-Leipzig, 1935).
[K] K. Kuratowski, *Topologie* I (Warszawa-Lwów, 1933).
[S, FM] W. Sierpiński, papers published in [FM].

CHAPTER I

1. Braces enclosing a number of letters designate a set consisting of the elements designated by these letters.
2. A Monteiro, Portug. Math., vol. 2 (1914), 58.
3. See example in § 5.
4. It can be shown that (1), (2), (3), and (4) are equivalent to (1) and the single property: $E_1 + \bar{E}_1 + \bar{E}_2 + \bar{\bar{E}}_2 \subset \overline{E_1 + E_2}$.
5. P. Szymański, Mathematica, vol. 17 (1941), 65-84.
6. Cf. S. Saks, Wiad. Matem., vol. 28 (1924/5), 17-22.
7. A. Khintchine, [FM], vol. 4 (1923), 165.
8. B. Knaster and K. Kuratowski, [FM], vol. 2 (1921), 241-244.
9. [S, FM], vol. 2 (1921), 82.
10. See, e.g., H. Tietze, *Über Analysis Situs* (Hamburg, 1923), 2.
11. We use the notation $< a, b >$ to indicate an interval of real numbers with two or one or no end-points.
12. By a linear set we shall mean throughout a set of real numbers or of points on a straight line, and by a plane set a set of pairs of real numbers or points in a plane with the usual definition of neighbourhoods.
13. K. Kuratowski, [FM], vol. 2 (1921), 158.
14. Cf. S. Saks, (FM], vol. 5 (1924), 291.
15. The cardinal $\bar{\bar{E}}$ of the set E is not to be confused with the closure of $\bar{E}$.
16. We call such a sequence descending.

CHAPTER II

1. Cf. P. Urysohn, Biul. Pol. Ak. Um. (1923), 13.

2. Spaces in which the family of all closed sets satisfy conditions 1, 2, 3 were investigated by Hausdorff, [FM], vol. 25 (1935), 486.

3. With condition γ changed, see § **35**.

4. [K], 15.

5. [AH], 37. The conditions satisfied by closure, as given by these authors, are somewhat weaker (hence their spaces are more general); for they replace condition II by conditions: $E \subset \bar{E}$ for all $E \subset K$ and $\bar{O} = O$, as originally given by Kuratowski in [FM], vol. 3 (1922), 182.

6. F. Hausdorff, [FM], vol. 25 (1933), 489.

7. $\bar{\alpha}$ denotes the cardinal number of a set whose ordinal number is α.

8. L. Scheeffer, Acta Math., vol. 5 (1884), and W. H. and G. C. Young, *The theory of sets of points* (Cambridge, 1906), 52 (Th. 18).

9. See B. Knaster and K. Kuratowski, [FM], vol. 2 (1921), 212.

10. *Ibid.*, 210.

11. *Ibid.*, 211.

CHAPTER III

1. A. Appert, *Proprietés des espaces abstraits les plus généraux*, Actualités Scient. et Industr., vol. 146 (1934), 84. Such spaces were already known to Urysohn and were mentioned by him (without proof) in a paper published in Biul. Pol. Ak. Um. (1923), 16.

2. [S, FM], vol. 33(1945), 299.

3. For a generalization of these theorems see W. Sierpiński, Biul. Pol. Ak. Um. (1921), 62-65.

4. Compare § **9** for the definition of a set contained in a (V)space and open in a subset of that space; also Theorem 39.

5. See e.g., [K], 156, Theorem VIII (a generalization of the Bolzano-Weierstrass theorem).

CHAPTER IV

1. Cf. [H], 213.

2. See e.g., [AH], 59, 67, 58, 68. Alexandroff and Hopf define topological spaces slightly differently so that sets consisting of single elements need not be closed.

3. M. Fréchet, R.C., Cir. Matem. Palermo, vol. 22 (1906); see also Bull. Sc. Math., vol. 42 (1918).

4. For, if $\mathbf{m}$ is the cardinal of the set of all different topological types appearing in K, then $\mathbf{m} \leqslant 2^{\mathbf{c}} = \mathbf{mc}$; consequently $\mathbf{m} > \mathbf{c}$ and so $\mathbf{mc} = \mathbf{m}$ which gives $\mathbf{m} = 2^{\mathbf{c}}$.

5. The expression "closed and compact" may be replaced both times by the expression "compact-in-itself," by the corollary to Theorem 53.

6. It should be noted that this part of the proof applies to general topological spaces.

7. We are obviously assuming here the axiom of choice without which it cannot be shown that every infinite set contains an infinite sequence of its elements.

CHAPTER V

1. See [AH], 31 (example 2) and 68-69.

2. See, e.g., Sierpiński, *Introduction to general topology* (Toronto, 1934), 66-67.

3. See [S, FM], vol. 2 (1921), 244.

4. P. Urysohn, Biul. Pol. Ak. Um. (1923), 16.

CHAPTER VI.

1. Some authors (e.g., Menger) write simply ab instead of $\rho(a, b)$ which, in the theory of sets, is not confusing, provided that a and b are neither sets nor numbers; others (e.g., Kuratowski) write $|a - b|$ instead of $\rho(a, b)$.

2. It is sufficient to postulate condition 3 only for the case that a, b, and c are all different. For, since $\rho \geqslant 0$, 3 is always satisfied when $a = b$ or $b = c$ and, if 2 is assumed, also when $a = c$.

Lindenbaum has observed that conditions 1, 2, and 3 may be replaced by 2 and the condition

$$\rho(a, c) \leqslant \rho(a, b) + \rho(c, b)$$

for any set of three elements a, b, and c of M (see [FM], vol. 8 (1926), 211).

3. K. Menger, Math. Ann., vol. 100 (1928), 115; Jber. Dtsch. Math. V., vol. 40 (1931), 202; Proc. Ak. Amst., vol. 30 (1927), 710.

4. E. W. Chittenden, Trans. Amer. Math. Soc., vol. 18 (1917), 161-166; see also Fréchet's (E)spaces in his book, *Les espaces abstraits* (Paris, 1928), 213-214.

5. With neighbourhoods as defined in § **48**.

6. See, e.g., K. Menger, C. R. Paris, vol. 202 (1936), 1007.

7. [AH], 28; also examples 5, 6, and 7, 29-30.

8. G. Birkoff, [FM], vol. 26 (1936), 156 ff.

9. H. Ribeiro, Portug. Math., vol. 4 (1943-5), 21-40.

10. K. Menger, Jber. Dtsch. Math. V., vol. 40 (1931), 210, and Math. Zeit., vol. 33 (1931), 396.

11. K. Menger, [FM], vol. 25 (1935), 445; S. Golab, [FM], vol. 31 (1938), 67.

12. L. Bieberbach, Sber. Preuss. Akad. Wiss., (1929), 612.

13. K. Kuratowski, [FM], vol. 6 (1924), 243.

14. D. König and S. Valkó, [FM], vol. 8 (1926), 131. A particular case of this theorem for $n = 2^s$ $(s = 1, 2, \ldots)$ was proved earlier in a much simpler manner by S. Banach and A. Tarski in [FM], vol. 6 (1924), 254, Cor. 12.

15. A. Lindenbaum, [FM], vol. 8 (1926), 217.

16. S. Mazurkiewicz and W. Sierpiński, C. R. Paris, vol. 158 (1914), 618.

17. I.e., E is the result of all such finite operations applied to the point $(0, 0)$.

18. [S, FM], vol. 34 (1947), 9.

19. J. von Neumann, [FM], vol. 11 (1928), 230-238.

20. S. Ruziewicz, [FM], vol. 5 (1924), 92.

21. W. Sierpiński, Atti Accad. Sci. Torino, vol. 75 (1940), 571-574, and [FM], vol. 33 (1945), 123.

22. In the sense of § **48**.

23. Cf. L. Bieberbach, Sber. Preuss. Akad. Wiss., (1929), 612.

24. See S. Banach and A. Tarski, [FM], vol. 6 (1924), 251, Th. 8.

25. *Ibid.*, 252, Cor. 9.

26. *Ibid.*, 247-248.

27. *Ibid.*, 264.

28. A. Tarski, *On the equivalence of polygons* (Polish) Przeg. Matem.-Fiz., vol. 2 (1924).

29. D. Hilbert, *Grundlagen der Geometrie*, 5th ed. (Leipzig and Berlin, 1922), Chap. IV.

30. S. Banach, *Sur le problème de mesure*, [FM], vol. 6 (1924), 7-33.

31. See F. Enriques, *Fragen der Elementargeometrie* (Leipzig and Berlin, 1911), vol. 1, 183.

32. M. Dehn, Math. Ann., vol. 55 (1920), 474; also F. Enriques, *loc. cit.*, 193.

33. To avoid the axiom of choice, it may be assumed that $\phi(p) = 1/n$, where n is the smallest natural number for which $S(p, 1/n) \cdot Q = 0$. Similarly for $\phi(q)$.

34. [K], 158, Theorem VIII.

35. [S, FM], vol. 34 (1947), 155.

36. P. Urysohn, [FM], vol. 9 (1927), 119.

37. [S, FM], vol. 21 (1933), 107.

38. *Ibid.*, 111.

39. N. Lusin, *Leçons sur les ensembles analytiques* (Paris, 1930), 294; also [FM], vol. 10 (1927), 62. Another example was given in [S, FM], vol. 18 (1932), 191.

40. See A. Tarski, [FM], vol. 30 (1938), 222, Cor. 1.17, and vol. 31 (1938), 63, Cor. 3.14. An elementary and direct proof of this theorem was given by the author in Actas Acad. Nac. Ciencias Lima, vol. II (1946), 113.

41. Another example was given by H. Hadwiger in Portug. Math., vol. 6 (1947), 47-48.

42. See H. Hadwiger, *ibid.*, 46-47.

43. A. Lindenbaum, [FM], vol. 8 (1926), 218, and H. Freudenthal and W. Hurewicz, [FM], vol. 26 (1936), 121, Theorem VI.

44. K. Menger, Math. Ann., vol. 100 (1928), 116; Amer. J. of Math., vol. 53 (1931), 721; Proc. Ak. Amst., vol. 30 (1927), 710.

45. K. Menger, Jber. Dtsch. Math. V., vol. 40 (1931), 210.

46. See e.g., [AH], 35.

47. Cf. [AH], 31, example 1; also 81, n. 2.

48. S. Banach, [FM], vol. 6 (1924), 239.

49. [S, FM], vol. 13 (1929), 117.

50. K. Kuratowski, [FM], vol. 8 (1926), 201.

51. F. Hausdorff, [FM], vol. 30 (1938), 41; also [FM], vol. 16 (1930), 353.

52. [S, FM], vol. 14 (1929), 123.

53. S. Ruziewicz, [FM], vol. 15 (1930), 95.

54. K. Kunugui, C. R. Paris, vol. 187 (1928), 876.

55. [S, FM], vol. 13 (1929), 277.

56. K. Kunugui, *loc. cit.*, 876.

57. K. Kunugui, C. R. Paris, vol. 188 (1929), 297.

58. This is known as *Urysohn's Metrisation Theorem*—see [A, H], 88, Th. VIII.

59. See Appendix, p. 270.

60. K. Kuratowski and W. Sierpiński, [FM], vol. 8 (1926), 200.

61. [S, FM], vol. 33 (1945), 124, Th. 5.

62. P. Urysohn, C. R. Paris, vol. 180 (1925), 83; also Bull. Sc. Math., 2nd ser. vol. 51 (1927), 1-38.

63. [B], 187. A simpler proof was given in [S, FM], vol. 33 (1945), 115-119.

64. [K], 79.

65. K. Kuratowski and W. Sierpiński, [FM], vol. 8 (1926), 197.

66. The problem of the existence of continuous curves filling a square has been extensively studied by many mathematicians. The first example of such a curve was given by Peano (Math. Ann., vol. 36 (1889/90), 157). The equation of the Peano curve was given by Cesàro (Bull. des Sc. Math., 2nd ser., vol. 21 (1897), 257) and the geometrical interpretation of the curve by A. Schoenflies and E. H. Moore. Another continuous curve filling a square was constructed by Hilbert (Math. Ann., vol. 38 (1891), 459 and Prace Mat.-Fiz., vol. 5 (1894), 13). Papers by the author on curves filling a square may be found in Prace Mat.-Fiz., vol. 23 (1912), 193-219 and in Biul. Pol. Ak. Um. (1912), 462-478. Equations of a continuous curve filling a square were also given by I. J. Shoenberg in Bull. Amer. Math. Soc., vol. 44 (1938), 519.

67. See [S, FM], vol. 1 (1920), 44-60; [H_1], 205-207.

68. H. Hahn, Jber. Dtsch. Math. V., vol. 23 (1914), 319; Wiener Ber., vol. 123 (1914), 2433. S. Mazurkiewicz, Sprawoz. Tow. Nauk. Wąrsaw, vol. 6 (1913), 305; [FM], vol. 1 (1920), 191. Cf. [H_1], 207.

69. B. Knaster, [FM], vol. 3 (1922), 274.

70. [S, FM], vol. 26 (1936), 46. Another example was given by Kuratowski in [FM], vol. 2 (1921), 158-160.

71. For a proof see [S, FM], vol. 26 (1936), 45-46.

72. *Ibid.*, 47.

73. *Ibid.*, 47-48.

74. *Ibid.*, 49.

75. M. Kondô, [FM], vol. 31 (1938), 1-30. Also Proc. Imp. Acad. Japan, vol. 14 (1937), 59.

76. W. Sierpiński, C. R. I. Congrès des Mathématiciens des pays Slaves (Warszawa, 1929), 53.

77. [S, FM], vol. 26 (1936), 48.

78. Z. Waraszkiewicz, [FM], vol. 23 (1934), 186 (corollary to (v); cf. *ibid.*, n. 20).

79. W. Sierpiński, Portug. Math., vol. 5 (1946), 193.

80. This theorem was communicated to the author by Kuratowski.

81. [B], 187.

82. Cf. [S, FM], vol. 33 (1945), 115; also Atti R. Accad. Sc. Torino, vol. 75 (1940).

83. P. Urysohn, C. R. Paris, vol. 180 (1925), 83; also Bull. Sc. Math., 2nd ser., vol. 51 (1927), 1-38.

84. [S, FM], vol. 9 (1926), 189-192.

85. Cf. S. Mazurkiewicz, *On the dimensional type of a hyperspace continuum* (Polish), Sprawoz. Tow. Nauk. Warsaw, vol. 24 (1931).

86. F. Hausdorff, [FM], vol. 26 (1936), 248.

87. N. Lusin, Isvestja Akad. Nauk. U.S.S.R., Ser. Math., vol. 11 (1947), 403-410.

89. $\bar{E}$ here indicates the cardinal number of the set E.

89. Cf. F. Hausdorff, *loc. cit.*, 245, where he refers to the set B as lying close to the family R; also N. Lusin, *loc. cit.*, 404, where he refers to B as being tangent to the family R.

90. Two ordered sets are said to be co-final if from a certain stage on in each set the remainders have the same ordinal type.

91. F. Hausdorff, [FM], vol. 26 (1936), 243 and 247; a proof of the existence of such sequences, which is not essentially different from Hausdorff's, was also given by Lusin, *loc. cit.*, 407-409.

92. [S, FM], vol. 33 (1945), 271. See also "A Weaker Form of the Problem of the Continuum," N. Lusin, Ann. Scu. Norm. Sup. Pisa, vol. 2 (1933), 271.

93. F. Hausdorff, Math. Zeit., vol. 5 (1919), 307. He introduced a different notation in [H_1], 178.

94. This could not be asserted unless $\xi_n \geqslant 2$. Note that property 8 is not true for $\alpha = 2$. For instance, it can be shown that the open interval (0, 1) (which is an F_σ) is not the sum of a countable aggregate of disjoint closed sets.

95. For a generalization of this corollary see [S, FM], vol. 23 (1934), 292-303.

96. Cf. D. Montgomery, [FM], vol. 25 (1935), 527-528.

97. Where ϕ denotes any transfinite ordinal number.

98. Another proof of Theorem 87 was given by Kuratowski, [FM], vol. 25 (1935), 534 ff.

99. K. Zarankiewicz, Wiad. Matem., vol. 30 (1927/8), 117.

100. This is a theorem of Banach; see [FM], vol. 16 (1930), 395; also [B], 13, Theorem 1.

101. W. Sierpiński, Hypothèse du continue, Monog. Matematyczne, vol. IV (Warszawa-Lwów, 1934), 176-177.

102. *Ibid.*, 159.

103. W. Sierpiński, *Functions representable analytically* (Polish) (Lwów-Warszawa-Kraków, 1925), 4.

CHAPTER VII

1. W. Niemytzki and A. Tychonoff, [FM], vol. 12 (1928), 118.

2. Hausdorff calls this theorem "*Zweiter Durchschnittssatz*"—see [H_1], 130.

3. For any two real numbers a and b we have

$$a^2 = (a - b + b)^2 \leqslant 2(a - b)^2 + 2b^2.$$

4. Cf. E. Čech, *Point sets* (Czech) (Prague, 1936), 103, Theorem 17.1.4

5. *Ibid.*, 106, Theorem 17.2.5.

6. See, e.g., [H_1], 214, III; also [S, FM], vol. 11 (1928), 203-205.

7. W. Sierpiński, Sprawoz. Tow. Nauk. Warsaw, vol. 21 (1928), 131-134.

8. F. Hausdorff, [FM], vol. 23 (1934), 279; for euclidean spaces see [S, FM], vol. 16 (1930), 173.

9. M. Lavrentieff, [FM], vol. 6 (1924), 149.

10. W. Sierpiński, C. R. Paris, vol. 178 (1923), 545.

11. This theorem was first proved for euclidean spaces by Mazurkiewicz, Biul. Pol. Ak. Um. (1916), 490-496. Another proof was given in [S, FM], vol. 8 (1926), 135.

12. W. Sierpiński, C. R. Paris, vol. 171 (1920), 24.

13. K. Menger, Jber. Dtsch. Math. V., vol. 37 (1928), 224 and 305, n. 6.

14. K. Menger, *loc cit.*, 222-223; cf. Brouwer's definition of finite sets in Math. Ann., vol. 93 (1925), 245.

15. W. Sierpiński, Sprawoz. Tow. Nauk. Warsaw, vol. 22 (1929), 163.
16. [S, FM], vol. 11 (1928), 16.
17. W. Sierpiński, Actas Acad. Nac. Ciencias Lima, vol. 10 (1947), 17-23.
18. W. Sierpiński, Biul. Pol. Ak. Um. (1927), 697.
19. *Ibid.*, 701; [S, FM], vol. 11 (1928), 294.
20. W. Hurewicz, [FM], vol. 12 (1928), 101.
21. [S, FM], vol. 14 (1929), 345.
22. P. Alexandroff, [FM], vol. 5 (1924), 164; M. Lavrentieff, [FM], vol. 6 (1924), 154.
23. S. Mazurkiewicz, [FM], vol. 10 (1927), 172.
24. W. Sierpiński, *Functions representable analytically*, 13 (Theorem 8).
25. N. Lusin, [FM], vol. 10 (1927), 12-15; W. Sierpiński, *ibid.*, 169-171.
26. W. Sierpiński, Sprawoz. Tow. Nauk. Warsaw, vol. 22 (1929), 1.
27. W. Sierpiński, Mathematica, vol. 5 (1931), 51.
28. [S, FM], vol. 6 (1924), 163.
29. S. Mazurkiewicz and W. Sierpiński, [FM], vol. 6 (1924), 163 and 166; K. Kuratowski, [FM], vol. 17 (1931), 261.
30. [S, FM], vol. 17 (1931), 77.
31. K. Kuratowski and E. Szpilrajn, [FM], vol. 18 (1932), 168.
32. N. Lusin, *Leçons sur les ensembles analytiques*, 255-259; also S. Braun, [FM], vol. 20 (1933), 168 and 172 (Theorem 9).
33. K. Kunugui, J. Fac. Sc. Hokkaido, vol. 8 (1940), 88, Theorem 11.
34. K. Kunugui, *ibid.*, 99, Cor. 1; P. Novikoff, C. R. Acad. Sc. U.S.S.R., vol. 23 (1939), 864-865.
35. W. Sierpiński, Mathematica, vol. 5 (1931), 51.
36. W. Sierpiński, *ibid.*, 51 and 52; also [FM], vol. 17 (1931), 30.
37. K. Kuratowski and E. Szpilrajn, [FM], vol. 18 (1932), 161-169.
38. This theorem was obtained by Souslin in 1916; see N. Lusin, C. R. Paris, vol. 164 (1917); also [FM], vol. 10 (1927), 25.
39. Condition (85) does not come into consideration for $k = 1$.
40. F. Hausdorff, Math. Ann., vol. 77 (1916), 430. See also [FM], vol. 5 (1924), 166.
41. P. Alexandroff, C. R. Paris, vol. 162 (1916).
42. This proof is due to Lusin. For a generalization of the Theorem see [S, FM], vol. 21 (1933), 265.
43. See [S, FM], vol. 25 (1935), 29-32.
44. K. Kuratowski, C. R. Paris, vol. 176 (1923), 229; [S, FM], vol.6 (1924), 39.
45. [S, FM], vol. 34 (1947), 66-68.
46. For $k = 1$, condition (116) is irrelevant.
47. N. Lusin, [FM], vol. 10 (1927), 60. For a generalization of this theorem see [S, FM], vol. 21 (1933), 271 (Theorems VIII and VIII-a).

48. It is not known, however, whether or not every continuous image of the complement of an analytic set is also the intersection of $\aleph_1$ Borel sets.

49. For a generalization of this formula, see K. Kuratowski, and A. Tarski, [FM], vol. 17 (1931), 243; also [K], 10 (3).

50. [S, FM], vol. 11 (1928), 126; also vol. 13 (1929), 239.

51. The general concept of a plane universal set with respect to a given family of linear sets is due to Lebesgue (J. de Math., Sér. 6, vol. I (1905), 207). It was studied more closely by Lusin (C. R. Paris, vol. 181 (1925), 95; Leçons sur les ensembles analitiques, 164 and 190). Also [S, FM], vol. 14 (1929), 82; O. Nikodym, [FM], vol. 14 (1929), 145; [K], 172; L. V. Kantorovitch, J. Soc. Phys.-Math. Leningrad, vol. II (1929), 13-21.

52. [S, FM], vol. 7 (1925), 198-200.

53. For $m = 3$, cf. T. Ważewski, [FM], vol. 4 (1923), 214-245.

54. W. Hurewicz, [FM], vol. 15 (1930), 17.

55. N. Lusin, [FM], vol. 10 (1927), 76 ff.

56. W. Sierpiński, Mathematica, vol. 12 (1936), 31-36. Lusin obtained doubly universal sets from universal sets by means of a curve filling the plane see C. R. Paris, vol. 189 (1929), 392.

INDEX

(Numbers refer to pages)